Preface

Nanotechnology is design, fabrication and application of nanostructures or nanomaterials, and the fundamental understanding of the relationships between physical properties or phenomena and material dimensions. Nanotechnology deals with materials or structures in nanometer scales, typically ranging from subnanometers to several hundred nanometers. One nanometer is 10^{-3} micrometer or 10^{-9} meter. Nanotechnology is a new field or a new scientific domain. Similar to quantum mechanics, on nanometer scale, materials or structures may possess new physical properties or exhibit new physical phenomena. Some of these properties are already known. For example, band gaps of semiconductors can be tuned by varying material dimension. There may be many more unique physical properties not known to us yet. These new physical properties or phenomena will not only satisfy everlasting human curiosity, but also promise new advancement in technology. For example, ultra-strong ultra-light multifunctional materials may be made from hierarchical nanostructures. Nanotechnology also promises the possibility of creating nanostructures of metastable phases with non-conventional properties including superconductivity and magnetism. Yet another very important aspect of nanotechnology is the miniaturization of current and new instruments, sensors and machines that will greatly impact the world we live in. Examples of possible miniaturization are: computers with infinitely great power that compute algorithms to mimic human brains, biosensors that warn us at the early stage of the onset of disease and preferably at the molecular level and target specific drugs that automatically attack the diseased cells on site, nanorobots that can repair internal damage and remove chemical toxins in human bodies, and nanoscaled electronics that constantly monitor our local environment.

Nanotechnology has an extremely broad range of potential applications from nanoscale electronics and optics, to nanobiological systems and nanomedicine, to new materials, and therefore it requires formation of and contribution from multidisciplinary teams of physicists, chemists, materials scientists, engineers, molecular biologists, pharmacologists and others to work together on (i) synthesis and processing of nanomaterials and nanostructures, (ii) understanding the physical properties related to the nanometer scale, (iii) design and fabrication of nano-devices or devices with nanomaterials as building blocks, and (iv) design and construction of novel tools for characterization of nanostructures and nanomaterials.

Synthesis and processing of nanomaterials and nanostructures are the essential aspect of nanotechnology. Studies on new physical properties and applications of nanomaterials and nanostructures are possible only when nanostructured materials are made available with desired size, morphology, crystal and microstructure and chemical composition. Work on the fabrication and processing of nanomaterials and nanostructures started long time ago, far earlier than nanotechnology emerged as a new scientific field. Such research has been drastically intensified in the last decade, resulting in overwhelming literatures in many journals across different disciplines. The research on nanotechnology is evolving and expanding very rapidly. That makes it impossible for a book of this volume to cover all the aspects of the field. The readers will readily find that this book has been focused primarily on inorganic materials. However, efforts were made to include the relevant organic materials such as self-assembled monolayers and Langmuir–Blodgett films. Of course, in the synthesis and processing of nanomaterials, organic materials often play an indispensable role, such as surfactants in the synthesis of ordered mesoporous materials, and capping polymers in the synthesis of monodispersed nanoparticles. The aim of this book is to summarize the fundamentals and established techniques of synthesis and processing of nanomaterials and nanostructures so as to provide readers a systematic and coherent picture about synthesis and processing of nanomaterials. In addition, the last two chapters of the book have been devoted to characterization, properties and applications of nanomaterials and nanostructures. This book would serve as a general introduction to people just entering the field, and also for experts seeking for information in other subfields. This is not a handbook with quick recipes for synthesis and processing of nanomaterials; it has been the intention of the author that this book is to be a tutorial and not a comprehensive review. Therefore, this book is well suited as a textbook for upper-level undergraduate, graduate and professional short courses.

The contents benefit greatly from the interaction between the author and colleagues or students. Most of the subjects covered in this book have been taught as a short course in 2002 SPIE annual meeting and a regular graduate course at the University of Washington. The precious feedback from the students who attended the classes have been incorporated into this book. I am grateful in particular to Mary Shang who took care of the figures and went through all the details to clarify many points in the presentation. I am also indebted to Ying Wang and Steven Limmer who got all the copyright permission and helped in preparing some of the figures. Our department chair, Prof. Raj Bordia, is acknowledged for having graciously reduced my department committee work. Editors, Mr. Stanford Chong and Ms. Lakshmi Narayanan at World Scientific Publishing and Imperial College Press are acknowledged for initiating and editing the book, respectively. The writing of this book started coincidently with the birth of my son, Doran, and finished when he turned a year old. It has been exciting to see both the little boy and the manuscript grow at the same time. This book would not have been possible without the dedication of my mother-in-law, who left Hong Kong to come to the States to take care of our son. My wife, Yuk Lan Li, deserves special thanks for her support and understanding.

Guozhong Cao
Seattle, WA
August 2003

Contents

Preface — v

1. Introduction — 1
  1.1. Introduction — 1
  1.2. Emergence of Nanotechnology — 4
  1.3. Bottom-Up and Top-Down Approaches — 7
  1.4. Challenges in Nanotechnology — 10
  1.5. Scope of the Book — 11
  References — 14

2. Physical Chemistry of Solid Surfaces — 15
  2.1. Introduction — 15
  2.2. Surface Energy — 17
  2.3. Chemical Potential as a Function of Surface Curvature — 26
  2.4. Electrostatic Stabilization — 32
    2.4.1. Surface charge density — 32
    2.4.2. Electric potential at the proximity of solid surface — 33
    2.4.3. Van der Waals attraction potential — 36
    2.4.4. Interactions between two particles: DLVO theory — 38
  2.5. Steric Stabilization — 42
    2.5.1. Solvent and polymer — 43
    2.5.2. Interactions between polymer layers — 45
    2.5.3. Mixed steric and electric interactions — 47
  2.6. Summary — 48
  References — 48

3. Zero-Dimensional Nanostructures: Nanoparticles — 51
  3.1. Introduction — 51
  3.2. Nanoparticles through Homogeneous Nucleation — 53
    3.2.1. Fundamentals of homogeneous nucleation — 53
    3.2.2. Subsequent growth of nuclei — 58
      3.2.2.1. Growth controlled by diffusion — 59
      3.2.2.2. Growth controlled by surface process — 59
    3.2.3. Synthesis of metallic nanoparticles — 63
      3.2.3.1. Influences of reduction reagents — 67
      3.2.3.2. Influences by other factors — 69
      3.2.3.3. Influences of polymer stabilizer — 72
    3.2.4. Synthesis of semiconductor nanoparticles — 74
    3.2.5. Synthesis of oxide nanoparticles — 81
      3.2.5.1. Introduction to sol-gel processing — 82
      3.2.5.2. Forced hydrolysis — 85
      3.2.5.3. Controlled release of ions — 87
    3.2.6. Vapor phase reactions — 88
    3.2.7. Solid state phase segregation — 89
  3.3. Nanoparticles through Heterogeneous Nucleation — 93
    3.3.1. Fundamentals of heterogeneous nucleation — 93
    3.3.2. Synthesis of nanoparticles — 95
  3.4. Kinetically Confined Synthesis of Nanoparticles — 96
    3.4.1. Synthesis inside micelles or using microemulsions — 96
    3.4.2. Aerosol synthesis — 98
    3.4.3. Growth termination — 99
    3.4.4. Spray pyrolysis — 100
    3.4.5. Template-based synthesis — 101
  3.5. Epitaxial Core-Shell Nanoparticles — 101
  3.6. Summary — 104
  References — 105

4. One-Dimensional Nanostructures: Nanowires and Nanorods — 110
  4.1. Introduction — 110
  4.2. Spontaneous Growth — 111
    4.2.1. Evaporation (dissolution)–condensation growth — 112
      4.2.1.1. Fundamentals of evaporation (dissolution)–condensation growth — 112
      4.2.1.2. Evaporation–condensation growth — 119
      4.2.1.3. Dissolution–condensation growth — 123

Contents

   4.2.2. Vapor (or solution)–liquid–solid
     (VLS or SLS) growth   127
     4.2.2.1. Fundamental aspects of VLS and
       SLS growth   127
     4.2.2.2. VLS growth of various nanowires   131
     4.2.2.3. Control of the size of nanowires   134
     4.2.2.4. Precursors and catalysts   138
     4.2.2.5. SLS growth   140
   4.2.3. Stress-induced recrystallization   142
  4.3. Template-Based Synthesis   143
   4.3.1. Electrochemical deposition   144
   4.3.2. Electrophoretic deposition   151
   4.3.3. Template filling   157
     4.3.3.1. Colloidal dispersion filling   158
     4.3.3.2. Melt and solution filling   160
     4.3.3.3. Chemical vapor deposition   161
     4.3.3.4. Deposition by centrifugation   161
   4.3.4. Converting through chemical reactions   162
  4.4. Electrospinning   164
  4.5. Lithography   165
  4.6. Summary   168
  References   168

5. Two-Dimensional Nanostructures: Thin Films   **173**
  5.1. Introduction   173
  5.2. Fundamentals of Film Growth   174
  5.3. Vacuum Science   178
  5.4. Physical Vapor Deposition (PVD)   182
   5.4.1. Evaporation   183
   5.4.2. Molecular beam epitaxy (MBE)   185
   5.4.3. Sputtering   186
   5.4.4. Comparison of evaporation and sputtering   188
  5.5. Chemical Vapor Deposition (CVD)   189
   5.5.1. Typical chemical reactions   189
   5.5.2. Reaction kinetics   190
   5.5.3. Transport phenomena   191
   5.5.4. CVD methods   194
   5.5.5. Diamond films by CVD   197
  5.6. Atomic Layer Deposition (ALD)   199
  5.7. Superlattices   204
  5.8. Self-Assembly   205

5.8.1. Monolayers of organosilicon or alkylsilane derivatives	208
5.8.2. Monolayers of alkanethiols and sulfides	210
5.8.3. Monolayers of carboxylic acids, amines and alcohols	212
5.9. Langmuir–Blodgett Films	213
5.10. Electrochemical Deposition	218
5.11. Sol-Gel Films	219
5.12. Summary	223
References	224

6. Special Nanomaterials — 229

6.1. Introduction	229
6.2. Carbon Fullerenes and Nanotubes	230
6.2.1. Carbon fullerenes	230
6.2.2. Fullerene-derived crystals	232
6.2.3. Carbon nanotubes	232
6.3. Micro and Mesoporous Materials	238
6.3.1. Ordered mesoporous structures	239
6.3.2. Random mesoporous structures	245
6.3.3. Crystalline microporous materials: zeolites	249
6.4. Core-Shell Structures	257
6.4.1. Metal-oxide structures	257
6.4.2. Metal–polymer structures	260
6.4.3. Oxide–polymer structures	261
6.5. Organic–Inorganic Hybrids	263
6.5.1. Class I hybrids	263
6.5.2. Class II hybrids	264
6.6. Intercalation Compounds	266
6.7. Nanocomposites and Nanograined Materials	267
6.8. Summary	268
References	269

7. Nanostructures Fabricated by Physical Techniques — 277

7.1. Introduction	277
7.2. Lithography	278
7.2.1. Photolithography	279
7.2.2. Phase-shifting photolithography	283
7.2.3. Electron beam lithography	284
7.2.4. X-ray lithography	287
7.2.5. Focused ion beam (FIB) lithography	288

Contents

7.2.6. Neutral atomic beam lithography	290
7.3. Nanomanipulation and Nanolithography	291
7.3.1. Scanning tunneling microscopy (STM)	292
7.3.2. Atomic force microscopy (AFM)	294
7.3.3. Near-field scanning optical microscopy (NSOM)	296
7.3.4. Nanomanipulation	298
7.3.5. Nanolithography	303
7.4. Soft Lithography	308
7.4.1. Microcontact printing	308
7.4.2. Molding	310
7.4.3. Nanoimprint	310
7.4.4. Dip-pen nanolithography	313
7.5. Assembly of Nanoparticles and Nanowires	314
7.5.1. Capillary forces	315
7.5.2. Dispersion interactions	316
7.5.3. Shear force assisted assembly	318
7.5.4. Electric-field assisted assembly	318
7.5.5. Covalently linked assembly	319
7.5.6. Gravitational field assisted assembly	319
7.5.7. Template-assisted assembly	319
7.6. Other Methods for Microfabrication	321
7.7. Summary	321
References	322
8. Characterization and Properties of Nanomaterials	**329**
8.1. Introduction	329
8.2. Structural Characterization	330
8.2.1. X-ray diffraction (XRD)	331
8.2.2. Small angle X-ray scattering (SAXS)	333
8.2.3. Scanning electron microscopy (SEM)	336
8.2.4. Transmission electron microscopy (TEM)	338
8.2.5. Scanning probe microscopy (SPM)	340
8.2.6. Gas adsorption	343
8.3. Chemical Characterization	344
8.3.1. Optical spectroscopy	345
8.3.2. Electron spectroscopy	349
8.3.3. Ionic spectrometry	350
8.4. Physical Properties of Nanomaterials	352
8.4.1. Melting points and lattice constants	353
8.4.2. Mechanical properties	357
8.4.3. Optical properties	362

	8.4.3.1. Surface plasmon resonance	362
	8.4.3.2. Quantum size effects	367
8.4.4.	Electrical conductivity	371
	8.4.4.1. Surface scattering	371
	8.4.4.2. Change of electronic structure	374
	8.4.4.3. Quantum transport	375
	8.4.4.4. Effect of microstructure	379
8.4.5.	Ferroelectrics and dielectrics	380
8.4.6.	Superparamagnetism	382
8.5. Summary		384
References		384

9. Applications of Nanomaterials — 391

9.1.	Introduction	391
9.2.	Molecular Electronics and Nanoelectronics	392
9.3.	Nanobots	394
9.4.	Biological Applications of Nanoparticles	396
9.5.	Catalysis by Gold Nanoparticles	397
9.6.	Band Gap Engineered Quantum Devices	399
	9.6.1. Quantum well devices	399
	9.6.2. Quantum dot devices	401
9.7.	Nanomechanics	402
9.8.	Carbon Nanotube Emitters	404
9.9.	Photoelectrochemical Cells	406
9.10.	Photonic Crystals and Plasmon Waveguides	409
	9.10.1. Photonic crystals	409
	9.10.2. Plasmon waveguides	411
9.11.	Summary	411
References		412

Appendix

1. Periodic Table of the Elements	419
2. The International System of Units	420
3. List of Fundamental Physical Constants	421
4. The 14 Three-Dimensional Lattice Types	422
5. The Electromagnetic Spectrum	423
6. The Greek Alphabet	424

Index — 425

Chapter 1

Introduction

1.1. Introduction

Nanotechnology deals with small structures or small-sized materials. The typical dimension spans from subnanometer to several hundred nanometers. A nanometer (nm) is one billionth of a meter, or 10^{-9} m. Figure 1.1 gives a partial list of zero-dimensional nanostructures with their typical ranges of dimensions.[1,2] One nanometer is approximately the length equivalent to 10 hydrogen or 5 silicon atoms aligned in a line. Small features permit more functionality in a given space, but nanotechnology is not only a simple continuation of miniaturization from micron meter scale down to nanometer scale. Materials in the micrometer scale mostly exhibit physical properties the same as that of bulk form; however, materials in the nanometer scale may exhibit physical properties distinctively different from that of bulk. Materials in this size range exhibit some remarkable specific properties; a transition from atoms or molecules to bulk form takes place in this size range. For example, crystals in the nanometer scale have a low melting point (the difference can be as large as 1000°C) and reduced lattice constants, since the number of surface atoms or ions becomes a significant fraction of the total number of atoms or ions and the surface energy plays a significant role in the thermal stability. Crystal structures stable at elevated temperatures are stable at much lower

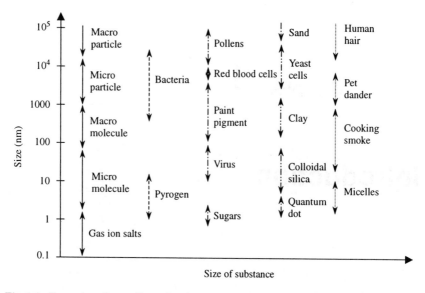

Fig. 1.1. Examples of zero-dimensional nanostructures or nanomaterials with their typical ranges of dimension.

temperatures in nanometer sizes, so ferroelectrics and ferromagnetics may lose their ferroelectricity and ferromagnetism when the materials are shrunk to the nanometer scale. Bulk semiconductors become insulators when the characteristic dimension is sufficiently small (in a couple of nanometers). Although bulk gold does not exhibit catalytic properties, Au nanocrystal demonstrates to be an excellent low temperature catalyst.

Currently there are a lot of different opinions about what exactly is nanotechnology. For example, some people consider the study of microstructures of materials using electron microscopy and the growth and characterization of thin films as nanotechnology. Other people consider a bottom-up approach in materials synthesis and fabrication, such as self-assembly or biomineralization to form hierarchical structures like abalone shell, is nanotechnology. Drug delivery, e.g. by putting drugs inside carbon nanotubes, is considered as nanotechnology. Micro-electromechanical systems (MEMS) and lab-on-a-chip are considered as nanotechnology. More futuristic or science fiction-like opinions are that nanotechnology means something very ambitious and startlingly new, such as miniature submarines in the bloodstream, smart self-replication nanorobots monitoring our body, space elevators made of nanotubes and the colonization of space. There are many other definitions that people working in nanotechnology use to define the field. These definitions are true to certain specific research fields, but none of them covers the full spectrum of nanotechnology. The many diverse definitions of nanotechnology reflect the fact that nanotechnology covers a broad spectrum of research field and requires true interdisciplinary and multidisciplinary efforts.

In general, nanotechnology can be understood as a technology of design, fabrication and applications of nanostructures and nanomaterials. Nanotechnology also includes fundamental understanding of physical properties and phenomena of nanomaterials and nanostructures. Study on fundamental relationships between physical properties and phenomena and material dimensions in the nanometer scale, is also referred to as nanoscience. In the United States, nanotechnology has been defined as being "concerned with materials and systems whose structures and components exhibit novel and significantly improved physical, chemical and biological properties, phenomena and processes due to their nanoscale size".[3]

In order to explore novel physical properties and phenomena and realize potential applications of nanostructures and nanomaterials, the ability to fabricate and process nanomaterials and nanostructures is the first corner stone in nanotechnology. Nanostructured materials are those with at least one dimension falling in nanometer scale, and include nanoparticles (including quantum dots, when exhibiting quantum effects), nanorods and nanowires, thin films, and bulk materials made of nanoscale building blocks or consisting of nanoscale structures. Many technologies have been explored to fabricate nanostructures and nanomaterials. These technical approaches can be grouped in several ways. One way is to group them according to the growth media:

(1) Vapor phase growth, including laser reaction pyrolysis for nanoparticle synthesis and atomic layer deposition (ALD) for thin film deposition.
(2) Liquid phase growth, including colloidal processing for the formation of nanoparticles and self assembly of monolayers.
(3) Solid phase formation, including phase segregation to make metallic particles in glass matrix and two-photon induced polymerization for the fabrication of three-dimensional photonic crystals.
(4) Hybrid growth, including vapor–liquid–solid (VLS) growth of nanowires.

Another way is to group the techniques according to the form of products:

(1) Nanoparticles by means of colloidal processing, flame combustion and phase segregation.
(2) Nanorods or nanowires by template-based electroplating, solution–liquid–solid growth (SLS), and spontaneous anisotropic growth.
(3) Thin films by molecular beam epitaxy (MBE) and atomic layer deposition (ALD).

(4) Nanostructured bulk materials, for example, photonic bandgap crystals by self-assembly of nanosized particles.

There are many other ways to group different fabrication and processing techniques such as top-down and bottom-up approaches, spontaneous and forced processes. Top-down is in general an extension of lithography. The concept and practice of a bottom-up approach in material science and chemistry are not new either. Synthesis of large polymer molecules is a typical bottom-up approach, in which individual building blocks (monomers) are assembled to a large molecule or polymerized into bulk material. Crystal growth is another bottom-up approach, where growth species either atoms, or ions or molecules orderly assemble into desired crystal structure on the growth surface.

1.2. Emergence of Nanotechnology

Nanotechnology is new, but research on nanometer scale is not new at all. The study of biological systems and the engineering of many materials such as colloidal dispersions, metallic quantum dots, and catalysts have been in the nanometer regime for centuries. For example, the Chinese are known to use Au nanoparticles as an inorganic dye to introduce red color into their ceramic porcelains more than thousand years ago.[4,5] Use of colloidal gold has a long history, though a comprehensive study on the preparation and properties of colloidal gold was first published in the middle of the 19th century.[6] Colloidal dispersion of gold prepared by Faraday in 1857,[7] was stable for almost a century before being destroyed during World War II.[6] Medical applications of colloidal gold present another example. Colloidal gold was, and is still, used for treatment of arthritis. A number of diseases were diagnosed by the interaction of colloidal gold with spinal fluids obtained from the patient.[8] What has changed recently is an explosion in our ability to image, engineer and manipulate systems in the nanometer scale. What is really new about nanotechnology is the combination of our ability to see and manipulate matter on the nanoscale and our understanding of atomic scale interactions.

Although study on materials in the nanometer scale can be traced back for centuries, the current fever of nanotechnology is at least partly driven by the ever shrinking of devices in the semiconductor industry and supported by the availability of characterization and manipulation techniques at the nanometer level. The continued decrease in device dimensions has followed the well-known Moore's law predicted in 1965 and illustrated in Fig. 1.2.[9] The figure shows that the dimension of a device

Introduction

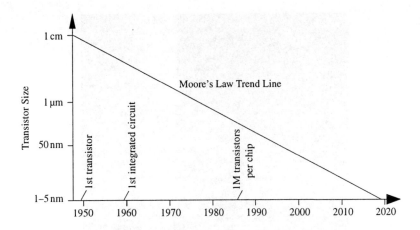

Fig. 1.2. "Moore's Law" plot of transistor size versus year. The trend line illustrates the fact that the transistor size has decreased by a factor of 2 every 18 months since 1950.

halves approximately every eighteen months and today's transistors have well fallen in the nanometer range. Figure 1.3 shows the original centimeter scale contact transistor made by Bardeen, Brattain, and Shockley on 23 December 1947 at AT&T Bell Lab.[10] Figure 1.4 shows an electronic device that is based on a single Au nanoparticle bridging two molecular monolayers for electrical studies.[11] Many scientists are currently working on molecular and nanoscaled electronics, which are constructed using single molecules or molecular monolayers.[12–14] Although the current devices operate far below fundamental limits imposed by thermodynamics and quantum mechanics,[15] a number of challenges in transistor design have already arisen from materials limitations and device physics.[16] For example, the off-currents in a metal-oxide semiconductor field-effect transistor (MOSFET) increase exponentially with device scaling. Power dissipation and overheating of chips have also become a serious issue in further reduction of device sizes. The continued size shrinkage of transistors will sooner or later meet with the limitations of the materials' fundamentals. For example, the widening of the band gap of semiconductors occurs when the size of the materials reaches de Broglie's wavelength.

Miniaturization is not necessarily limited to semiconductor-based electronics, though simple miniaturization already brings us significant excitement.[17] Promising applications of nanotechnology in the practice of medicine, often referred to as nanomedicine, have attracted a lot of attention and have become a fast growing field. One of the attractive applications in nanomedicine is the creation of nanoscale devices for improved therapy and diagnostics. Such nanoscale devices are known as nanorobots

Fig. 1.3. The original contact transistor made by Bardeen, Brattain, and Shockley on December 23, 1947 at AT&T Bell Lab. [M. Riordan and L. Hoddeson, *Crystal Fire*, W.W. Norton and Company, New York, 1997.]

or more simply as nanobots.[18] These nanobots have the potential to serve as vehicles for delivery of therapeutic agents, detectors or guardians against early disease and perhaps repair of metabolic or genetic defects. Studies in nanotechnology are not limited to miniaturization of devices. Materials in nanometer scale may exhibit unique physical properties and have been explored for various applications. For example, gold nanoparticles have found many potential applications using surface chemistry and its uniform size. Au nanoparticles can function as carrier vehicles to accommodate multiple functionalities through attaching various functional organic molecules or bio-components.[19] Bandgap engineered quantum devices, such as lasers and heterojunction bipolar transistors, have been developed with unusual electronic transport and optical effects.[20] The discovery of synthetic materials, such as carbon fullerenes,[21] carbon nanotubes,[22] and ordered mesoporous materials,[23] has further fueled the research in nanotechnology and nanomaterials.

Introduction

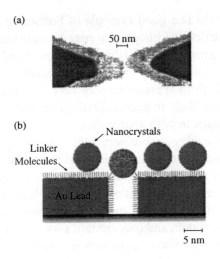

Fig. 1.4. (a) Field emission SEM image of an Au lead-structure before the nanocrystals are introduced. The light gray region is formed by the angle evaporation, and is ~10 nm thick. The darker region is from a normal angle evaporation and is ~70 nm thick. (b) Schematic cross-section of nanocrystals bound via a bifunctional linker molecule to the leads. Transport between the leads occurs through the mottled nanocrystal bridging the gap. [D.L. Klein, P.L. McEuen, J.E. Bowen Katari, R. Roth, and A.P. Alivisatos, *Appl. Phys. Lett.* **68**, 2574 (1996).]

The invention and development of scanning tunneling microscopy (STM) in the early 1980s[24] and subsequently other scanning probe microscopy (SPM) such as atomic force microscopy (AFM),[25] have opened up new possibilities for the characterization, measurement and manipulation of nanostructures and nanomaterials. Combining with other well-developed characterization and measurement techniques such as transmission electron microscopy (TEM), it is possible to study and manipulate the nanostructures and nanomaterials to a great detail and often down to the atomic level. Nanotechnology is already all around us if you know where to look.[26] This technology is not new, it is the combination of existing technologies and our new found ability to observe and manipulate at the atomic scale, this makes nanotechnology so compelling from scientific, business and political viewpoints.

1.3. Bottom-Up and Top-Down Approaches

Obviously there are two approaches to the synthesis of nanomaterials and the fabrication of nanostructures: top-down and bottom-up. Attrition or milling is a typical top-down method in making nanoparticles, whereas the colloidal dispersion is a good example of bottom-up approach in the synthesis of nanoparticles. Lithography may be considered as a hybrid approach, since the growth of thin films is bottom-up whereas etching is top-down, while nanolithography and nanomanipulation are commonly a bottom-up approach. Both approaches play very important roles in modern industry and most likely in nanotechnology as well. There are advantages and disadvantages in both approaches.

Among others, the biggest problem with top-down approach is the imperfection of the surface structure. It is well known that the conventional top-down techniques such as lithography can cause significant crystallographic damage to the processed patterns,[27] and additional defects may be introduced even during the etching steps.[28] For example, nanowires made by lithography is not smooth and may contain a lot of impurities and structural defects on surface. Such imperfections would have a significant impact on physical properties and surface chemistry of nanostructures and nanomaterials, since the surface over volume ratio in nanostructures and nanomaterials is very large. The surface imperfection would result in a reduced conductivity due to inelastic surface scattering, which in turn would lead to the generation of excessive heat and thus impose extra challenges to the device design and fabrication. Regardless of the surface imperfections and other defects that top-down approaches may introduce, they will continue to play an important role in the synthesis and fabrication of nanostructures and nanomaterials.

Bottom-up approach is often emphasized in nanotechnology literature, though bottom-up is nothing new in materials synthesis. Typical material synthesis is to build atom by atom on a very large scale, and has been in industrial use for over a century. Examples include the production of salt and nitrate in chemical industry, the growth of single crystals and deposition of films in electronic industry. For most materials, there is no difference in physical properties of materials regardless of the synthesis routes, provided that chemical composition, crystallinity, and microstructure of the material in question are identical. Of course, different synthesis and processing approaches often result in appreciable differences in chemical composition, crystallinity, and microstructure of the material due to kinetic reasons. Consequently, the material exhibits different physical properties.

Bottom-up approach refers to the build-up of a material from the bottom: atom-by-atom, molecule-by-molecule, or cluster-by-cluster. In organic chemistry and/or polymer science, we know polymers are synthesized by connecting individual monomers together. In crystal growth, growth species, such as atoms, ions and molecules, after impinging onto the growth surface, assemble into crystal structure one after another.

Introduction

Although the bottom-up approach is nothing new, it plays an important role in the fabrication and processing of nanostructures and nanomaterials. There are several reasons for this. When structures fall into a nanometer scale, there is little choice for a top-down approach. All the tools we have possessed are too big to deal with such tiny subjects.

Bottom-up approach also promises a better chance to obtain nanostructures with less defects, more homogeneous chemical composition, and better short and long range ordering. This is because the bottom-up approach is driven mainly by the reduction of Gibbs free energy, so that nanostructures and nanomaterials such produced are in a state closer to a thermodynamic equilibrium state. On the contrary, top-down approach most likely introduces internal stress, in addition to surface defects and contaminations.

Figure 1.5 shows a miniature bull fabricated by a technique called two-photon polymerization[29] whereas Fig. 1.6 shows a "molecular person", consisting of 14 carbon monoxide molecules arranged on a metal surface fabricated and imaged by STM.[30] These two figures show what the current technology or nanotechnology is capable of, and new capabilities are

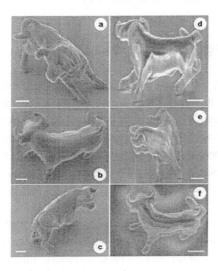

Fig. 1.5. Miniature bulls were fabricated by two-photon polymerization. A titanium sapphire laser operating in mode-lock at 76 MHz and 780 nm with a 150-femtosecond pulse width was used as an exposure source. The laser was focused by an objective lens of high numerical aperture (~1.4). (a–c) Bull sculpture produced by raster scanning; the process took 180 min. (d–f) The surface of the bull was defined by two-photon absorption (TPA; that is, surface-profile scanning) and was then solidified internally by illumination under a mercury lamp, reducing the TPA-scanning time to 13 min. Scale bars, 2 μm. [K. Kawata, H.B. Sun, T. Tanaka, and K. Takada, *Nature* **412**, 697 (2001).]

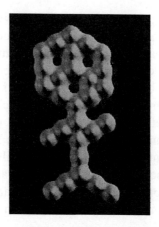

Fig. 1.6. A molecular person consisting of 14 carbon monoxide molecules arranged on a metal surface fabricated and imaged by scanning tunneling microscopy. [P. Zeppenfeld & D.M. Eigler, *New Scientist* **129**, 20 (23 February 1991), and http://www.almaden.ibm.com/vis/stm/atomo.html]

being developed and the existing techniques are being further improved pushing the current limit to a smaller size.

1.4. Challenges in Nanotechnology

Although many of the fundamentals have long been established in different fields such as in physics, chemistry, materials science and device science and technology, and research on nanotechnology is based on these established fundamentals and technologies, researchers in the field face many new challenges that are unique to nanostructures and nanomaterials. Challenges in nanotechnology include the integration of nanostructures and nanomaterials into or with macroscopic systems that can interface with people.

Challenges include the building and demonstration of novel tools to study at the nanometer level what is being manifested at the macro level. The small size and complexity of nanoscale structures make the development of new measurement technologies more challenging than ever. New measurement techniques need to be developed at the nanometer scale and may require new innovations in metrological technology. Measurements of physical properties of nanomaterials require extremely sensitive instrumentation, while the noise level must be kept very low. Although material properties such as electrical conductivity, dielectric constant, tensile strength, are independent of dimensions and weight of the material in question, in practice, system properties are measured experimentally. For example, electrical conductance, capacitance and tensile stress are measured and used to calculate electrical conductivity, dielectric constant and tensile strength. As the dimensions of materials shrink from centimeter or millimeter scale to nanometer scale, the system properties would change accordingly, and mostly decrease with the reducing dimensions of the sample materials. Such a decrease can easily be as much as 6 orders of magnitude as sample size reduces from centimeter to nanometer scale.

Other challenges arise in the nanometer scale, but are not found in the macro level. For example, doping in semiconductors has been a very well established process. However, random doping fluctuations become extremely important at nanometer scale, since the fluctuation of doping concentration would be no longer tolerable in the nanometer scale. With a typical doping concentration of $10^{18}/cm^3$, there will be just one dopant atom in a device of $10 \times 10 \times 10\,nm^3$ in size. Any distribution fluctuation of dopants will result in a totally different functionality of device in such a size range. Making the situation further complicated is the location of the dopant atoms. Surface atom would certainly behave differently from the centered atom. The challenge will be not only to achieve reproducible and uniform distribution of dopant atoms in the nanometer scale, but also to precisely control the location of dopant atoms. To meet such a challenge, the ability to monitor and manipulate the material processing in the atomic level is crucial. Furthermore, doping itself also imposes another challenge in nanotechnology, since the self-purification of nanomaterials makes doping very difficult.

For the fabrication and processing of nanomaterials and nanostructures, the following challenges must be met:

(1) Overcome the huge surface energy, a result of enormous surface area or large surface to volume ratio.
(2) Ensure all nanomaterials with desired size, uniform size distribution, morphology, crystallinity, chemical composition, and microstructure, that altogether result in desired physical properties.
(3) Prevent nanomaterials and nanostructures from coarsening through either Ostwald ripening or agglomeration as time evolutes.

1.5. Scope of the Book

The aim of this book is to summarize the fundamentals and technical approaches in synthesis, fabrication and processing of nanostructures and nanomaterials so as to provide the readers a systematic and coherent picture of the field. Therefore, this book would serve as a general introduction to people just entering the field and for experts seeking for information in other sub-fields. It has been the intention of the author that this book is intended to be tutorial and not a comprehensive review. The research on nanotechnology is evolving and expanding very rapidly. That makes it impossible for a book to cover all the aspects of the nanotechnology field. Furthermore, this book has been primarily focused on inorganic materials, although, efforts have been made to include the relevant organic materials such as self-assembled monolayers and Langmuir–Blodgett films as part of Chapter 5. Of course, in the synthesis, fabrication and processing of nanostructures and nanomaterials, organic materials often play an indispensable role, such as surfactants in the synthesis of ordered mesoporous materials, and capping polymers in the synthesis of monodispersed nanoparticles.

In the synthesis, fabrication and processing of nanostructures and nanomaterials, one of the great challenges is to deal with the large surface to volume ratio and the resulting surface energy. Therefore, the entire chapter, Chapter 2, has been devoted to the discussion on the physical chemistry of solid surface prior to introducing various synthesis techniques for various nanostructures and nanomaterials. A good understanding of the surface properties of solids is essential for the understanding of the fabrication and process of nanostructures and nanomaterials.

Chapter 3 is focused on the synthesis and processing of zero-dimensional nanostructures including nanoparticles and heteroepitaxial core-shell structures. In this chapter, the fundamentals of homogeneous and heterogeneous nucleation as well as the continued growth immediately following the initial nucleation will be discussed in detail. Particular attention will be paid to the fundamentals for the control of particle size, size distribution and chemical composition. Various methods for the synthesis of nanoparticles and heleroepitaxial core-shell structures are reviewed.

The formation of one-dimensional nanostructures is the subject of Chapter 4. One-dimensional nanostructures include nanorods, nanowires and nanotubules. In this chapter, we discuss spontaneous anisotropic growth, catalyst induced anisotropic growth such as vapor–liquid–solid growth, and nanolithography. Essential fundamentals are discussed first, prior to the discussion of the details of various techniques used in the synthesis of one-dimensional nanostructures.

Chapter 5 is on the formation of two-dimensional structure, i.e. thin films. Since there are relatively abundant information on the deposition of thin (less than 100 nm) and thick (above 100 nm here) films, the discussion in this chapter has been kept as brief as possible. The focus has been

Introduction

mainly on the less extensively covered subjects on conventional thin film books: atomic layer deposition and self-assembled monolayers. These two techniques are extremely important in making very thin films, and are capable of making films less than 1 nm in thickness.

Chapter 6 discusses the synthesis of various special nanomaterials. The coverage in this chapter is somewhat different from other chapters. Here we have also included some brief introduction to those special nanomaterials. Carbon fullerenes and nanotubes have been discussed first with a brief introduction to what are carbon fullerenes and nanotubes including their crystal structure and some physical properties. Mesoporous materials were discussed next. In this section, three types of mesoporous materials were included — ordered mesoporous materials with surfactant templating, random structured mesoporous materials and zeolites. Another group of special nanomaterials discussed in this chapter is the core-shell structures. Organic–inorganic hybrid materials and intercalation compounds have been discussed in this chapter as well.

In Chapter 7, various physical techniques for the fabrication of nanostructures are discussed. A variety of lithography methods using light, electron beams, focused ion beams, neutral atoms and X-rays were discussed first. Nanomanipulation and nanolithography were discussed with a brief introduction of scanning tunneling microscopy (STM) and atomic force microscopy (AFM). Then soft lithography for the fabrication of nanostructures was discussed.

Chapter 8 is the characterization and properties of nanomaterials. Most commonly used structural and chemical characterization methods have been reviewed in the beginning. The structural characterization methods include X-ray diffraction and small (XRD) angle X-ray scattering (SAXS), scanning and transmission electron microscopy (SEM/TEM), and various scanning probe microscopy (SPM) with emphasis on STM and AFM. Chemical characterization methods include electron spectroscopy, ion spectroscopy and optical spectroscopy. Physical properties of nanomaterials include melting points, lattice constants, mechanical properties, optical properties, electrical conduction, ferroelectrics and dielectrics and superparamagnetism.

Chapter 9 gives some examples of applications of nanostructures and nanomaterials. Examples include nanoscale and molecular electronics, catalysis of gold nanocrystals, nanobots, nanoparticles as biomolecular probes, bandgap engineered quantum devices, nanomechanics, carbon nanotube emitters, photoelectrochemical cells and photonic crystals and plasmon devices.

References

1. *Microscopy and Histology Catalog*, Polysciences, Warrington, PA, 1993–1994.
2. N. Itoh, in *Functional Thin Films and Functional Materials: New Concepts and Technologies*, ed. D.L. Shi, Tsinghua University Press and Springer-Verlag, Berlin, p. 1, 2003.
3. *National Nanotechnology Initiative 2000 Leading to the Next Industrial Revolution*, A Report by the Interagency Working Group on Nanoscience, Engineering and Technology (Washington, DC: Committee on Technology, National Science and Technology Council) http://www/nano.gov.
4. J. Ayers, in *Ceramics of the World: From 4000 BC to the Present*, eds. L. Camusso and S. Bortone, Abrams, New York, p. 284, 1992.
5. H. Zhao and Y. Ning, *Gold Bull.* **33**, 103 (2000).
6. J. Turkevich, *Gold Bull.* **18**, 86 (1985).
7. M. Faraday, *Phil. Trans.* **147**, 145 (1857).
8. J. Turkevich, *Gold Bull.* **18**, 86 (1985).
9. B.E. Deal, *Interface* **6**, 18 (1976).
10. M. Riordan and L. Hoddeson, *Crystal Fire*, W.W. Norton and Company, New York, 1997.
11. D.L. Klein, P.L. McEuen, J.E. Bowen Katari, R. Roth, and A.P. Alivisatos, *Appl. Phys. Lett.* **68**, 2574 (1996).
12. M.A. Reed, C. Zhou, C.J. Muller, T.P. Burgin, and J.M. Tour, *Science* **278**, 252 (1997).
13. R.F. Service, *Science* **293**, 782 (2001).
14. J.H. Schön, H. Meng, and Z. Bao, *Science* **294**, 2138 (2001).
15. J.D. Meindl, Q. Chen, and J.A. Davis, *Science* **293**, 2044 (2001).
16. M. Lundstrom, *Science* **299**, 210 (2003).
17. R.P. Feynman, *J. Microelectromechan. Syst.* **1**, 1 (1992).
18. C.A. Haberzettl, *Nanotechnology* **13**, R9 (2002).
19. D.L. Feldheim and C.D. Keating, *Chem. Soc. Rev.* **27**, 1 (1998).
20. F. Capasso, *Science* **235**, 172 (1987).
21. W. Krastchmer, L.D. Lamb, K. Fostiropoulos, and D.R. Huffman, *Nature* **347**, 354 (1990).
22. S. Iijima, *Nature* **354**, 56 (1991).
23. C.T. Kresge, M.E. Leonowicz, W.J. Roth, J.C. Vartulli, and J.S. Beck, *Nature* **359**, 710 (1992).
24. G. Binnig, H. Rohrer, C. Gerber, and E. Weibel, *Phys. Rev. Lett.* **49**, 57 (1982).
25. G. Binnig, C.F. Quate, and Ch. Gerber, *Phys. Rev. Lett.* **56**, 930 (1986).
26. T. Harper, *Nanotechnology* **14**, 1 (2003).
27. B. Das, S. Subramanium, and M.R. Melloch, *Semicond. Sci. Technol.* **8**, 1347 (1993).
28. C. Vieu, F. Carcenac, A. Pepin, Y. Chen, M. Mejias, L. Lebib, L. Manin Ferlazzo, L. Couraud, and H. Launois, *Appl. Surf. Sci.* **164**, 111 (2000).
29. K. Kawata, H.B. Sun, T. Tanaka, and K. Takada, *Nature* **412**, 697 (2001).
30. P. Zeppenfeld and D.M. Eigler, *New Scientist* **129**, 20 (February 23, 1991) and http://www.almaden.ibm.com/vis/stm/atomo.html.

Chapter 2

Physical Chemistry of Solid Surfaces

2.1. Introduction

Nanostructures and nanomaterials possess a large fraction of surface atoms per unit volume. The ratio of surface atoms to interior atoms changes dramatically if one successively divides a macroscopic object into smaller parts. For example, for a cube of iron of 1 cm^3, the percentage of surface atoms would be only 10^{-5}%. When the cube is divided into smaller cubes with an edge of 10 nm, the percentage of the surface atoms would increase to 10%. In a cube of iron of 1 nm^3, every atom would be a surface atom. Figure 2.1 shows the percentage of surface atoms changes with the palladium cluster diameter.[1] Such a dramatic increase in the ratio of surface atoms to interior atoms in nanostructures and nanomaterials might illustrate why changes in the size range of nanometers are expected to lead to great changes in the physical and chemical properties of the materials.

The total surface energy increases with the overall surface area, which is in turn strongly dependent on the dimension of material. Table 2.1 indicates how the specific surface area and total surface energy of 1 g of sodium chloride vary with particle size.[2] The calculation was done based on the following assumptions: surface energy of 2×10^{-5} J/cm^2 and edge energy of 3×10^{-13} J/cm, and the original 1 g cube was successively divided into smaller cubes. It should be noted that the specific surface area

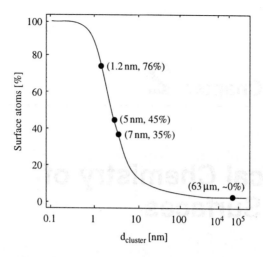

Fig. 2.1. The percentage of surface atoms changes with the palladium cluster diameter. [C. Nützenadel, A. Züttel, D. Chartouni, G. Schmid, and L. Schlapbach, *Eur. Phys. J.* **D8**, 245 (2000).]

Table 2.1. Variation of surface energy with particle size.[22]

Side (cm)	Total surface area (cm^2)	Total edge (cm)	Surface energy (J/g)	Edge energy (J/g)
0.77	3.6	9.3	7.2×10^{-5}	2.8×10^{-12}
0.1	28	550	5.6×10^{-4}	1.7×10^{-10}
0.01	280	5.5×10^4	5.6×10^{-3}	1.7×10^{-8}
0.001	2.8×10^3	5.5×10^6	5.6×10^{-2}	1.7×10^{-6}
10^{-4} (1 μm)	2.8×10^4	5.5×10^8	0.56	1.7×10^{-4}
10^{-7} (1 nm)	2.8×10^7	5.5×10^{14}	560	170

and, thus, the total surface energy are negligible when cubes are large, but become significant for very small particles. When the particles change from centimeter size to nanometer size, the surface area and the surface energy increase seven orders of magnitude.

Due to the vast surface area, all nanostructured materials possess a huge surface energy and, thus, are thermodynamically unstable or metastable. One of the great challenges in fabrication and processing of nanomaterials is to overcome the surface energy, and to prevent the nanostructures or nanomaterials from growth in size, driven by the reduction of overall surface energy. In order to produce and stabilize nanostructures and nanomaterials, it is essential to have a good understanding of surface energy and surface physical chemistry of solid surfaces. In this chapter, the origin of

Physical Chemistry of Solid Surfaces

the surface energy will be reviewed first, followed with detailed discussion of the possible mechanisms for a system or material to reduce the overall surface energy. Then attention will be focused on the chemical potentials as a function of surface curvature and its implications. Finally, two mechanisms to prevent the agglomeration of nanomaterials will be discussed.

2.2. Surface Energy

Atoms or molecules on a solid surface possess fewer nearest neighbors or coordination numbers, and thus have dangling or unsatisfied bonds exposed to the surface. Because of the dangling bonds on the surface, surface atoms or molecules are under an inwardly directed force and the bond distance between the surface atoms or molecules and the sub-surface atoms or molecules, is smaller than that between interior atoms or molecules. When solid particles are very small, such a decrease in bond length between the surface atoms and interior atoms becomes significant and the lattice constants of the entire solid particles show an appreciable reduction.[3] The extra energy possessed by the surface atoms is described as surface energy, surface free energy or surface tension. Surface energy, γ, by definition, is the energy required to create a unit area of "new" surface:

$$\gamma = \left(\frac{\partial G}{\partial A}\right)_{n_i,T,P} \tag{2.1}$$

where A is the surface area. Let us consider separating a rectangular solid material into two pieces as illustrated in Fig. 2.2. On the newly created surfaces, each atom is located in an asymmetric environment and will move towards the interior due to breaking of bonds at the surface. An extra force is required to pull the surface atoms back to its original position. Such a surface is ideal and also called singular surface. For each atom on such a singular surface, the energy required to get it back to its original position will be equal to the number of broken bonds, N_b, multiplying by

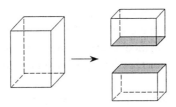

Fig. 2.2. Schematic showing two new surfaces being created by breaking a rectangular solid into two pieces.

half of the bond strength, ε. Therefore, the surface energy is given by:

$$\gamma = \frac{1}{2} N_b \varepsilon \rho_a \qquad (2.2)$$

where ρ_a is the surface atomic density, the number of atoms per unit area on the new surface. This crude model ignores interactions owing to higher order neighbors, assumes that the value of ε is the same for surface and bulk atoms, and does not include entropic or pressure–volume contributions. This relation only gives a rough estimation of the true surface energy of a solid surface, and is only applicable to solids with rigid structure where no surface relaxation occurs. When there is an appreciable surface relaxation, such as the surface atoms moving inwardly, or there is a surface restructuring, surface energy will be lower than that estimated by the above equation. In spite of the overly simplified assumptions used in Eq. (2.2), it does provide some general guidance. Let us take an elemental crystal with a face-centered cubic (FCC) structure having a lattice constant of a, as an example to illustrate the surface energy on various facets. Each atom in such a FCC crystal has a coordination number of 12. Each surface atom on {100} facets would have four broken chemical bonds, and the surface energy of {100} surface can be calculated using Eq. (2.2) and Fig. 2.3A:

$$\gamma_{\{100\}} = \frac{1}{2} \frac{2}{a^2} \cdot 4 \cdot \varepsilon = \frac{4\varepsilon}{a^2} \qquad (2.3)$$

Similarly, each atom on {110} surface has 5 broken chemical bonds and {111} has 3. The surface energies of {110} and {111} surfaces are given, calculating from Figs. 2.3B and 2.3C:

$$\gamma_{\{110\}} = \frac{5}{\sqrt{2}} \cdot \frac{\varepsilon}{a^2} \qquad (2.4)$$

$$\gamma_{\{111\}} = 2\sqrt{3} \cdot \frac{\varepsilon}{a^2} \qquad (2.5)$$

The readers can easily figure out the fact that low index facets have low surface energy according to Eq. (2.2). Thermodynamics tells us that any material or system is stable only when it is in a state with the lowest Gibbs free energy. Therefore, there is a strong tendency for a solid or a liquid to minimize the total surface energy. There are a variety of mechanisms to reduce the overall surface energy. The various mechanisms can be grouped into atomic or surface level, individual structures and the overall system.

For a given surface with a fixed surface area, the surface energy can be reduced through (i) surface relaxation, the surface atoms or ions shift inwardly which occur more readily in liquid phase than in solid surface due

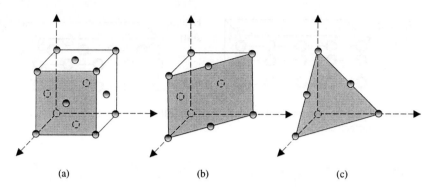

Fig. 2.3. Schematic representing low index faces of a face-centered cubic (fcc) crystal structure: (a) {100}, (b) {110} and (c) {111}.

to rigid structure in solids, (ii) surface restructuring through combining surface dangling bonds into strained new chemical bonds, (iii) surface adsorption through chemical or physical adsorption of terminal chemical species onto the surface by forming chemical bonds or weak attraction forces such as electrostatic or van der Waals forces, and (iv) composition segregation or impurity enrichment on the surface through solid-state diffusion.

Let us take the surface atoms on an atomic flat {100} surface as an example, assuming the crystal has a simple cubic structure and each atom has a coordination number of six. The surface atoms are linked with one atom directly beneath and four other surrounding surface atoms. It is reasonable to consider each chemical bond acting as an attractive force; all the surface atoms are under the influence of a net force pointing inwardly and perpendicular to the surface. Understandably, under such a force, the distance between the surface atomic layer and the subsurface atomic layer would be smaller than that inside the bulk, though the structure of the surface atomic layer remains unchanged. In addition, the distance between the atomic layers under the surface would also be reduced. Such surface relaxation has been well established.[4–7] Furthermore, the surface atoms may also shift laterally relative to the subsurface atomic layer. Figure 2.4 schematically depicts such surface atomic shift or relaxation. For bulk materials, such a reduction in the lattice dimension is too small to exhibit any appreciable influence on the overall crystal lattice constant and, therefore, can be ignored. However, such an inward or lateral shift of surface atoms would result in a reduction of the surface energy. Such a surface relaxation becomes more pronounced in less rigid crystals, and can result in a noticeable reduction of bond length in nanoparticles.[3]

If a surface atom has more than one broken bonds, surface restructuring is a possible mechanism to reduce the surface energy.[8–11] The broken

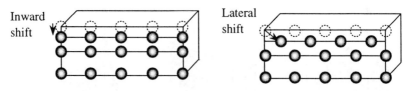

Fig. 2.4. Schematic showing surface atoms shifting either inwardly or laterally so as to reduce the surface energy.

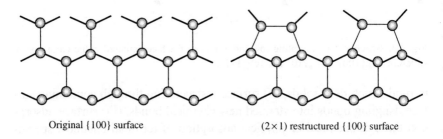

Original {100} surface   (2×1) restructured {100} surface

Fig. 2.5. Schematic illustrating the (2 × 1) restructure of silicon {100} surface.

bonds from neighboring surface atoms combine to form a highly strained bond. For example, such surface restructuring is found in the {100} surface of silicon crystals.[12] Surface energy of {100} faces in diamond and silicon crystals before restructuring is higher than of both {111} and {110} faces. However, restructured {100} faces have the lowest surface energy among three low indices faces,[13–15] and such surface restructuring can have a significant impact on the crystal growth.[16–19] Figure 2.5 shows the original {100} surface and 2 × 1 restructured {100} surface of diamond crystal.

Another way to reduce the surface energy is chemical and physical adsorption on solid surfaces, which can effectively lower the surface energy.[20–23] For example, the surface of diamond is terminated with hydrogen and that of silicon is covered with hydroxyl groups before restructuring as schematically shown in Fig. 2.6. These are considered as chemical adsorption. Yet another approach to reduce the surface energy is composition segregation or enrichment of impurities on the surfaces. Although composition segregation, such as enrichment of surfactants on the surface of a liquid is an effective way to reduce the surface energy, it is not common in a solid surface. In bulk solids, composition segregation is not significant, since the activation energy required for solid-state diffusion is high and the diffusion distance is large. In nanostructures and nanomaterials, however, phase segregation may play a significant role in the reduction of surface energy, considering the great impact of surface energy and the

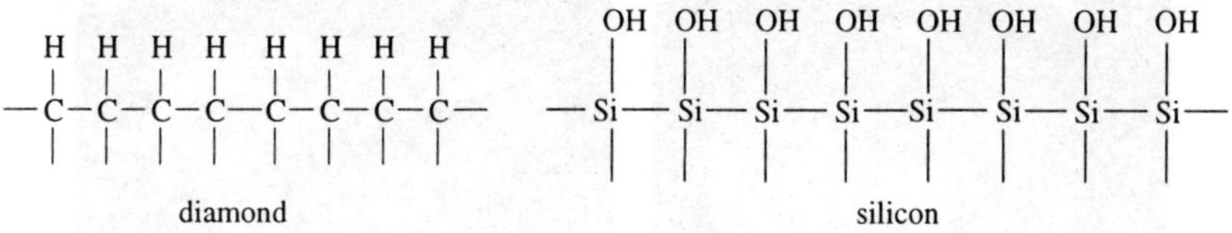

```
    H  H  H  H  H  H  H  H              OH  OH  OH  OH  OH  OH  OH  OH
    |  |  |  |  |  |  |  |              |   |   |   |   |   |   |   |
   —C—C—C—C—C—C—C—C—          —Si— Si— Si— Si—Si— Si— Si— Si—
    |  |  |  |  |  |  |  |              |   |   |   |   |   |   |   |
              diamond                              silicon
```

Fig. 2.6. Schematic showing the surface of diamond is covered with hydrogen and that of silicon is covered with hydroxyl groups through chemisorption before restructuring.

short diffusion distance. Although there is no direct experimental evidence to show the impact of composition segregation on the reduction of surface energy in nanostructured materials, the difficulty in doping nanomaterials and the ease in getting near perfect crystal structure in nanomaterials are indicative that the impurities and defects are readily to be repelled from the interior to the surface of nanostructures and nanomaterials.

At the individual nanostructure level, there are two approaches to the reduction of the total surface energy. One is to reduce the overall surface area, assuming the material is entirely isotropic. Water on a hydrophobic surface always balls up and forms a spherical droplet in free form to minimize the overall surface area. The same is also found for a glass. When heating a piece of glass to temperatures above its glass transition point, sharp corners will round up. For liquid and amorphous solids, they have isotropic microstructure and, thus, isotropic surface energy. For such materials, reduction of the overall surface area is the way to reduce the overall surface energy. However, for a crystalline solid, different crystal facets possess different surface energy. Therefore, a crystalline particle normally forms facets, instead of having a spherical shape, which in general, possesses a surface energy higher than a faceted particle. The thermodynamically equilibrium shape of a given crystal can be determined by considering the surface energies of all facets, since there is a minimal surface energy when a group of surfaces is combined in a certain pattern.

In spite of the overly simplified assumptions used in the derivation of Eq. (2.2), one can use it to estimate the surface energy of various facets of a given crystal. For example, {111} surfaces in a monatomic FCC crystal have the lowest surface energy, followed by {110} and {100}. It is also easy to find that crystal surfaces with low Miller indices in general have a lower surface energy than that with high Miller indices. It does explain why a crystal is often surrounded by low index surfaces. Figure 2.7 gives some typical images of crystals with equilibrium facets.

Wulff plot is often used to determine the shape or the surfaces of an equilibrium crystal.[24,25] For an equilibrium crystal, i.e. the total surface energy reaches minimum, there exists a point in the interior such that its

Fig. 2.7. Examples of single crystals with thermodynamic equilibrium shape. (Top-left) Sodium chloride, (top-right) silver, (bottom-left) silver, and (bottom-right) gold. Gold particles are formed at 1000°C and some facets have gone through roughening transition.

perpendicular distance, h_i, from the ith face is proportional to the surface energy, γ_i:

$$\gamma_i = C h_i \qquad (2.6)$$

where C is a constant. For a given crystal, C is the same for all the surfaces. A Wulff plot can be constructed with the following steps:

(1) Given a set of surface energies for the various crystal faces, draw a set of vectors from a common point of length proportional to the surface energy and direction normal to that the crystal face.
(2) Construct the set of faces normal to each vectors and positioned at its end, and
(3) Find a geometric figure whose sides are made up entirely from a particular set of such faces that do not interest any of the other planes.

Figure 2.8 gives a conformation for a hypothetical two-dimensional crystal to illustrate how the equilibrium shape of a crystal is obtained using the Wulff construction described above.[2] It should be reemphasized that the geometric figure determined by the Wulff plot represents the ideal situation, i.e. the crystal reaches the minimal surface energy level

Physical Chemistry of Solid Surfaces

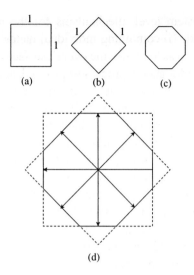

Fig. 2.8. Conformation for a hypothetical two-dimensional crystal. (a) (10) plane, (b) (11) plane, (c) shape given by the Wulff construction, and (d) Wulff construction considering only (10) and (11) planes. [A.W. Adamson and A.P. Gast, *Physical Chemistry of Surfaces*, 6th edition, John Wiley & Sons, New York, 1997.]

thermodynamically. In practice, however, geometric figure of a crystal is also determined by kinetic factors, which in turn are dependent on the processing or crystal growth conditions. The kinetic factors explain the fact that for the same crystal, different morphologies are obtained when the processing conditions are changed.[26]

Furthermore, it should be noted that not all crystals grown under equilibrium conditions form equilibrium facets as predicted by Wulff plots. The equilibrium crystal surfaces may not be smooth, and difference in surface energy of various crystal facets may disappear.[27] Such a transition is called surface roughening or roughening transition. Below roughening temperature, a crystal is faceted. Above the roughening temperature, the thermal motion predominates and the difference in surface energy among various crystal facets becomes negligible. As a result, a crystal does not form facets above the roughening temperature. Such a physical property can be understood by considering such a solid surface above the roughening temperature as a liquid surface.[28] Crystals grown at temperatures above the surface roughening temperature do not form facets. Examples include silicon crystals grown by Czochraski method.[29,30] Kinetic factors may also prevent the formation of facets. As will be seen in the next chapter, most nanoparticles grown by solution methods at elevated temperatures are spherical in shape and do not form any facets.

At the overall system level, mechanisms for the reduction of overall surface energy include (i) combining individual nanostructures together to form large structures so as to reduce the overall surface area, if large enough activation is available for such a process to proceed, and (ii) agglomeration of individual nanostructures without altering the individual nanostructures. Specific mechanisms of combining individual nanostructures into large structures include (i) sintering, in which individual structures merge together and (ii) Ostwald ripening, in which relatively large structures grow at the expense of smaller ones. In general, sintering is negligible at low temperatures including room temperature, and becomes important only when materials are heated to elevated temperatures, typically 70% of the melting point of the material in question. Ostwald ripening occurs at a wide range of temperatures, and proceeds at relatively low temperatures when nanostructures are dispersed and have an appreciable solubility in a solvent.

Sintering is a process that must be prevented in the fabrication and processing of nanomaterials. Fortunately, sintering becomes significant only at high temperatures. However, considering the small dimensions of nanomaterials and, thus, the extremely high surface energy, sintering can become a serious issue when nanomaterials are brought to moderate temperatures. Sintering is a complex process and involves solid-state diffusion, evaporation–condensation or dissolution–precipitation, viscous flow and dislocation creep. Solid-state diffusion can be further divided into three categories: surface diffusion, volume diffusion and cross grain-boundary diffusion. Surface diffusion requires the smallest activation energy, and thus is a predominant process at relatively low temperatures, whereas cross grain boundary diffusion demands the highest activation energy and, thus, becomes significant only at high temperatures. At moderate temperatures, volume diffusion dominates the sintering process, resulting in densification and removal of pores in bulk materials. Although three solid-state diffusion processes result in markedly different microstructures, they all result in a reduction of overall surface or interface energy. Evaporation–condensation is important when nanomaterials have an appreciable vapor pressure at the processing temperature. Dissolution–precipitation occurs when the solid is dispersed in a liquid in which the solid is partially soluble. Viscous flow occurs when the material is amorphous and the temperature is above the glass transition point. Creep dislocation is important particularly when the material is under a mechanical stress. To preserve nanostructures during the synthesis and processing of nanomaterials and for various practical applications of nanomaterials, sintering must be avoided. A variety of mechanisms have been explored to promote sintering by the ceramic and powder metallurgy research community. A simple reverse engineering of sintering process may offer many possible approaches to prevent nanomaterials from sintering. For detailed discussion and further information on sintering, the readers are suggested to consult ceramic processing and powder metallurgy books.[31–33]

In general, sintering can be considered as a process to replace solid–vapor surface by solid–solid interface through reshaping the nanostructures in such a way that individual nanostructures are packed such that there is no gap among solid nanostructures. Ostwald ripening takes a radically different approach, in which two individual nanostructures become a single one. A large one grows at the expense of the smaller one until the latter disappears completely. Details of Ostwald ripening will be discussed further in the next section. The product of sintering is a polycrystalline material, whereas Ostwald ripening results in a single uniform structure. Figure 2.9 shows schematically the two different processes, though the results of both processes are similar, i.e. a reduction of total surface energy. Macroscopically, the reduction of total surface energy is the driving force for both sintering and Ostwald ripening. Microscopically, the differential surface energy of surfaces with different surface curvature is the true driving force for the mass transport during sintering or Ostwald ripening. In the next section, we will discuss the dependence of chemical potential on the surface curvature.

In addition to combining the individual nanostructures together to form large structures through sintering or Ostwald ripening, agglomeration is another way to reduce the overall surface energy. In agglomerates, many nanostructures are associated with one another through chemical bonds and physical attraction forces at interfaces. Once formed, agglomerates are very difficult to destroy. The smaller the individual nanostructures are, the stronger they are associated with one another, and the more difficult to separate. For practical applications of nanomaterials, the formation

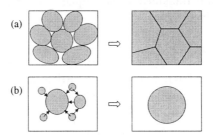

Fig. 2.9. Schematic showing sintering and Ostwald ripening processes. (a) Sintering is to combine individual particles to a bulk with solid interfaces to connect each other (b) Ostwald ripening is to merge smaller particles into a larger particle. Both processes reduce the solid–gas surface area.

of agglomerates should be prevented. Later in this chapter, two common methods of preventing the formation of agglomerates are discussed in detail.

So far, we have discussed the origin of surface energy and various possible mechanisms for a system to minimize its overall surface energy. In the next section, we will discuss the influences of surface curvature on surface energy. It will become clear that for a given material, concave surfaces have much lower surface energy than convex surfaces. Such differences are reflected in their respective equilibrium vapor pressure and solubility, and thus their stabilities.

2.3. Chemical Potential as a Function of Surface Curvature

As discussed in previous sections, the properties of surface atoms or molecules are different from that of interior atoms or molecules, due to fewer bonds linking to their nearest neighbor atoms or molecules as compared with their interior counterparty. Further, the chemical potential is also dependent on the radius of curvature of a surface. To understand the relationship between chemical potential and surface curvature, let us consider transferring material from an infinite flat surface to a spherical solid particle as illustrated in Fig. 2.10. As a result of transferring of dn atoms from a flat solid surface to a particle with a radius of R, the volume change of spherical particle, dV, is equal to the atomic volume, Ω, times dn, that is:

$$dV = 4\pi R^2 \, dR = \Omega \, dn \qquad (2.7)$$

The work per atom transferred, $\Delta\mu$, equals to the change of chemical potential, and is given by:

$$\Delta\mu = \mu_c - \mu_\infty = \gamma \frac{dA}{dn} = \gamma \, 8\pi R \, dR \, \frac{\Omega}{dV} \qquad (2.8)$$

where μ_c is the chemical potential on the particle surface, whereas μ_∞ is the chemical potential on the flat surface. Combining with Eq. (2.7), we have

$$\Delta\mu = 2\gamma \frac{\Omega}{R} \qquad (2.9)$$

This equation is also known as Young–Laplace equation, and describes the chemical potential of an atom in a spherical surface with respect to a flat reference surface. This equation can be readily generalized for any type of

Physical Chemistry of Solid Surfaces

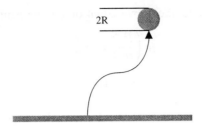

Fig. 2.10. Transport of *n* atoms from the flat surface of a semi-infinite reference solid to the curved surface of a solid sphere.

curved surfaces. It is known[34] that any curved surface can be described by two principal radii of curvature, R_1 and R_2, so we have:

$$\Delta\mu = \gamma \Omega \left(\frac{1}{R_1} + \frac{1}{R_2}\right) \tag{2.10}$$

For a convex surface, the curvature is positive, and thus the chemical potential of an atom on such a surface is higher than that on a flat surface. Mass transfer from a flat surface to a convex surface results in an increase in surface chemical potential. It is obvious that when mass is transferred from a flat surface to a concave surface, the chemical potential decreases. Thermodynamically, an atom on a convex surface possesses the highest chemical potential, whereas an atom on a concave surface has the lowest chemical potential. Such a relationship is also reflected by the difference in vapor pressure and solubility of a solid. Assuming the vapor of solid phase obeys the ideal gas law, for the flat surface one can easily arrive at:

$$\mu_v - \mu_\infty = -kT \ln P_\infty \tag{2.11}$$

where μ_v is the chemical potential of a vapor atom, k, the Boltzmann constant, P_∞, the equilibrium vapor pressure of flat solid surface, and T, temperature. Similarly, for a curved surface we have:

$$\mu_v - \mu_c = -kT \ln P_c \tag{2.12}$$

where P_c is the equilibrium vapor pressure of the curved solid surface. Combining Eqs. (2.11) and (2.12), we have:

$$\mu_c - \mu_\infty = \Delta\mu = kT \ln\left(\frac{P_c}{P_\infty}\right) \tag{2.13}$$

Combining with Eq. 2.10 and rearranging it, we have:

$$\ln\left(\frac{P_c}{P_\infty}\right) = \gamma \Omega \frac{R_1^{-1} + R_2^{-1}}{kT} \tag{2.14}$$

For a spherical particle, the above equation can be simplified as:

$$\ln\left(\frac{P_c}{P_\infty}\right) = \frac{2\gamma\Omega}{kRT} \qquad (2.15)$$

The above equation is also generally and commonly referred to as the Kelvin equation and has been verified experimentally.[35,36] The same relation can be derived for the dependence of the solubility on surface curvature:

$$\ln\left(\frac{S_c}{S_\infty}\right) = \gamma\Omega\,\frac{R_1^{-1}+R_2^{-1}}{kT} \qquad (2.16)$$

where S_c is the solubility of a curved solid surface, S_∞ is the solubility of a flat surface. This equation is also known as the Gibbs–Thompson relation.[37] Figure 2.11 shows the dependence of solubility of silica as a

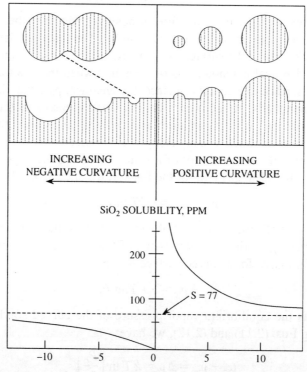

Fig. 2.11. Variation in solubility of silica with radius of curvature of surface. The positive radii of curvature are shown in cross-section as particles and projections from a planar surface; negative radii are shown as depressions or holes in the surface, and in the crevice between two particles. [R.K. Iler, *The Chemistry of Silica*, Wiley, New York, 1979.]

Physical Chemistry of Solid Surfaces

function of surface curvature.[38] The vapor pressure of small particles is notably higher than that of the bulk material[39–42] and Fig. 2.12 shows the vapor pressure of a number of liquids as a function of droplet radius.[41]

When two particles with different radii, assuming $R_1 \gg R_2$, are put into a solvent, each particle will develop an equilibrium with the surrounding solvent. According to Eq. (2.16), solubility of the smaller particle will be larger than that of the larger particle. Consequently, there would be a net diffusion of solute from proximity of the small particle to proximity of the large particle. To maintain the equilibrium, solute will deposit onto the surface of the large particle, whereas the small particle has to continue dissolving so as to compensate for the amount of solute diffused away. As a result, the small particle gets smaller, whereas the large particle gets

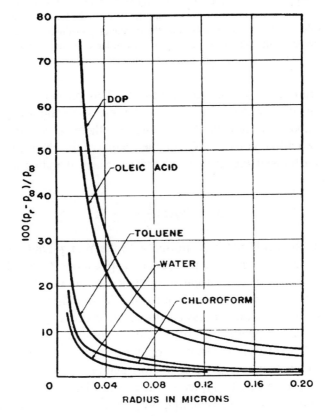

Fig. 2.12. Vapor pressure of a number of liquids as a function of droplet radius. [V.K. La Mer and R. Gruen, *Trans. Faraday Soc.* **48**, 410 (1952).]

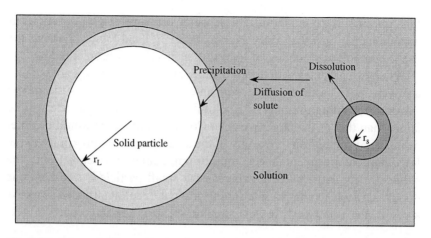

Fig. 2.13. Schematic illustrating the Ostwald ripening processing. Smaller particle has a larger solubility or vapor pressure due to its larger curvature, whereas the larger particle possesses a smaller solubility or vapor pressure. To maintain the local concentration equilibrium, smaller particle would dissolve into the surrounding medium; solute at proximity of smaller particle diffuses away; solute at proximity of larger particle would deposit. The process would continue till disappearance of the smaller particle.

larger. Figure 2.13 depicts such a process. This phenomenon is called Ostwald ripening, which occurs also in the forms of solid-state diffusion and of evaporation–condensation. Assuming there is no other change between two different particles, then the change of the chemical potential of an atom transferring from a spherical surface of radius R_1 to R_2 is given by:

$$\Delta\mu = 2\gamma\Omega \left(\frac{1}{R_1} - \frac{1}{R_2}\right) \quad (2.17)$$

This equation should not be confused with the Young–Laplace equation [Eq. (2.9)]. Depending on the process and applications, Ostwald ripening can have either positive or negative influence on the resulting materials. Ostwald ripening can either widen or narrow the size distribution, depending on the control of the process conditions. In processing of many materials, Ostwald ripening is often not desirable. In sintering of polycrystalline materials, Ostwald ripening results in abnormal grain growth, leading to inhomogeneous microstructure and inferior mechanical properties of the products. Typically one or a few large grains grow at the expense of a number of surrounding small grains, resulting in inhomogeneous microstructure. However, Ostwald ripening has been explored in the synthesis of nanoparticles. More specifically, Ostwald ripening has been used to narrow the size distribution of nanoparticles by eliminating small particles. The situation here is very different. Many relatively large particles grow at the expense of a relatively small number of smaller particles. The result is the elimination of smaller particles, and thus the size distribution of nanoparticles becomes narrower. Ostwald ripening can be promoted by varying processing temperatures. In the synthesis of nanoparticles from solution, after the initial nucleation and subsequent growth, the temperature is raised, and thus the solubility of solid in solvent increases to promote Ostwald ripening. As a result, the concentration of solid in solvent falls below the equilibrium solubility of small nanoparticles, and the small particles dissolve into the solvent. As dissolution of a nanoparticle proceeds, the nanoparticle becomes smaller and has higher solubility. It is clear that once a nanoparticle starts dissolving into the solvent, the dissolution process stops only when the nanoparticle is dissolved completely. On the other hand, the concentration of solid in solvent is still higher than the equilibrium solubility of larger particles and, thus, these large particles would continue to grow. Such a growth process would stop when the concentration of solid in the solvent equals the equilibrium solubility of these relatively large nanoparticles.

The reduction of overall surface energy is the driving force for the surface restructuring, formation of faceted crystals, sintering and Ostwald ripening. These are the reduction mechanisms for individual surface, individual nanostructures and the overall system. The system can have another mechanism to reduce the overall surface energy, in addition to sintering and Ostwald ripening. This is agglomeration. When small nanostructures form agglomerates, it is very difficult to disperse them. In nanostructure fabrication and processing, it is very important to overcome the huge total surface energy to create the desired nanostructures. It is equally important to prevent the nanostructures from agglomeration. As the dimension of nanostructured materials reduces, van der Waals attraction force between nanostructured materials becomes increasingly important. Without appropriate stabilization mechanisms applied, the nanostructured materials are most likely and readily to form agglomerates. The following sections are devoted to the stabilization mechanisms for the prevention of agglomeration of individual nanostructures. Although the discussion will be focused mainly on nanoparticles, the same principles are applicable to other nanostructures, such as nanorods and nanofibrils. There are two major stabilization mechanisms widely used: electrostatic stabilization and steric stabilization. Two mechanisms have some distinct differences. For example, a system using electrostatic stabilization is kinetically stable, whereas steric stabilization makes the system thermodynamically stable.

2.4. Electrostatic Stabilization

2.4.1. *Surface charge density*

When a solid emerges in a polar solvent or an electrolyte solution, a surface charge will be developed through one or more of the following mechanisms:

(1) Preferential adsorption of ions
(2) Dissociation of surface charged species
(3) Isomorphic substitution of ions
(4) Accumulation or depletion of electrons at the surface
(5) Physical adsorption of charged species onto the surface.

For a given solid surface in a given liquid medium, a fixed surface electrical charge density or electrode potential, E, will be established, which is given by the Nernst equation:

$$E = E_o + \frac{R_g T}{n_i F} \ln a_i \quad (2.18)$$

where E_o is the standard electrode potential when the concentration of ions is unity, n_i is the valence state of ions, a_i is the activity of ions, R_g is the gas constant and T is temperature, and F is the Faraday's constant. Equation (2.18) clearly indicates that the surface potential of a solid varies with the concentration of the ions in the surrounding solution, and can be either positive or negative. Electrochemistry of metals will be discussed further in Sec. 4.3.1 in Chapter 4. The focus of the discussion here will be on non-conductive materials or dielectrics, more specifically on oxides.

The surface charge in oxides is mainly derived from preferential dissolution or deposition of ions. Ions adsorbed on the solid surface determine the surface charge, and thus are referred to as charge determining ions, also known as co-ions or coions. In the oxide systems, typical charge determining ions are protons and hydroxyl groups and their concentrations are described by pH (pH = $-$log [H$^+$]). As the concentration of charge determining ions varies, the surface charge density changes from positive to negative or vice versa. The concentration of charge determining ions corresponding to a neutral or zero-charged surface is defined as a point of zero charge (p.z.c.) or zero-point charge (z.p.c.). For the sake of clarity and consistence, in the rest of this book, we will use the term of point of zero charge or p.z.c. only. Table 2.2 gives a list of some p.z.c. values of selected oxides.[43–45] At pH > p.z.c., the oxide surface is negatively charged, since the surface is covered with hydroxyl groups, OH$^-$, which

Table 2.2. A list of p.z.c. of some common oxides in water.[45]

Solids	p.z.c.
WO_3	0.5
V_2O_5	1–2
δ-MnO_2	1.5
β-MnO_2	7.3
SiO_2	2.5
SiO_2 (quartz)	3.7
TiO_2	6
TiO_2 (calcined)	3.2
SnO_2	4.5
Al-O-Si	6
ZrO_2	6.7
FeOOH	6.7
Fe_2O_3	8.6
ZnO	8
Cr_2O_3	8.4
Al_2O_3	9
MgO	12

is the electrical determining ion. At pH < p.z.c., H^+ is the charge determining ions and the surface is positively charged. The surface charge density or surface potential, E in volt, can then be simply related to the pH and the Nernst equation [Eq. (2.18)] can be written as[45]:

$$E = \frac{2.303\ R_g T\ [(\text{p.z.c.}) - \text{pH}]}{F} \quad (2.19)$$

At room temperature, the above equation can be further simplified:

$$E \approx 0.06\ [(\text{p.z.c.}) - \text{pH}] \quad (2.20)$$

2.4.2. *Electric potential at the proximity of solid surface*

When a surface charge density of a solid surface is established, there will be an electrostatic force between the solid surface and the charged species in the proximity to segregate positive and negatively charged species. However, there also exist Brownian motion and entropic force, which homogenize the distribution of various species in the solution. In the solution, there always exist both surface charge determining ions and counter ions, which have charge opposite to that of the determining ions. Although charge neutrality is maintained in a system, distributions of charge determining ions and counter ions in the proximity of the solid surface are inhomogeneous and very different. The distributions of both ions are mainly controlled by a combination of the following forces:

(1) Coulombic force or electrostatic force,
(2) Entropic force or dispersion,
(3) Brownian motion.

The combined result is that the concentration of counter ions is the highest near the solid surface and decreases as the distance from the surface increases, whereas the concentration of determining ions changes in the opposite manner. Such inhomogeneous distributions of ions in the proximity of the solid surface lead to the formation of so-called double layer structure, which is schematically illustrated in Fig. 2.14. The double layer consists of two layers, Stern layer and Gouy layer (also called diffuse double layer), and the two layers are separated by the Helmholtz plane.[46] Between the solid surface and the Helmholtz plane is the Stern layer, where the electric potential drops linearly through the tightly bound layer of solvent and counter ions. Beyond the Helmholtz plane until the counter ions reach average concentration in the solvent is the Gouy layer or diffuse double layer. In the Gouy layer, the counter ions diffuse freely and the

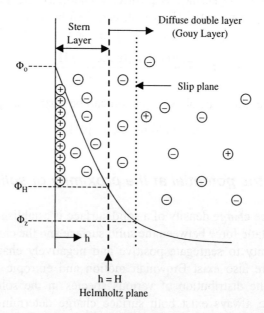

Fig. 2.14. Schematic illustrating electrical double layer structure and the electric potential near the solid surface with both Stern and Gouy layers indicated. Surface charge is assumed to be positive.

electric potential does not reduce linearly. The electric potential drops approximately following:

$$E \propto e^{-\kappa(h-H)} \quad (2.21)$$

where $h \geq H$, which is the thickness of the Stern layer, $1/\kappa$ is known as the Debye–Hückel screening strength and is also used to describe the thickness of double layer, and κ is given by

$$\kappa = \sqrt{\frac{F^2 \Sigma_i C_i Z_i^2}{\varepsilon_r \varepsilon_0 R_g T}} \quad (2.22)$$

where F is Faraday's constant, ε_0 is the permittivity of vacuum, ε_r is the dielectric constant of the solvent, and C_i and Z_i are the concentration and valence of the counter ions of type i. This equation clearly indicates that the electric potential at the proximity of solid surface decreases with increased concentration and valence state of counter ions, and increases with an increased dielectric constant of the solvent exponentially. Higher concentration and valence state of counter ions would result in a reduced thickness of both Stern layer and Gouy layer.[47,48] In theory, the Gouy diffusion layer would end at a point where the electric potential reaches zero, which would be the case only when the distance from the solid surface is infinite. However, in practice, double layer thickness is typically of approximately 10 nm or larger.

Although the above discussion has been focused on a flat solid surface in an electrolyte solution, the concepts are applicable to curved surfaces as well, assuming that the surface is smooth and thus the surface charge is distributed uniformly. For a smooth curved surface, the surface charge density is constant, so that the electric potential in the surrounding solution can be described using Eqs. (2.21) and (2.22). Such assumptions are certainly valid for spherical particles, when particles are dispersed in an electrolyte solution and the distance between any two particles are large enough so that the charge distribution on particle surface is not influenced by other particles. Interactions between particles are complex. One of the interactions between particles is directly associated with the surface charge and the electric potential adjacent to the interface. The electrostatic repulsion between two particles arises from the electric surface charges, which are attenuated to a varied extent by the double layers. When two particles are far apart, there will be no overlap of two double layers and electrostatic repulsion between two particles is zero. However, when two particles approach one another, double layer overlaps and a repulsive force develops. An electrostatic repulsion between two equally sized spherical particles of radius r, and separated by a distance S, is given by[46]:

$$\Phi_R = 2\pi \, \varepsilon_r \varepsilon_0 \, rE^2 \, e^{-\kappa S} \quad (2.23)$$

2.4.3. Van der Waals attraction potential

When particles are small, typically in micrometers or less, and are dispersed in a solvent, van der Waals attraction force and Brownian motion play important roles, whereas the influence of gravity becomes negligible. For the sake of simplicity, we will refer these particles to as nanoparticles, though particles in micrometer size behave the same and are also included in the discussion here. Furthermore, we will limit our discussion on spherical nanoparticles. Van der Waals force is a weak force and becomes significant only at a very short distance. Brownian motion ensures that the nanoparticles are colliding with each other all the time. The combination of van der Waals attraction force and Brownian motion would result in the formation of agglomeration of the nanoparticles.

Van der Waals interaction between two nanoparticles is the sum of the molecular interaction for all pairs of molecules composed of one molecule in each particle, as well as to all pairs of molecules with one molecule in a particle and one in the surrounding medium such as solvent. Integration of all the van der Waals interactions between two molecules over two spherical particles of radius, r, separated by a distance, S, as illustrated in Fig. 2.15 gives the total interaction energy or attraction potential[46]:

$$\Phi_A = -A/6 \left\{ \frac{2r^2}{S^2 + 4rS} + \frac{2r^2}{S^2 + 4rS + 4r^2} + \ln\left(\frac{S^2 + 4rS}{S^2 + 4rS + 4r^2}\right) \right\} \quad (2.24)$$

where the negative sign represents the attraction nature of the interaction between two particles, and A is a positive constant termed the Hamaker constant, which has a magnitude on the order of 10^{-19} to 10^{-20} J, and depends on the polarization properties of the molecules in the two particles and in the medium which separates them. Table 2.3 listed some Hamaker constants for a few common materials.[45] Equation (2.24) can be simplified under various boundary conditions. For example, when the separation distance between two equal sized spherical particles are significantly smaller than the particle radius, i.e. $S/r \ll 1$, the simplest expression of

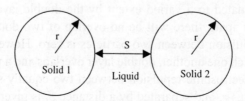

Fig. 2.15. Pair of particles used to derive the van der Waals interaction.

Table 2.3. Hamaker constants for some common materials.[45]

Materials	$A_i(10^{-20}$ J)
Metals	16.2–45.5
Gold	45.3
Oxides	10.5–15.5
Al_2O_3	15.4
MgO	10.5
SiO_2 (fused)	6.5
SiO_2 (quartz)	8.8
Ionic crystals	6.3–15.3
CaF_2	7.2
Calcite	10.1
Polymers	6.15–6.6
Polyvinyl chloride	10.82
Polyethylene oxide	7.51
Water	4.35
Acetone	4.20
Carbon tetrachloride	4.78
Chlorobenzene	5.89
Ethyl acetate	4.17
Hexane	4.32
Toluene	5.40

the van der Waals attraction could be obtained:

$$\Phi_A = \frac{-A\ r}{12\ S} \tag{2.25}$$

Other simplified expressions of the van der Waals attraction potential are summarized in Table 2.4.[46] From this table, it is noticed that the van der Waals attraction potential between two particles are different from that between two flat surfaces. Furthermore, it should be noted that the interaction between two molecules are significantly different from that between two particles. Van der Waals interaction energy between two molecules can be simply represented by:

$$\Phi_A \propto -S^{-6} \tag{2.26}$$

Although the nature of the attraction energy between two particles is the same as that between two molecules, integration of all the interaction between molecules from two particles and from medium results in a totally different dependence of force on distance. The attraction force between two particles decay much slowly and extends over distances of nanometers. As a result, a barrier potential must be developed to prevent agglomeration. Two methods are widely applied to prevent agglomeration of particles: electrostatic repulsion and steric exclusion.

Table 2.4. Simple formulas for the van der Waals attraction between two particles.[46]

Particles	Φ_A
Two spheres of equal radius, r*	$-Ar/12S$
Two spheres of unequal radii, r_1 and r_2*	$-A\,r_1 r_2/6S(r_1+r_2)$
Two parallel plates with thickness of δ, interaction per unit area	$-A/12\pi[S^{-2}+(2\delta+S)^{-2}+(\delta+S)^{-2}]$
Two blocks, interaction per unit area	$-A/12\pi S^2$

* r, r_1 and $r_2 \gg S$

2.4.4. Interactions between two particles: DLVO theory

The total interaction between two particles, which are electrostatic stabilized, is the combination of van der Waals attraction and electrostatic repulsion:

$$\Phi = \Phi_A + \Phi_R \qquad (2.27)$$

The electrostatic stabilization of particles in a suspension is successfully described by the DLVO theory, named after Derjaguin, Landau, Verwey and Overbeek. The interaction between two particles in a suspension is considered as the combination of van der Waals attraction potential and the electric repulsion potential. There are some important assumptions in the DLVO theory:

(1) Infinite flat solid surface,
(2) Uniform surface charge density,
(3) No redistribution of surface charge, i.e. the surface electric potential remains constant,
(4) No change of concentration profiles of both counter ions and surface charge determining ions, i.e. the electric potential remains unchanged, and
(5) Solvent exerts influences via dielectric constant only, i.e. no chemical reactions between the particles and solvent.

It is very clear that some of the assumptions are far from the real picture of two particles dispersed in a suspension. For example, the surface of particles is not infinitely flat, and the surface charge density is most likely to change when two charged particles get very close to each other. However, in spite of the assumptions, the DLVO theory works very well in explaining the interactions between two approaching particles, which are electrically charged, and thus is widely accepted in the research community of colloidal science.

Physical Chemistry of Solid Surfaces

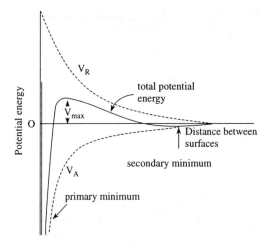

Fig. 2.16. Schematic of DLVO potential: V_A = attractive van der Waals potential, V_R = repulsive electrostatic potential.

Figure 2.16 shows the van der Waals attraction potential, electric repulsion potential, and the combination of the two opposite potentials as a function of distance from the surface of a spherical particle.[47] At a distance far from the solid surface, both van der Waals attraction potential and electrostatic repulsion potential reduce to zero. Near the surface is a deep minimum in the potential energy produced by the van der Waals attraction. A maximum is located a little farther away from the surface, as the electric repulsion potential dominates the van der Waals attraction potential. The maximum is also known as repulsive barrier. If the barrier is greater than $\sim 10\,kT$, where k is Boltzmann constant, the collisions of two particles produced by Brownian motion will not overcome the barrier and agglomeration will not occur. Since the electric potential is dependent on the concentration and valence state of counter ions as given in Eqs. (2.21) and (2.22) and the van der Waals attraction potential is almost independent of the concentration and valence state of counter ions, the overall potential is strongly influenced by the concentration and valence state of counter ions. An increase in concentration and valence state of counter ions results in a faster decay of the electric potential as schematically illustrated in Fig. 2.17.[49] As a result, the repulsive barrier is reduced and its position is pushed towards the particle surface. The secondary minimum in Fig. 2.17 is not necessary to exist in all situations, and it is present only when the concentration of counter ions is higher enough. If secondary minimum is established, particles are likely to be associated with each other, which is known as flocculation.

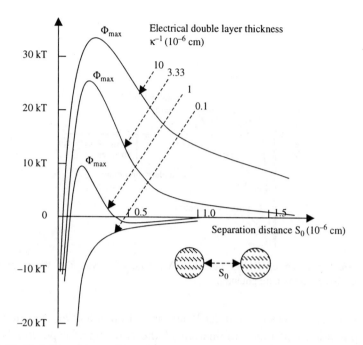

Fig. 2.17. Variation of the total interaction energy Φ between two spherical particles, as a function of the closest separation distance S_0 between their surfaces, for different double layer thickness κ^{-1} obtained with different monovalent electrolyte concentrations. The electrolyte concentration is C (mol.L^{-1}) = 10^{-15} κ^2 (cm^{-1}). [J.T.G. Overbeek, *J. Colloid Interf. Sci.* **58**, 408 (1977).]

When two particles are far apart or the distance between the surfaces of two particles is larger than the combined thickness of two electric double layers of two particles, there would be no overlap of diffusion double layers, and thus there would be no interaction between two particles (Fig. 2.18(a)). However, when two particles move closer and the two electric double layers overlap, a repulsion force is developed. As the distance reduces, the repulsion increases and reaches the maximum when the distance between two particle surfaces equals the distance between the repulsive barrier and the surface (Fig. 2.18(b)). Such a repulsion force can be understood in two ways. One is that the repulsion derives from the overlap of electric potentials of two particles. It should be noted that the repulsion is not directly due to the surface charge on solid particles, instead it is the interaction between two double layers. The other is the osmotic flow. When two particles approach one another, the concentrations of ions between two particles where two double layers overlap, increase significantly, since each double layer would retain its original concentration profile. As a result, the original equilibrium concentration

Physical Chemistry of Solid Surfaces

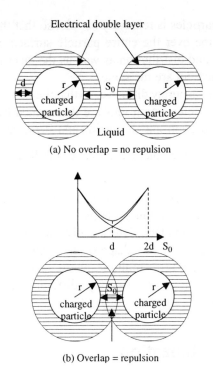

Fig. 2.18. Schematic illustrating the conditions for the occurrence of electrostatic repulsion between two particles.

profiles of counter ions and surface charge determining ions are destroyed. To restore the original equilibrium concentration profiles, more solvent needs to flow into the region where the two double layers overlap. Such an osmotic flow of solvent effectively repels two particles apart, and the osmotic force disappears only when the distance between the two particles equals to or becomes larger than the sum of the thickness of the two double layers.

Although many important assumptions of the DLVO theory are not satisfied in the really colloidal systems, in which small particles are dispersed in a diffusive medium, the DLVO theory is still valid and has been widely applied in practice, as far as the following conditions are met:

(1) Dispersion is very dilute, so that the charge density and distribution on each particle surface and the electric potential in the proximity next to each particle surface are not interfered by other particles.
(2) No other force is present besides van der Waals force and electrostatic potential, i.e. the gravity is negligible or the particle is significantly small, and there exist no other forces, such as magnetic field.

(3) Geometry of particles is relatively simple, so that the surface properties are the same over the entire particle surface, and, thus surface charge density and distribution as well as the electric potential in the surrounding medium are the same.
(4) The double layer is purely diffusive, so that the distributions of counter ions and charge determining ions are determined by all three forces: electrostatic force, entropic dispersion and Brownian motion.

However, it should be noted that electrostatic stabilization is limited by the following facts:

(1) Electrostatic stabilization is a kinetic stabilization method.
(2) It is only applicable to dilute systems.
(3) It is not applicable to electrolyte sensitive systems.
(4) It is almost not possible to redisperse the agglomerated particles.
(5) It is difficult to apply to multiple phase systems, since in a given condition, different solids develop different surface charge and electric potential.

2.5. Steric Stabilization

Steric stabilization, also called polymeric stabilization is a method widely used in stabilization of colloidal dispersions and thoroughly discussed in literature,[50–52] though it is less well understood as compared with electrostatic stabilization method. Polymeric stabilization does offer several advantages over electrostatic stabilization:

(1) It is a thermodynamic stabilization method, so that the particles are always redispersible.
(2) A very high concentration can be accommodated, and the dispersion medium can be completely depleted.
(3) It is not electrolyte sensitive.
(4) It is suitable to multiple phase systems.

In this section, we will briefly summarize the essential concepts of polymeric stabilization. Compared to electrostatic stabilization mechanism, polymeric stabilization offers an additional advantage in the synthesis of nanoparticles, particularly when narrow size distribution is required. Polymer layer adsorbed on the surface of nanoparticles serves as a diffusion barrier to the growth species, resulting in a diffusion-limited growth in the subsequent growth of nuclei. As will be discussed in detail in the next chapter, diffusion-limited growth would reduce the size distribution

Physical Chemistry of Solid Surfaces

of the initial nuclei, leading to monosized nanoparticles. The dual functionalities of polymeric layer on the surface of nanoparticles explain the fact that steric stabilization is widely used in the synthesis of nanoparticles.

2.5.1. Solvent and polymer

Solvents can be grouped into aqueous solvent, which is water, H_2O, and non-aqueous solvents or organic solvents. Solvents can also been categorized into protic solvent, which can exchange protons and examples of which include: methanol, CH_3OH, and ethanol, C_2H_5OH and aprotic solvent, which cannot exchange protons, such as benzene, C_6H_6. Table 2.5 gives some examples of typical protic and aprotic solvents.[53]

Not all polymers are dissolvable into solvents and those non-solvable polymers will not be discussed in this chapter, since they cannot be used for the steric stabilization. When a solvable polymer dissolves into a solvent, the polymer interacts with the solvent. Such interaction varies with the system as well as temperature. When polymer in a solvent tends to expand to reduce the overall Gibbs free energy of the system, such a solvent is called a "good solvent". When polymer in a solvent tends to coil up or collapse to reduce the Gibbs free energy, the solvent is considered to be a "poor solvent".

Table 2.5. List of some solvents with their dielectric constants.

Solvent	Formula	Dielectric constant	Type
Acetone	C_3H_6O	20.7	Aprotic
Acetic acid	$C_2H_4O_2$	6.2	Protic
Ammonia	NH_3	16.9	Protic
Benzene	C_6H_6	2.3	Aprotic
Chloroform	$CHCl_3$	4.8	Aprotic
Dimethylsulfoxide	$(CH_3)_2SO$	45.0	Aprotic
Dioxanne	$C_4H_8O_2$	2.2	Aprotic
Water	H_2O	78.5	Protic
Methanol	CH_3OH	32.6	Protic
Ethanol	C_2H_5OH	24.3	Protic
Formamide	CH_3ON	110.0	Protic
Dimethylformamide	C_3H_7NO	36.7	Aprotic
Nitrobenzene	$C_6H_5NO_2$	34.8	Aprotic
Tetrahydrofuran	C_4H_8O	7.3	Aprotic
Carbon tetrachloride	CCl_4	2.2	Aprotic
Diethyl ether	$C_4H_{10}O$	4.3	Aprotic
Pyridine	C_5H_5N	14.2	Aprotic

For a given system, i.e. a given polymer in a given solvent, whether the solvent is a "good" or "poor" solvent is dependent on the temperature. At high temperatures, polymer expands, whereas at low temperatures, polymer collapses. The temperature, at which a poor solvent transfers to a good solvent, is the Flory–Huggins theta temperature, or simply the θ temperature. At $T = θ$, the solvent is considered to be at the theta state, at which the Gibbs free energy does not change whether the polymer expands or collapses.

Depending on the interaction between polymer and solid surface, a polymer can be grouped into:

(1) Anchored polymer, which irreversibly binds to solid surface by one end only, and typically are diblock polymer (Fig. 2.19(a)),
(2) Adsorbing polymer, which adsorbs weakly at random points along the polymer backbone (Fig. 2.19(b)),
(3) Non-adsorbing polymer, which does not attach to solid surface and thus does not contribute to polymer stabilization, and so is not discussed further in this chapter.

The interaction between a polymer and solid surface is limited to adsorption of polymer molecules onto the surface of solid. The adsorption can be either by forming chemical bonds between surface ions or atoms on the solid and polymer molecules or by weak physical adsorption. Furthermore, there is no restriction whether one or multiple bonds are formed between solid and polymer. No other interactions such as chemical reactions or further polymerization between polymer and solvent or between polymers are considered for the current discussion.

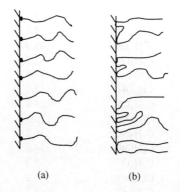

(a) (b)

Fig. 2.19. Schematic of different polymers according to the interaction between polymers and solid surface: (a) anchored polymer and (b) absorbing polymer.

2.5.2. Interactions between polymer layers

First let us consider two solid particles covered with terminally anchored polymers as schematically illustrated in Fig. 2.20(a). When two particles approach one another, the attached polymers interact only when the separation distance, H, between the surfaces of two particles is less than twice the thickness, L, of polymer layers. Beyond this distance, there is no interaction between two particles and their polymer layers on surfaces. However, when the distance reduces to less than $2L$, but is still larger than L, there will be interactions between solvent and polymer and between two polymer layers. But there is no direct interaction between the polymer layer of one particle and the solid surface of the opposite particle. In a good solvent, in which polymer expands, if the coverage of polymer on the solid surface is not complete, particularly less than 50% coverage, when the concentration of polymer in the solvent is insufficient, two polymer layers tend to interpenetrate so as to reduce the available space between polymers. Such an interpenetration of two polymer layers of two approaching particles would result in a reduction of the freedom of polymers, which leads to a reduction of entropy, i.e. $\Delta S < 0$. As a result, the Gibbs free energy of the system would increase, assuming the change of enthalpy due to the interpenetration of two polymer layers negligible, i.e. $\Delta H \approx 0$, according to:

$$\Delta G = \Delta H - T\Delta S > 0 \qquad (2.28)$$

So two particles repel one another and the distance between two particles must be equal to or larger than twice the thickness of polymer layers. When the coverage of polymer is high, particularly approaching 100%, there would be no interpenetration. As a result, the two polymer layers will be compressed, leading to the coil up of polymers in both layers. The overall Gibbs free energy increases, and repels two particles apart. When the distance between the surfaces of two particles is less than the thickness of polymer layers, a further reduction of the distance would force polymers to coil up and result in an increase in the Gibbs free energy. Figure 2.20(b) sketches the Gibbs free energy as a function of the distance between two particles, and shows that the overall energy is always positive and increases with a decreasing distance when H is smaller than $2L$.

The situation is rather different in a poor solvent, with a low coverage of polymer on the solid surface. With a low coverage, when the distance between two particles is less than twice the thickness of polymer layers but larger than the thickness of single polymer layer, i.e. $L < H < 2L$, polymers adsorbed onto the surface of one particle surface tend to penetrate

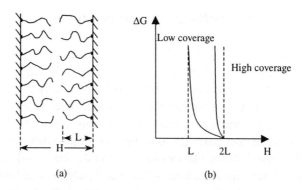

Fig. 2.20. Schematic of interactions between polymer layers: (a) the schematic of two approaching polymer layers, and (b) the Gibbs free energy as a function of the distance between two particles.

into the polymer layer of the approach particle. Such interpenetration of two polymer layers will promote further coil up of polymers, and result in a reduction of the overall Gibbs free energy. Two particles tend to associate with one another. However, with a high coverage, similar to polymer in a good solvent, there would be no penetration and the reduction in distance results in a compressive force, leading to an increase in the overall free energy. When the distance between two particles is less than the thickness of the polymer layer, a reduction in distance always produces a repulsive force and an increase in the overall Gibbs free energy. Figure 2.21 summarizes the dependence of free energy as a function of distance between two particles. Regardless of the difference in coverage and solvent, two particles covered with polymer layers are prevented from agglomeration by the space exclusion or steric stabilization.

Next, let us look at the adsorbing polymers. The situation of adsorbing polymers is more complicated due to the following two reasons. First, polymer originally attached to the solid surface of one particle may interact with and adsorb onto another particle surface, and thus form bridges between two particles, when two particles approach to a sufficiently close distance between each other. Second, given sufficient time, attached polymer can desorb from the surface and migrate out of the polymer layer.

When polymer has a strong adsorption and forms a full coverage, the interaction between two polymer layers produces a purely repulsive force and results in an increased free energy, when the distance between two particles reduces below twice the thickness of polymer layer. This is the same as that of anchored polymer at full coverage. When only a partial coverage is achieved, the nature of solvent can have a significant influence on the interaction between two particles. In a good solvent, two partially

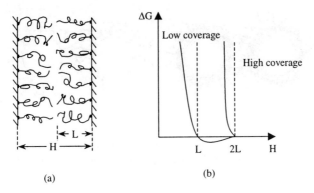

Fig. 2.21. Schematic of interactions between polymer layers: (a) the schematic of two approaching polymer layers and (b) the Gibbs free energy as a function of the distance between two particles.

covered polymer layers interpenetrate each other, resulting in a reduced space and more ordered polymer arrangement. As a result, the entropy reduces and the Gibbs free energy increases. However, in a poor solvent, interpenetration promotes further coil up of polymers, leads to increased entropy, and thus results in a reduced free energy. This interaction force of adsorbing polymer layers in a poor solvent is very similar to that of anchored polymer layers with partial coverage in poor solvent; however, the process involved is significantly different due to multiple adsorption sites at both surfaces. It is always the case that a repulsive force develops and repels two particles away from each other, when the distance is less than the thickness of polymer layer.

The physical basis for the steric stabilization is (i) a volume restriction effect arising from the decrease in possible configurations in the region between the two surfaces when two particles approach one another, and (ii) an osmotic effect due to the relatively high concentration of adsorbed polymeric molecules in the region between the two particles.

2.5.3. *Mixed steric and electric interactions*

Steric stabilization can be combined with electrostatic stabilization, which is also referred to as electrosteric stabilization and is sketched in Fig. 2.22.[50] When polymers are attached to a charged particle surface, a polymer layer would form as discussed above. In addition, an electric potential adjacent to the solid surface would retain. When two particles approach each other, both electrostatic repulsion and steric restriction would prevent agglomeration.

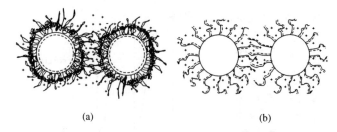

(a) (b)

Fig. 2.22. Schematic representation of electrosteric stabilization: (a) charged particles with nonionic polymers and (b) polyelectrolytes attached to uncharged particles.

2.6. Summary

This chapter has discussed the origins of the surface energy of solids, the various mechanisms for a material to reduce its surface energy, the influences of the surface curvature on the chemical potential, and the two mechanisms for the stabilization of nanoparticles from agglomeration. All the concepts and theories discussed in this chapter have been well established in surface science and materials fields. However, the impact of surface energy on nanostructures and nanomaterials would be far more significant, due to the huge surface area involved in nanostructures and nanomaterials. A good understanding of these fundamentals is not only important for the fabrication and processing of nanomaterials, it is equally important to the applications of nanomaterials.

References

1. C. Nützenadel, A. Züttel, D. Chartouni, G. Schmid, and L. Schlapbach, *Eur. Phys. J.* **D8**, 245 (2000).
2. A.W. Adamson and A.P. Gast, *Physical Chemistry of Surfaces*, 6th edition, John Wiley & Sons, New York, 1997.
3. A.N. Goldstein, C.M. Echer, and A.P. Alivisatos, *Science* **256**, 1425 (1992).
4. M.A. Van Hove, W.H. Weinberg, and C.M. Chan, *Low-Energy Electron Diffraction*, Springer-Verlag, Berlin, 1986.
5. M.W. Finnis and V. Heine, *J. Phys.* **F4**, L37 (1974).
6. U. Landman, R.N. Hill, and M. Mosteller, *Phys. Rev.* **B21**, 448 (1980).
7. D.L. Adams, H.B. Nielsen, J.N. Andersen, I. Stengsgaard, R. Friedenhans'l, and J.E. Sorensen, *Phys. Rev. Lett.* **49**, 669 (1982).
8. C.M. Chan, M.A. Van Hove, and E.D. Williams, *Surface Sci.* **91**, 440 (1980).
9. M.A. Van Hove, R.J. Koestner, P.C. Stair, J.P. Birberian, L.L. Kesmodell, I. Bartos, and G.A. Somorjai, *Surface Sci.* **103**, 189 (1981).

10. I.K. Robinson, Y. Kuk, and L.C. Feldman, *Phys. Rev.* **B29**, 4762 (1984).
11. R.M. Tromp, R.J. Hamers, and J.E. Demuth, *Phys. Rev.* **B34**, 5343 (1986).
12. G. Binnig, H. Rohrer, Ch. Gerber, and E. Weibel, *Phys. Rev. Lett.* **50**, 120 (1983).
13. R. Schlier and H. Farnsworth, *J. Chem. Phys.* **30**, 917 (1959).
14. R.M. Tromp, R.J. Hames, and J.E. Demuth, *Phys. Rev. Lett.* **55**, 1303 (1985).
15. R.M. Tromp, R.J. Hames, and J.E. Demuth, *Phys. Rev. Lett.* **B24**, 5343 (1986).
16. J.M. Jasinski, B.S. Meyerson, and B.A. Scott, *Annu. Rev. Phys. Chem.* **38**, 109 (1987).
17. M. McEllistrem, M. Allgeier, and J.J. Boland, *Science* **279**, 545 (1998).
18. Z. Zhang, F. Wu, and M.G. Lagally, *Annu. Rev. Mater. Sci.* **27**, 525 (1997).
19. T. Tsuno, T. Imai, Y. Nishibayashi, K. Hamada, and N. Fujimori, *Jpn. J. Appl. Phys.* **30**, 1063 (1991).
20. C.J. Davisson and L.H. Germer, *Phys. Rev.* **29**, 908 (1927).
21. K. Christmann, R.J. Behm, G. Ertl, M.A. Van Hove, and W.H. Weinberg, *J. Chem. Phys.* **70**, 4168 (1979).
22. H.D. Shih, F. Jona, D.W. Jepsen, and P.M. Marcus, *Surface Sci.* **60**, 445 (1976).
23. J.M. MacLaren, J.B. Pendry, P.J. Rous, D.K. Saldin, G.A. Somorjai, M.A. Van Hove, and D.D. Vvedensky, eds., *Surface Crystallography Information Service*, Reidel Publishing, Dordrecht, 1987.
24. C. Herring, *Structure and Properties of Solid Surfaces*, University of Chicago, Chicago, IL, 1952.
25. W.W. Mullins, *Metal Surfaces: Structure Energetics and Kinetics*, The American Society for Metals, Metals Park, OH, 1963.
26. E. Matijevi, *Annu. Rev. Mater. Sci.* **15**, 483 (1985).
27. H.N.V. Temperley, *Proc. Cambridge Phil. Soc.* **48**, 683 (1952).
28. W.K. Burton and N. Cabrera, *Disc. Faraday Soc.* **5**, 33 (1949).
29. G.K. Teal, *IEEE Trans. Electron Dev.* **ED-23**, 621 (1976).
30. W. Zuhlehner and D. Huber, *Czochralski Grown Silicon, in Crystals 8*, Springer-Verlag, Berlin, 1982.
31. W.D. Kingery, H.W. Bowen, and D.R. Uhlmann, *Introduction to Ceramics*, 2nd edition, Wiley, New York, 1976.
32. J.S. Reed, *Introduction to Principles of Ceramic Processing*, Wiley, New York, 1988.
33. E.P. DeGarmo, J.T. Black, and R.A. Kohner, *Materials and Processes in Manufacturing*, Macmillan, New York, 1988.
34. A.W. Adamson, *Physical Chemistry of Surfaces*, Wiley, New York, 1976.
35. L.R. Fisher and J.N. Israelachvili, *J. Colloid Interf. Sci.* **80**, 528 (1981).
36. J.C. Melrose, *Langmuir* **5**, 290 (1989).
37. R.W. Vook, *Int. Metals Rev.* **27**, 209 (1982).
38. R.K. Iler, *The Chemistry of Silica: Solubility, Polymerization, Colloid and Surface Properties, and Biochemistry*, John Wiley & Sons, New York, 1979.
39. J.R. Sambles, L.M. Skinner, and N.D. Lisgarten, *Proc. R. Soc.* **A324**, 339 (1971).
40. N.D. Lisgarten, J.R. Sambles, and L.M. Skinner, *Contemp. Phys.* **12**, 575 (1971).
41. V.K. La Mer and R. Gruen, *Trans. Faraday Soc.* **48**, 410 (1952).
42. F. Piuz and J-P. Borel, *Phys. Status Solid.* **A14**, 129 (1972).
43. R.J. Hunter, *Zeta Potential in Colloid Science*, Academic Press, New York, 1981.
44. G.A. Parks, *Chem. Rev.* **65**, 177 (1965).
45. A.C. Pierre, *Introduction to Sol-Gel Processing*, Kluwer, Norwell, MA, 1998.
46. P.C. Hiemenz, *Principles of Colloid and Surface Chemistry*, Marcel Dekker, New York, 1977.

47. G.D. Parfitt, in *Dispersion of Powders in Liquids with Special Reference to Pigments*, ed. G.D. Parfitt, Applied Science, London, p.1, 1981.
48. C.J. Brinker and G.W. Scherer, *Sol-Gel Science: The Physics and Chemistry of Sol-Gel Processing*, Academic Press, San Diego, CA, 1990.
49. J.T.G. Overbeek, *J. Colloid Interf. Sci.* **58**, 408 (1977).
50. D.H. Napper, *Polymeric Stabilization of Colloidal Dispersions*, Academic Press, New York, 1983.
51. W.B. Russel, D.A. Saville, and W.R. Schowalter, *Colloidal Dispersions*, Cambridge University Press, Cambridge, 1991.
52. P. Somasundaran, B. Markovic, S. Krishnakumar, and X. Yu, in *Handbook of Surface and Colloid Chemistry*, ed. K.S. Birdi, CRC Press, Boca Raton, FL, p. 559, 1997.
53. J.J. Lagowski, *The Chemistry of Non-aqueous Systems*, Vols. 1–4, Academic Press, New York, 1965, 1967, 1970, 1976.

Chapter 3

Zero-Dimensional Nanostructures: Nanoparticles

3.1. Introduction

Many techniques, including both top-down and bottom-up approaches, have been developed and applied for the synthesis of nanoparticles. Top-down approaches include milling or attrition, repeated quenching and lithography. Attrition can produce nanoparticles ranging from a couple of tens to several hundreds nanometers in diameter. However, nanoparticles produced by attrition have a relatively broad size distribution and varied particle shape or geometry. In addition, they may contain a significant amount of impurities from the milling medium and defects resulting from milling. Such prepared nanoparticles are commonly used in the fabrication of nanocomposites and nanograined bulk materials, which require much lower sintering temperatures. In nanocomposites and nanograined bulk materials, defects may be annealed during sintering, size distribution, particle shape, and a small amount of impurities are relatively insensitive for their applications. Repeated thermal cycling may also break a bulk material into small pieces, if the material has very small thermal conductivity but a large volume change as a function of temperature. A big volume change associated with phase transition can be effectively utilized in this approach. Although very fine particles can be produced, this process is difficult to design and control so as to produce desired particle size and shape. It is also limited to materials with very poor thermal conductivity but a large volume change. Lithography, which will be discussed in Chapter 7, is another method to make small particles.[1,2]

Bottom-up approaches are far more popular in the synthesis of nanoparticles and many methods have been developed. For example, nanoparticles are synthesized by homogeneous nucleation from liquid or vapor, or by heterogeneous nucleation on substrates. Nanoparticles or quantum dots can also be prepared by phase segregation through annealing appropriately designed solid materials at elevated temperatures. Nanoparticles can be synthesized by confining chemical reactions, nucleation and growth processes in a small space such as micelles. Various synthesis methods or techniques can be grouped into two categories: thermodynamic equilibrium approach and kinetic approach. In the thermodynamic approach, synthesis process consists of (i) generation of supersaturation, (ii) nucleation, and (iii) subsequent growth. In the kinetic approach, formation of nanoparticles is achieved by either limiting the amount of precursors available for the growth such as used in molecular beam epitaxy, or confining the process in a limited space such as aerosol synthesis or micelle synthesis. In this chapter, the attention will be focused mainly on the synthesis of nanoparticles through thermodynamically equilibrium approach. However, some typical kinetic approaches such as microemulsion, aerosol pyrolysis and template-based deposition, will be highlighted as well. For the thermodynamic equilibrium approach, this chapter will take the solution synthesis of nanoparticles as an example to illustrate the fundamental requirements and consideration, however the fundamentals and principles are applicable to other systems without or with minimal modification.

For the fabrication of nanoparticles, a small size is not the only requirement. For any practical application, the processing conditions need to be controlled in such a way that resulting nanoparticles have the following characteristics: (i) identical size of all particles (also called monosized or with uniform size distribution), (ii) identical shape or morphology, (iii) identical chemical composition and crystal structure that are desired among different particles and within individual particles, such as core and surface composition must be the same, and (iv) individually dispersed or monodispersed, i.e. no agglomeration. If agglomeration does occur, nanoparticles should be readily redispersible.

Nanoparticles discussed in this chapter include single crystal, polycrystalline and amorphous particles with all possible morphologies, such as spheres, cubes and platelets. In general, the characteristic dimension of the particles is not larger than several hundred nanometers, mostly less than a couple of hundred nanometers. Some other terminologies are commonly used in the literature to describe some specific subgroups of nanoparticles. If the nanoparticles are single crystalline, they are often referred to as nanocrystals. When the characteristic dimension of the nanoparticles is sufficiently small and quantum effects are observed, quantum dots are the common term used to describe such nanoparticles.

3.2. Nanoparticles through Homogeneous Nucleation

For the formation of nanoparticles by homogeneous nucleation, a supersaturation of growth species must be created. A reduction in temperature of an equilibrium mixture, such as a saturated solution would lead to supersaturation. Formation of metal quantum dots in glass matrix by annealing at moderate temperatures is a good example of this approach. Another method is to generate a supersaturation through *in situ* chemical reactions by converting highly soluble chemicals into less soluble chemicals. For example, semiconductor nanoparticles are commonly produced by pyrolysis of organometallic precursors. Nanoparticles can be synthesized through homogeneous nucleation in three mediums: liquid, gas and solid; however, the fundamentals of nucleation and subsequent growth processes are essentially the same.

Before discussing the detailed approaches for the synthesis of uniformly sized monodispersed nanoparticles, we will first review the fundamentals of homogeneous nucleation and subsequent growth. Solution synthesis of metallic, semiconductor, and oxide nanoparticles will then be discussed in detail. Vapor phase reaction and solid phase segregation for the formation of nanoparticles are also included in this section.

3.2.1. *Fundamentals of homogeneous nucleation*

When the concentration of a solute in a solvent exceeds its equilibrium solubility or temperature decreases below the phase transformation point, a new phase appears. Let us consider the case of homogeneous nucleation of a solid phase from a supersaturated solution, as an example. A solution with solute exceeding the solubility or supersaturation possesses a high Gibbs free energy; the overall energy of the system would be reduced by segregating solute from the solution. Figure 3.1 is a schematic showing the reduction of the overall Gibbs free energy of a supersaturated solution by forming a solid phase and maintaining an equilibrium concentration in the

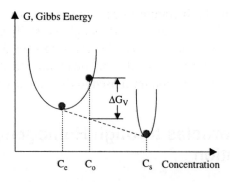

Fig. 3.1. Schematic showing the reduction of the overall Gibbs free energy of a supersaturated solution by forming a solid phase and maintaining an equilibrium concentration in the solution.

solution. This reduction of Gibbs free energy is the driving force for both nucleation and growth. The change of Gibbs free energy per unit volume of the solid phase, ΔG_v, is dependent on the concentration of the solute:

$$\Delta G_v = -\frac{kT}{\Omega}\ln(C/C_o) = -\frac{kT}{\Omega}\ln(1+\sigma) \qquad (3.1)$$

where C is the concentration of the solute, C_o is the equilibrium concentration or solubility, k is the Boltzmann constant, T is the temperature, Ω is the atomic volume, and σ is the supersaturation defined by $(C-C_o)/C_o$. Without supersaturation (i.e. $\sigma = 0$), ΔG_v is zero, and no nucleation would occur. When $C > C_o$, ΔG_v is negative and nucleation occurs spontaneously. Assuming a spherical nucleus with a radius of r, the change of Gibbs free energy or volume energy, $\Delta \mu_v$, can be described by:

$$\Delta \mu_v = \frac{4}{3}\pi r^3 \Delta G_v \qquad (3.2)$$

However, this energy reduction is counter balanced by the introduction of surface energy, accompanied with the formation of a new phase. This results in an increase in the surface energy, $\Delta \mu_s$, of the system:

$$\Delta \mu_s = 4\pi r^2 \gamma \qquad (3.3)$$

where γ is the surface energy per unit area. The total change of chemical potential for the formation of the nucleus, ΔG, is given by:

$$\Delta G = \Delta \mu_v + \Delta \mu_s = \frac{4}{3}\pi r^3 \Delta G_v + 4\pi r^2 \gamma \qquad (3.4)$$

Figure 3.2 schematically shows the change of volume free energy, $\Delta \mu_v$, surface free energy, $\Delta \mu_s$, and total free energy, ΔG, as functions of

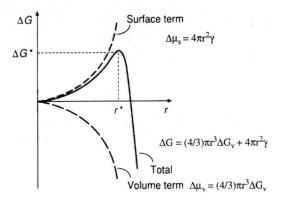

Fig. 3.2. Schematic illustrating the change of volume free energy, $\Delta\mu_v$, surface free energy, $\Delta\mu_s$, and total free energy, ΔG, as functions of nucleus' radius.

nucleus' radius. From this figure, one can easily see that the newly formed nucleus is stable only when its radius exceeds a critical size, r^*. A nucleus smaller than r^* will dissolve into the solution to reduce the overall free energy, whereas a nucleus larger than r^* is stable and continues to grow bigger. At the critical size $r = r^*$, $d\Delta G/dr = 0$ and the critical size, r^*, and critical energy, ΔG^*, are defined by:

$$r^* = -2\frac{\gamma}{\Delta G_v} \tag{3.5}$$

$$\Delta G^* = \frac{16\pi\gamma}{(3\Delta G_v)^2} \tag{3.6}$$

ΔG^* is the energy barrier that a nucleation process must overcome and r^* represents the minimum size of a stable spherical nucleus. The above discussion was based on a supersaturated solution; however, all the concepts can be generalized for a supersaturated vapor and a supercooled gas or liquid.

In the synthesis and preparation of nanoparticles or quantum dots by nucleation from supersaturated solution or vapor, this critical size represents the limit on how small nanoparticles can be synthesized. To reduce the critical size and free energy, one needs to increase the change of Gibbs free energy, ΔG_v, and reduce the surface energy of the new phase, γ. Equation (3.1) indicates that ΔG_v can be significantly increased by increasing the supersaturation, σ, for a given system. Figure 3.3 compares the critical sizes and critical free energy of three spherical nuclei with different values of supersaturation, which increases with a decreasing temperature. Temperature can also influence surface energy. Surface energy of the solid nucleus can change more significantly near the roughening temperature. Other possibilities include: (i) use of different solvent,

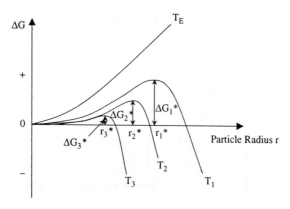

Fig. 3.3. The effect of temperature on the critical sizes and critical free energy of three spherical nuclei. Supersaturation increases with a decreasing temperature and surface energy also varies with temperature. $T_E > T_1 > T_2 > T_3$ with T_E being the equilibrium temperature.

(ii) additives in solution, and (iii) incorporation of impurities into solid phase, when other requirements are not compromised.

The rate of nucleation per unit volume and per unit time, R_N, is proportional to (i) the probability, P, that a thermodynamic fluctuation of critical free energy, ΔG^*, given by:

$$P = \exp\left(-\frac{\Delta G^*}{kT}\right) \qquad (3.7)$$

(ii) the number of growth species per unit volume, n, which can be used as nucleation centers (in homogeneous nucleation, it equals to the initial concentration, C_o), and (iii) the successful jump frequency of growth species, Γ, from one site to another, which is given by:

$$\Gamma = \frac{kT}{3\pi\lambda^3\eta} \qquad (3.8)$$

where λ is the diameter of the growth species and η is the viscosity of the solution. So the rate of nucleation R_N can be described by:

$$R_N = nP\Gamma = \left\{\frac{C_o kT}{3\pi\lambda^3\eta}\right\} \exp\left(-\frac{\Delta G^*}{kT}\right) \qquad (3.9)$$

This equation indicates that high initial concentration or supersaturation (so, a large number of nucleation sites), low viscosity and low critical energy barrier favor the formation of a large number of nuclei. For a given concentration of solute, a larger number of nuclei mean smaller sized nuclei.

Figure 3.4 schematically illustrated the processes of nucleation and subsequent growth.[3] When the concentration of solute increases as a function of time, no nucleation would occur even above the equilibrium solubility.

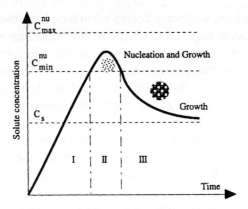

Fig. 3.4. Schematic illustrating the processes of nucleation and subsequent growth. [M. Haruta and B. Delmon, *J. Chim. Phys.* **83**, 859 (1986).]

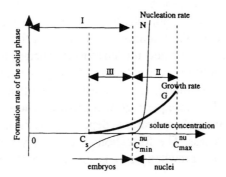

Fig. 3.5. Schematic showing, from a slightly different point of view, the relations between the nucleation and growth rates and the concentration of growth species. [M. Haruta and B. Delmon, *J. Chim. Phys.* **83**, 859 (1986).]

The nucleation occurs only when the supersaturation reaches a certain value above the solubility, which corresponds to the energy barrier defined by Eq. (3.6) for the formation of nuclei. After the initial nucleation, the concentration or supersaturation of the growth species decreases and the change of Gibbs free energy reduces. When the concentration decreases below this specific concentration, which corresponds to the critical energy, no more nuclei would form, whereas the growth will proceed until the concentration of growth species has attained the equilibrium concentration or solubility. Figure 3.5 schematically shows, from a slightly different point of view, the relations between the nucleation and growth rates and the concentration of growth species.[3] When the concentration of the growth species increases above the equilibrium concentration, initially there will be no nucleation. However, nucleation occurs when the concentration reaches the minimum saturation required to overcome the critical energy barrier, and the nucleation rate increases very rapidly as the concentration increases further. Although growth process cannot proceed when there is no nucleus, growth rate is above zero for a concentration above its equilibrium solubility. Once nuclei are formed, growth occurs simultaneously. Above the minimum concentration, nucleation and growth are inseparable processes; however, these two processes proceed at different speeds.

For the synthesis of nanoparticles with uniform size distribution, it is best if all nuclei are formed at the same time. In this case, all the nuclei are likely to have the same or similar size, since they are formed under the same conditions. In addition, all the nuclei will have the same subsequent growth. Consequently, monosized nanoparticles can be obtained. So it is obvious that it is highly desirable to have nucleation occur in a very short period of time. In practice, to achieve a sharp nucleation, the concentration of the growth species is increased abruptly to a very high supersaturation and then quickly brought below the minimum concentration for nucleation. Below this concentration, no more new nucleus forms, whereas the existing nuclei continue to grow until the concentration of the growth species reduces to the equilibrium concentration. The size distribution of nanoparticles can be further altered in the subsequent growth process. The size distribution of initial nuclei may increase or decrease depending on the kinetics of the subsequent growth process. The formation of uniformly sized nanoparticles can be achieved if the growth process is appropriately controlled.

3.2.2. Subsequent growth of nuclei

The size distribution of nanoparticles is dependent on the subsequent growth process of the nuclei. The growth process of the nuclei involves multi-steps and the major steps are (i) generation of growth species, (ii) diffusion of the growth species from bulk to the growth surface, (iii) adsorption of the growth species onto the growth surface, and (iv) surface growth through irreversible incorporation of growth species onto the solid surface. These steps can be further grouped into two processes. Supplying the growth species to the growth surface is termed as diffusion, which includes the generation, diffusion and adsorption of growth species onto the growth surface, whereas incorporation of growth species adsorbed on the growth surface into solid structure is denoted as growth. A diffusion-limited growth would result in a different size distribution of nanoparticles as compared with that by growth-limited process.

3.2.2.1. Growth controlled by diffusion

When the concentration of growth species reduces below the minimum concentration for nucleation, nucleation stops, whereas the growth continues. If the growth process is controlled by the diffusion of growth species from the bulk solution to the particle surface, the growth rate is given by[4]:

$$dr/dt = D(C-C_s)\frac{V_m}{r} \qquad (3.10)$$

where r is the radius of spherical nucleus, D is the diffusion coefficient of the growth species, C is the bulk concentration, C_s is the concentration on the surface of solid particles, and V_m is the molar volume of the nuclei as illustrated in Fig. 3.6. By solving this differential equation and assuming the initial size of nucleus, r_o, and the change of bulk concentration negligible, we have:

$$r^2 = 2D(C-C_s)V_m t + r_o^2 \qquad (3.11)$$

or

$$r^2 = k_D t + r_o^2 \qquad (3.12)$$

where $k_D = 2D(C-C_s)V_m$. For two particles with initial radius difference, δr_o, the radius difference, δr, decreases as time increases or particles grow bigger, according to:

$$\delta r = \frac{r_o \delta r_o}{r} \qquad (3.13)$$

Combining with Eq. (3.12), we have:

$$\delta r = \frac{r_o \delta r_o}{\sqrt{k_D t + r_o^2}} \qquad (3.14)$$

Both Eqs. (3.13) and (3.14) indicate that the radius difference decreases with increase of nuclear radius and prolonged growth time. The diffusion-controlled growth promotes the formation of uniformly sized particles.

3.2.2.2. Growth controlled by surface process

When the diffusion of growth species from the bulk to the growth surface is sufficiently rapid, i.e. the concentration on the surface is the same as that in the bulk as illustrated by a dash line also in Fig. 3.6, the growth rate is controlled by the surface process. There are two mechanisms for the surface processes: mononuclear growth and poly-nuclear growth. For the mononuclear growth, the growth proceeds layer by layer; the growth species are incorporated into one layer and proceeds to another layer only

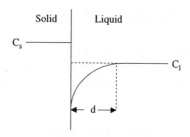

Fig. 3.6. Schematic diagram of the concentration profile of an alloy component or impurity distribution across the solid–liquid interface, showing the formation of a depletion boundary layer in the liquid phase.

after the growth of the previous layer is complete. There is a sufficient time for the growth species to diffuse on the surface. The growth rate is thus proportional to the surface area[4]:

$$\frac{dr}{dt} = k_m r^2 \tag{3.15}$$

where k_m is a proportionality constant, dependent on the concentration of growth species. The growth rate is given by solving the above equation:

$$\frac{1}{r} = \frac{1}{r_o} - k_m t \tag{3.16}$$

The radius difference increases with an increasing radius of the nuclei:

$$\delta r = r^2 \frac{\delta r_o}{r_o^2} \tag{3.17}$$

Substituting Eq. (3.16) into (3.17) yields:

$$\delta r = \frac{\delta r_o}{(1 - k_m r_o t)^2} \tag{3.18}$$

where $k_m r_o t < 1$. This boundary condition is derived from Eq. (3.16), and it means that the radius, is not infinitely large, i.e. $r < \infty$. Equation (3.18) shows that the radius difference increases with a prolonged growth time. Obviously, this growth mechanism does not favor the synthesis of monosized particles.

During poly-nuclear growth, which occurs when the surface concentration is very high, surface process is so fast that second layer growth proceeds before the first layer growth is complete. The growth rate of particles is independent of particle size or time,[5] i.e. the growth rate is constant:

$$\frac{dr}{dt} = k_p \tag{3.19}$$

Where k_p is a constant only dependent on temperature. Hence the particles grow linearly with time:

$$r = k_p t + r_o \tag{3.20}$$

The relative radius difference remains constant regardless of the growth time and the absolute particle size:

$$\delta r = \delta r_o \qquad (3.21)$$

It is worth noting that although the absolute radius difference remains unchanged, the relative radius difference would be inversely proportional to the particle radius and the growth time. As particles get bigger, the radius difference become smaller; so this growth mechanism also favors the synthesis of monosized particles.

Figures 3.7 and 3.8 schematically illustrate the radius difference as functions of particle size and growth time for all three mechanisms of subsequent growth discussed above. It is obvious that a diffusion controlled growth mechanism is required for the synthesis of monosized particles by homogeneous nucleation. Williams et al.[5] suggested that the growth of nanoparticles involve all three mechanisms. When the nuclei are small, monolayer growth mechanism may dominate, poly-nuclear growth may become predominant as the nuclei become bigger. Diffusion is predominant for the growth of relatively large particles. Of course, this would only be the case when no other procedures or measures were applied to prevent certain growth mechanisms. Different growth mechanisms can become predominant when favorable growth conditions are established. For example, when the supply of growth species is very slow due to a slow chemical reaction, the growth of nuclei would most likely be predominant by the diffusion-controlled process.

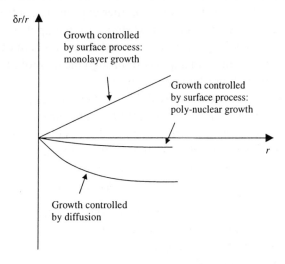

Fig. 3.7. Schematic illustrating the radius difference as functions of particle size for all three mechanisms of subsequent growth discussed above.

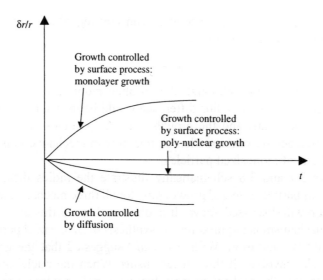

Fig. 3.8. Schematic illustrating the radius difference as functions of growth time for all three mechanisms of subsequent growth discussed above.

For the formation of monosized nanoparticles, diffusion-limited growth is desired. There are several ways to achieve diffusion-limited growth. For example, when the concentration of growth species is kept extremely low, diffusion distance would be very large and consequently diffusion could become the limiting step. Increasing the viscosity of solution is another possibility. Introduction of a diffusion barrier such as a monolayer on the surface of a growing particle is yet another approach. Controlled supply of growth species offers another method to manipulate the growth process. When growth species is generated through chemical reactions, the rate of reaction can be manipulated through the control of the concentration of by-product, reactant and catalyst.

In the following sections, we will discuss the synthesis of metal, semiconductor and oxide (including hydroxide) nanoparticles separately for the sake of clarity. First, we will focus our discussion on the synthesis of various types of nanoparticles through solution processes. Formation of nanoparticles dispersed in a solvent is the most common approach and offers several advantages, which include easiness of:

(1) stabilization of nanoparticles from agglomeration,
(2) extraction of nanoparticles from solvent,
(3) surface modification and application,
(4) processing control, and
(5) mass production.

3.2.3. Synthesis of metallic nanoparticles

Reduction of metal complexes in dilute solutions is the general method in the synthesis of metal colloidal dispersions, and a variety of methods have been developed to initiate and control the reduction reactions.[6–10] The formation of monosized metallic nanoparticles is achieved in most cases by a combination of a low concentration of solute and polymeric monolayer adhered onto the growth surfaces. Both a low concentration and a polymeric monolayer would hinder the diffusion of growth species from the surrounding solution to the growth surfaces, so that the diffusion process is likely to be the rate limiting step of subsequent growth of initial nuclei, resulting in the formation of uniformly sized nanoparticles.

In the synthesis of metallic nanoparticles, or more specifically speaking, metallic colloidal dispersion, various types of precursors, reduction reagents, other chemicals, and methods were used to promote or control the reduction reactions, the initial nucleation and the subsequent growth of initial nuclei. Table 3.1 briefly summarizes the precursors, reduction reagents and polymeric stabilizers commonly used in the production of metallic colloidal dispersions. The precursors include: elemental metals, inorganic salts and metal complexes, such as, Ni, Co, $HAuCl_4$, H_2PtCl_6, $RhCl_3$ and $PdCl_2$. Reduction reagents includes: sodium citrate, hydrogen peroxide, hydroxylamine hydrochloride, citric acid, carbon monoxide, phosphorus, hydrogen, formaldehyde, aqueous methanol, sodium carbonate and sodium hydroxide. Examples of polymeric stabilizers include polyvinyl alcohol (PVA) and sodium polyacrylate.

Colloidal gold has been studied extensively for a long time. In 1857 Faraday published a comprehensive study on the preparation and properties of colloidal gold.[11] A variety of methods have been developed for the synthesis of gold nanoparticles, and among them, sodium citrate reduction of chlorauric acid at 100°C was developed more than 50 years ago[12] and remains the most commonly used method. The classical (or standard) experimental conditions are as follows. Chlorauric acid dissolves into water to make 20 ml very dilute solution of $\sim 2.5 \times 10^{-4}$ M. Then 1 ml 0.5% sodium citrate is added into the boiling solution. The mixture is kept at 100°C till color changes, while maintaining the overall volume of the solution by adding water. Such prepared colloidal sol has excellent stability and uniform particle size of ~20 nm in diameter. It has been demonstrated that a large number of initial nuclei formed in the nucleation stage would result in a larger number of nanoparticles with smaller size and narrower size distribution. Figure 3.9 compares the size and size distribution of gold nanoparticles and the nucleation rates when the colloidal gold was prepared at different concentrations.[13]

Table 3.1. Summary of precursors, reduction reagents and polymer stabilizers.

Precursors	Formula
Metal anode	Pd, Ni, Co
Palladium chloride	$PdCl_2$
Hydrogen hexachloroplatinate IV	H_2PtCl_6
Potassium tetrachloroplatinate II	K_2PtCl_4
Silver nitrate	$AgNO_3$
Silver tetraoxylchlorate	$AgClO_4$
Chloroauric acid	$HAuCl_4$
Rhodium chloride	$RhCl_3$
Reduction Reagents	
Hydrogen	H_2
Sodium citrate	$Na_3C_6H_5O_7$
Hydroxylamine hydrochloride	$NH_4OH + HCl$
Citric acid	$C_6H_8O_7$
Carbon monoxide	CO
Phosphorus in ether	P
Methanol	CH_3OH
Hydrogen peroxide	H_2O_2
Sodium carbonate	Na_2CO_3
Sodium hydroxide	NaOH
Formaldehyde	HCHO
Sodium tetrahydroborate	$NaBH_4$
Ammonium ions	NH_4^-
Polymer stabilizers	
Poly(vinylpyrrolidone), PVP	
Polyvinylalcohol, PVA	
Polyethyleneimine	
Sodium polyphosphate	
Sodium polyacrylate	
Tetraalkylammonium halogenides	

Hirai and coworkers[14,15] prepared a colloidal dispersion of rhodium by refluxing a solution of rhodium chloride and PVA in a mixture of methanol and water at 79°C. The volume ratio of methanol to water was 1:1. Refluxing was carried out in argon or air for 0.2 to 16 hours. In this process, methanol was used as a reduction reagent and the reduction reaction was straightforward:

$$RhCl_3 + \frac{3}{2}CH_3OH \rightarrow Rh + \frac{3}{2}HCHO + 3HCl \qquad (3.22)$$

PVA was used as a polymer stabilizer and also served as a diffusion barrier. Rh nanoparticles prepared were found to have mean diameters

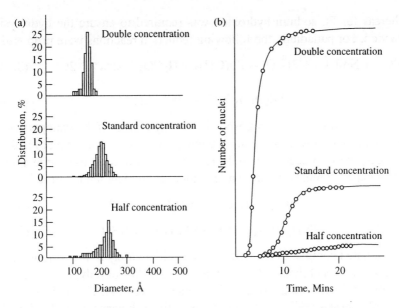

Fig. 3.9. (a) Particle size distribution curves of gold sol prepared at different concentrations (b) Nucleation rate curves for gold sols prepared at different concentrations. [J. Turkevich, *Gold Bull.* **18**, 86 (1985).]

ranging from 0.8 to 4 nm. However, a bimodal size distribution was found, with large particles of 4 nm and small ones of 0.8 nm. Increasing refluxing time was found to result in a decrease of small particles and an increase of large particles, which was attributed to Ostwald ripening.

Henglein et al.[16] studied and compared three different methods for the preparation of Pt nanoparticles: radiolysis, hydrogen reduction and citrate reduction. The γ-rays of ^{60}Co was used to generate hydrated electrons, hydrogen atoms and 1-hydroxylmethyl radicals. These radicals would subsequently reduce Pt^{2+} in K_2PtCl_4 to the zero-valence state, which formed Pt particles with a mean diameter of 1.8 nm. Citrate reduction of $PtCl_6^{2-}$ is also known as Turkevich method,[12,17,18] which was initially developed for the synthesis of uniformly sized gold nanoparticles. In this method, H_2PtCl_6 was mixed with sodium citrate and boiled for 1 hr, yielded Pt particles of 2.5 nm in diameter.

Hydrogen reduction of K_2PtCl_4 and $PdCl_2$ was developed by Rampino and Nord[19] and PVA was used to stabilize both Pt and Pd particles in the experiments. In this method, precursors in dilute aqueous solution were first hydrolyzed to form hydroxides prior to hydrogen reduction. For Pd, sodium carbonate was used as a catalyst to promote the hydrolysis reaction, whereas for Pt, sodium hydroxide was required to ensure the hydrolysis reaction. For palladium, the following reduction reactions were proposed:

$$PdCl_2 + Na_2CO_3 + 2H_2O \rightarrow Pd(OH)_2 + H_2CO_3 + 2Na^+ + 2Cl^- \quad (3.23)$$

$$Pd(OH)_2 + H_2 \rightarrow Pd + 2H_2O \quad (3.24)$$

Similar reactions were proposed for the synthesis of Pt nanoparticles. When no catalyst was used, during aging prior to the introduction of hydrogen gas, the Pt precursor complexes could be converted to a large extent into aquated complexes within a few hours at ambient temperature[20]:

$$PtCl_4^{2-} + H_2O \rightarrow Pt(H_2O)Cl_3^- + Cl^- \quad (3.25)$$

$$Pt(H_2O)Cl_3^- + H_2O \rightarrow Pt(H_2O)_2Cl_2 + Cl^- \quad (3.26)$$

The aquated complexes were then reduced by hydrogen. It was found that the polymeric stabilizer, either sodium polyacrylate or polyphosphate, had a strong influence on the rate of the reduction reaction. This indicates that the polymeric stabilizer may exert catalytic influences on reduction, in addition to their stabilization and diffusion barrier roles. Such prepared Pt particles have a mean diameter of 7.0 nm.

Various methods have been developed for the formation of silver nanoparticles. For example synthesis of Ag nanoparticles can be achieved by the UV illumination of aqueous solutions containing $AgClO_4$, acetone, 2-propanol and various polymer stabilizers.[31] UV illumination generates ketyl radicals via excitation of acetone and subsequent hydrogen atom abstraction from 2-propanol:

$$CH_3COCH_3^* + (CH_3)_2CHOH \rightarrow 2(CH_3)_2(OH)C\bullet \quad (3.27)$$

The ketyl radical may further undergo protolytic dissociation reaction:

$$(CH_3)_2(OH)C\bullet \Leftrightarrow (CH_3)_2OC\bullet^- + H^+ \quad (3.28)$$

Both the ketyl radical and radical anions react with and reduce silver ions to silver atoms:

$$(CH_3)_2(OH)C\bullet + Ag^+ \rightarrow (CH_3)_2CO + Ag + H^+ \quad (3.29)$$

$$(CH_3)_2OC\bullet^- + Ag^+ \rightarrow (CH_3)_2CO + Ag \quad (3.30)$$

Both reactions have a rather low reaction rate, and thus favor the production of monosized silver nanoparticles. With the presence of polyethyleneimine as polymer stabilizer, silver nanoparticles formed using the above photochemical reduction process have a mean size of 7 nm with a narrow size distribution.

Amorphous silver nanoparticles of ~20 nm were prepared by sonochemical reduction of an aqueous silver nitrate solution at a temperature of 10°C, in an atmosphere of argon and hydrogen.[21] The reaction was explained as follows. The ultrasound resulted in decomposition of water into hydrogen and hydroxyl radicals. Hydrogen radicals would reduce silver ions into silver atoms, which subsequently nucleate and grow to silver nanoclusters. Some hydroxyl radicals would combine to form an oxidant, hydrogen peroxide, which may oxidize silver nanoclusters to silver oxide, and the addition of hydrogen gas was to remove the hydrogen peroxides from the solution so as to prevent the oxidation of silver nanoparticles.[22]

Metallic nanoparticles can also be prepared by an electrochemical deposition method.[23,24] This synthesis employs a simple electrochemical cell containing only a metal anode and a metal or glassy carbon cathode. The electrolyte consists of organic solutions of tetraalkylammonium halogenides, which also serve as stabilizers for the produced metal nanoparticles. Upon application of an electric field, the anode undergoes oxidative dissolution forming metal ions, which would migrate toward the cathode. The reduction of metal ions by ammonium ions leads to the nucleation and subsequent growth of metallic nanoparticles in the solution. With this method, nanoparticles of Pd, Ni and Co with diameters ranging from 1.4 to 4.8 nm were produced. Furthermore, it was found that the current density has an appreciable influence on the size of metallic particles; increasing the current density results in a reduced particle size.[23]

3.2.3.1. Influences of reduction reagents

The size and size distribution of metallic colloids vary significantly with the types of reduction reagents used in the synthesis. In general, a strong reduction reaction promotes a fast reaction rate and favors the formation of smaller nanoparticles.[25,26] A weak reduction reagent induces a slow reaction rate and favors relatively larger particles. However, a slow reaction may result in either wider or narrower size distribution. If the slow reaction leads to continuous formation of new nuclei or secondary nuclei, a wide size distribution would be obtained. On the other hand, if no further nucleation or secondary nucleation occurs, a slow reduction reaction would lead to diffusion-limited growth, since the growth of the nuclei would be controlled by the availability of the zerovalent atoms. Consequently, a narrow size distribution would be obtained.

The influences of various reduction reagents on the size and size distribution of gold nanoparticles are summarized in Table 3.2.[27] Using the

Table 3.2. Comparison of average sizes of Au nanoparticles synthesized using various reduction reagents, all in nanometer.[27]

Reduction reagents	436 nm*	546 nm*	XRD#	SEM
Sodium citrate	29.1	28.6	17.5	17.6±0.6
Hydrogen peroxide	25.3	23.1	15.1	15.7±1.1
	31.0	31.3	18.7	19.7±2.6
Hydroxylamine hydrochloride			37.8	22.8±4.2
Citric acid	23.5	22.8		12.5±0.6
Carbon monoxide	9.1	7.4	9.0	5.0±0.5
	15.3	15.3	9.8	7.5±0.4
	18.9	18.3	13.1	12.2±0.5
Phosphorus			13.9	8.1±0.5
			21.0	15.5±1.7
			29.6	25.6±2.6
			36.9	35.8±9.7

* The particle sizes are determined using light scattering with the indicated wavelengths.
# The particle sizes are determined based on X-ray diffraction line broadening.

same reduction reagent, nanoparticle size can be varied by changing the synthesis conditions. In addition, it was found that the reduction reagents have noticeable influences on the morphology of the gold colloidal particles. Figure 3.10 shows electron micrographs of gold nanoparticles prepared with sodium citrate (a) and citric acid (b) as reduction reagents, respectively, under otherwise similar synthesis conditions.[27] Gold particles with spherical shape were obtained using sodium citrate or hydrogen peroxide as reduction reagents, whereas faceted gold particles were formed when hydroxylamine hydrochloride (cubical with {100} facets) and citric acid (trigons or very thin platelets of trigonal symmetry with {111} facets) were used as reduction reagents. Furthermore, concentration of the reduction reagents and pH value of the reagents have noticeable influences on the morphology of the grown gold nanoparticles. For example, lowering the pH value caused the {111} facets to develop at the expense of the {100} facets.

In preparation of transition metallic colloids, Reetz and Maase[28] found that the size of metallic colloids is strongly dependent on how strong a reduction reagent is, and stronger reducing reagents lead to smaller nanoparticles. For example, for the synthesis of Pd colloids from lead nitrate in THF, the particle size decreases in the following order:

$$r_{pivalate} \sim r_{acetate} > r_{glycolate} \gg r_{dichloroacetate} \qquad (3.31)$$

Figure 3.11 shows the particle size of Pd colloids as a function of peak potentials of reduction reagent, carboxylates, in which smaller peak potentials

Zero-Dimensional Nanostructures: Nanoparticles

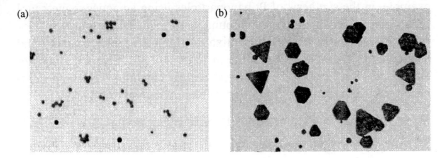

Fig. 3.10. SEM micrographs of gold nanoparticles prepared with sodium citrate (a) and citric acid (b) as reduction reagents, respectively, under otherwise similar synthesis conditions. [W.O. Miligan and R.H. Morriss, *J. Am. Chem. Soc.* **86**, 3461 (1964).]

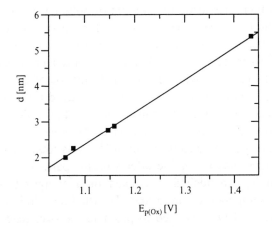

Fig. 3.11. The particle size of Pd colloids as a function of peak potentials of reduction reagent, carboxylates, in which smaller peak potentials mean stronger reduction reagents. [M.T. Reetz and M. Maase, *Adv. Mater.* **11**, 773 (1999).]

mean stronger reduction reagents.[28] Such an influence may be explained by the fact that stronger reduction reagent would generate an abrupt surge of the concentration of growth species, resulting in a very high supersaturation. Consequently, a large number of initial nuclei would form. For a given concentration of metal precursors, the formation of a larger number of nuclei would result in a smaller size of the grown nanoparticles.

3.2.3.2. Influences by other factors

In addition to the control by reduction reagents, the reduction reaction rate or the supply of the growth species can be influenced by other factors. For example, in the synthesis of Pt nanoparticles using an aqueous methanol reduction of H_2PtCl_6, Duff et al.[29] found that a high concentration of chloride ions present in the reaction mixture promoted monodispersity and near-spherical particle shape of the metallic colloids, favoring smoother and rounder surfaces, at the otherwise similar conditions. Such an influence could be understood from the two-step reduction reactions:

$$PtCl_6^{2-} + CH_3OH \rightarrow PtCl_4^{2-} + HCHO + 2H^+ + 2Cl^- \quad (3.32)$$

$$PtCl_4^{2-} + CH_3OH \rightarrow Pt + HCHO + 2H^+ + 4Cl^- \quad (3.33)$$

An increased concentration of chloride ions would favor slow reaction rates. Consequently, the supply of the growth species, i.e. zerovalent Pt atom, would be slow and, thus, favors diffusion-limited growth of initial Pt nuclei. Further, increasing the amount of polymer in the reaction mixture was found to increase the sphericity of the particles. It can be easily understood by considering the fact that increased amount of polymer produces steric resistance for the diffusion and consequently results in a diffusion controlled growth, which favors the formation of spherical particles.

A decreased reduction rate can also be achieved using a low concentration of reactant, which is illustrated by the following example. Nanosized silver particles were synthesized by reduction of silver nitrate using formaldehyde in aqueous solution.[30] It was found that the quantity of reducing agent had negligible effects on the particle size distribution; however, if only formaldehyde was used, the reaction rate would be too slow at room temperature due to low pH. Alkaline solution consisting of NaOH and/or Na_2CO_3 was used to promote the over reaction rate. The reaction between silver ions and reducing agent can be written as:

$$2Ag^+ + HCHO + 3OH^- \rightarrow 2Ag + HCOO^- + 2H_2O \quad (3.34)$$

$$Ag^+ + HCHO + OH^- \rightarrow Ag + HCOOH + \frac{1}{2}H_2 \quad (3.35)$$

The following reaction mechanism was proposed. First hydroxyl ions may undergo a nucleophilic addition reaction to formaldehyde producing hydride and formate ions, and then the hydride ions reduced silver ions to silver atoms.

When only NaOH was used, a higher pH was found to favor for higher reduction rate, and result in the formation of large silver precipitates, which settle at the bottom of solution. When a weak base of sodium carbonate was added to partially substitute NaOH, stable silver colloidal dispersions were obtained. The addition or substitution of sodium carbonate is to control the release of hydroxyl ions only when the pH became lower than certain value according to the following reaction:

$$Na_2CO_3 + 2H_2O \Leftrightarrow 2Na^+ + 2OH^- + H_2CO_3 \quad (3.36)$$

The concentration of hydroxyl ions would determine the rate of reactions 3.34 and 3.35, so as to control the production of silver atoms. Figure 3.12 shows the effect of the quantity of sodium carbonate on the average size of silver particles and the standard deviation of size distribution.[30] Well-dispersed crystalline silver particles of 7–20 nm in size and with spherical shape were obtained with a $Na_2CO_3/AgNO_3$ ratio ranging from 1 to 1.5. More Na_2CO_3 resulted in a higher pH or a higher concentration of hydroxyl ions, which would promote the reduction rates. A higher concentration of Na_2CO_3 would increase the concentration of hydroxyl ions and, thus, promote the reduction rate, resulting in the production of a large quantity of growth species and shift the growth away from diffusion limiting process. It should also be noted that during the synthesis, polyvinylpyrrolidone (PVP) or polyvinyl alcohol (PVA) was used to stabilize the grown silver nanoparticles. As discussed before, the presence of the polymeric layer would also serve as a diffusion barrier, which promotes the diffusion-limited growth, favoring a narrow size distribution. The influences of polymer stabilizers are discussed further in the next section.

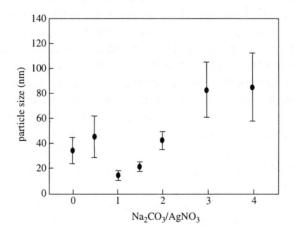

Fig. 3.12. Effect of $[Na_2CO_3]/[AgNO_3]$ ratio on silver average size and its standard deviation (other conditions: $[AgNO_3] = 0.005$ M, $[HCHO]/[AgNO_3] = 4$, $[NaOH]/[AgNO_3] = 1$, $PVP/[AgNO_3] = 9.27$). [K. Chou and C. Ren, *Mater. Chem. Phys.* **64**, 241 (2000).]

3.2.3.3. Influences of polymer stabilizer

Henglein[31] systematically studied the influences of various polymer stabilizers on the formation of silver colloidal dispersions. The polymer stabilizers studied were polyethyleneimine, sodium polyphosphate, sodium polyacrylate and poly(vinylpyrrolidone). Although polymer stabilizers are introduced primarily to form a monolayer on the surface of nanoparticles so as to prevent agglomeration of nanoparticles, the presence of such polymer stabilizers during the formation of nanoparticles can have various influences on the growth process of nanoparticles. Interaction between the surface of a solid particle and polymer stabilizer may vary significantly depending on the surface chemistry of solid, the polymer, solvent and temperature. A strong adsorption of polymer stabilizers would occupy the growth sites and thus reduce the growth rate of nanoparticles. A full coverage of polymer stabilizer would also hinder the diffusion of growth species from the surrounding solution to the surface of growing particle.

Polymer stabilizers may also interact with solute, catalyst, or solvent, and thus directly contribute to reaction. For example, Chou and Ren[30] reported that PVP is actually a weak acid and capable of combining with hydroxyl ions. As a result, the effective quantity of PVP as a stabilizer would be smaller than that was added. Polymer stabilizers have also been found to have catalytic effect on reduction reactions.[16] Furthermore, the pH of the solution would increase with an increasing concentration of PVP.

Ahmadi et al.[32] studied the influences of polymer stabilizer (also referred to as capping material), sodium polyacrylate, on the shape of colloidal platinum nanoparticles. Their results demonstrated that under the same experimental conditions and using the same polymer stabilizer, changing the ratio of the concentration of the capping material to that of Pt ions from 1 : 1 to 5 : 1 produced different shapes of Pt nanoparticles, with cubic particles corresponding to a ratio of 1 : 1 and tetrahedral particles to a 5 : 1 ratio. Obviously the different concentration ratio of capping material has determining influences on the growth rate of {111} and {100} facets of Pt nuclei. Figure 3.13 shows the different morphologies of Pt nanoparticles.[32]

It should also be noted that although polymer stabilizers play a very important role in the synthesis of metal nanoparticles, they can be prepared without using any polymer stabilizers.[21,33] Yin et al.[33] prepared silver nanoparticles through tollens process using a commercially available set of solution.[34] Without adding any stabilizing reagent, the as synthesized aqueous dispersion of silver nanoparticles of 20–30 nm in size was found to be stable for at least one year. The dispersion is likely to be stabilized

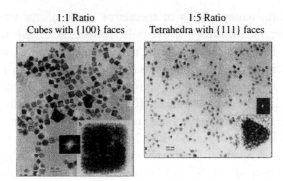

Fig. 3.13. Pt nanoparticles synthesized in colloidal solution and having different shapes (11 nm cubes on the left and ~7 nm tetrahedrons on the right). The potential use of these nanoparticles for different types of catalyses drives our research interest in these particles. [T.S. Ahmadi, Z.L. Wang, T.C. Green, A. Henglein, M.A. El-Sayed, *Science* **272**, 1924 (1996).]

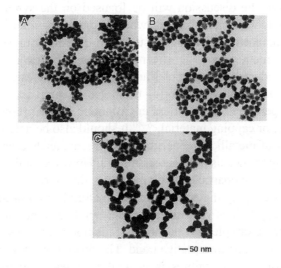

Fig. 3.14. TEM images of silver nanoparticles that were obtained as final products when the reactions were carried out under nitrogen at (A) 27, (B) 30, and (C) 35°C, respectively. The mean size of these silver nanoparticles changed from ~20, to ~30 and ~40 nm when the temperature was raised. [Y. Yin, Z. Li, Z. Zhong, B. Gates, Y. Xia, and S. Venkateswaran, *J. Mater. Chem.* **12**, 522 (2002).]

by electrostatic stabilization mechanism. However, the particle size is sensitively dependent on the synthesis temperature. A small variation of temperature would result in a significant change of diameters of metal nanoparticles. Figure 3.14 compares the silver nanoparticles synthesized under different temperatures.[33]

Furthermore, nanoparticles of metals or metal alloys were prepared through seeding nucleation. For example, Toneguzzo et al.[35] reported that polymetallic fine particles Co_xNi_{1-x} and $Fe_z[Co_xNi_{1-x}]_{1-z}$ were synthesized by precipitation from metallic precursors dissolved in 1,2-propanediol with an optimized amount of sodium hydroxide. The precursors used were tetrahydrated cobalt(II), nickel(II) acetate and tetrahydrated iron (II) chloride. The particle formation was initiated by adding a small amount of a solution of K_2PtCl_4 or $AgNO_3$ in 1,2-ethanediol. Pt or Ag is believed to act as nucleation agent. An increased concentration of Pt or Ag relative to the concentration of Co, Ni and Fe resulted in a reduced mean particle size, implying an increased number of particles.

3.2.4. Synthesis of semiconductor nanoparticles

In this section, the discussion will be focused on the synthesis of non-oxide semiconductor nanoparticles, whereas the formation of oxide semiconductor nanoparticles will be discussed in the following section, since the synthesis methods are significantly different from each other. Non-oxide semiconductor nanoparticles are commonly synthesized by pyrolysis of organometallic precursor(s) dissolved in anhydrate solvents at elevated temperatures in an airless environment in the presence of polymer stabilizer or capping material.[36–40] It should also be noted here that in the synthesis of metallic nanoparticles, polymers attached on the surface are commonly termed as polymer stabilizers. However, in the synthesis of semiconductor nanoparticles, polymers on the surface are generally referred to as capping materials. Capping materials are linked to the surface of nanocrystallites via either covalent bonds or other bonds such as dative bonds.[41] Examples are sulfur and transition metal ions and nitrogen lone pair of electrons form dative bond. The formation of monodispersed semiconductor nanocrystallites is generally achieved by the following approaches. First, temporally discrete nucleation is attained by a rapid increase in the reagent concentrations upon injection, resulting in an abrupt supersaturation. Second, Ostwald ripening during aging at increased temperatures promotes the growth of large particles at the expense of small ones, narrowing the size distribution. Third, size selective precipitation is applied to further enhance the size uniformity. It is noted that although organic molecules are used to stabilize the colloidal dispersion, similar to that in the formation of metallic colloidal dispersions, the organic monolayers on the surfaces of semiconductor nanoparticles play a relatively less significant role as a diffusion barrier during the subsequent growth of initial nuclei. This is simply because there is a less extent or negligible subsequent growth of initial nuclei due to the depletion of growth species and the drop of temperature at the nucleation stage.

Synthesis of CdE (E = S, Se, Te) semiconductor nanocrystallites reported by Murray et al.,[42] which is based on the earlier work by Steigerwald et al.[43,44] is used as an example to illustrate the general approach. Dimethylcadmium (Me_2Cd) was used as the Cd source and bis(trimethylsilyl) sulfide (($TMS)_2S$), trioctylphosphine selenide (TOPSe), and trioctylphosphine telluride (TOPTe) were used as S, Se and Te precursors, respectively. Mixed tri-*n*-octylphosphine (TOP) and tri-*n*-octylphosphine oxide (TOPO) solutions were used as solvents and capping materials, also known as coordinating solvents.

The procedure for the preparation of TOP/TOPO capped CdSe nanocrystallites is briefly outlined below.[42] Fifty grams of TOPO is dried and degassed in the reaction vessel by heating to ~200°C at ~1 torr for ~20 min, flushing periodically with argon. The temperature of the reaction flask is then stabilized at ~300°C under ~1 atm of argon. 1.00 mL of Me_2Cd is added to 25.0 mL of TOP in the dry box, and 10.0 mL of 1.0 M TOPSe stock solution is added to 15.0 mL of TOP. Two solutions are then combined and loaded into a syringe in the dry box. The heat is removed from the reaction vessel. The syringe containing the reagent mixture is quickly removed from the dry box and its content delivered to the vigorously stirring reaction flask in a single injection through a rubber septum. The rapid introduction of the reagent mixture produces a deep yellow/orange solution with an absorption feature at 440–460 nm. This is also accompanied by a sudden decrease in temperature to ~180°C. Heating is restored to the reaction flask and the temperature is gradually raised to and aged at 230–260°C. Depending on the aging time, CdSe nanoparticles with a series of sizes ranging from ~1.5 nm to 11.5 nm in diameter are prepared.

The above prepared colloidal dispersion is purified by cooling to ~60°C, slightly above the melting point of TOPO, and adding 20 mL of anhydrous methanol, which results in the reversible flocculation of the nanocrystallites. The flocculate is separated from the supernatant by centrifugation. Dispersion of the flocculation in 25 mL of anhydrous 1-butanol followed by further centrifugation results in an optically clear solution (more precisely speaking, a colloidal dispersion, but solution is a widely accepted term in the literature in this field) of nanocrystallites and a gray precipitate containing byproducts, consisting mostly of elemental Cd and Se, of the reaction. Addition of 25 mL of anhydrous methanol to the supernatant produces flocculation of the crystallites and removes excess TOP and TOPO. A final rinse of the flocculate with 50 mL of methanol and subsequent vacuum drying produces ~300 mg of free flowing TOP/TOPO capped CdSe nanocrystallite.

The purified nanocrystallites are subsequently dispersed in anhydrous 1-butanol forming an optically clear solution. Anhydrous methanol is then added drop wise to the dispersion until opalescence persists upon stirring or sonication. Separation of supernatant and flocculate by centrifugation produces a precipitate enriched with the largest crystallites in the sample. Dispersion of the precipitate in 1-butanol and size-selective precipitation with methanol is repeated until no further narrowing of the size distribution as indicated by sharpening of optical absorption spectrum.

Mixed phosphine and phosphine oxide solutions were found to be good solvents for the high temperature growth and annealing of CdSe crystallite.[45,46] The coordinating solvent plays a crucial role in controlling the growth process, stabilizing the resulting colloidal dispersion, and electronically passivating the semiconductor surface.

Injection of reagents into the hot reaction vessel results in a short burst of homogeneous nucleation due to an abrupt supersaturation and simultaneously a sharp drop in temperature associated with the introduction of room temperature precursor solution. The depletion of reagents through such nucleation prevents further nucleation and also largely hinders the subsequent growth of existing nuclei. Monodispersion is further achieved by gently reheating the solution to promote slow growth of initial nuclei. An increased temperature results in an increased solubility, and thus a reduced supersaturation of growth species in the solution. As a result, nuclei with small sizes may become unstable and dissolve back into the solution; dissolved species will then deposit onto the surfaces of large particles. This dissolution-growth process is also known as Ostwald ripening, in which large particles grow at the expense of small particles.[47] Such a growth process would result in the production of highly monodispersed colloidal dispersions from systems that may initially be polydispersed.[48] Lowering the synthesis temperature results in a wider size distribution with an increased amount of small particles. A lowered temperature would result in an increased supersaturation favoring continued nucleation with smaller sizes. An increased temperature will promote the growth of nanoparticles with a narrow size distribution.

Figure 3.15 shows the SEM images and optical absorption spectra of CdSe nanocrystallites in size ranging from ~1.2 nm to 11.5 nm and dispersed in hexane.[42] Figure 3.16 shows the X-ray powder diffraction spectra of CdSe crystallites ranging from ~1.2 to 11.5 nm in diameter, and indicates CdSe crystallites have a predominantly wurtzite crystal structure with the lattice spacing of the bulk material.[42] Finite size broadening in all

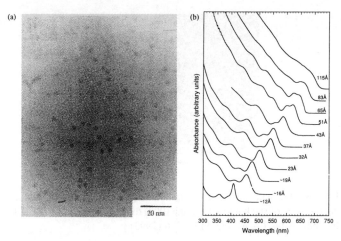

Fig. 3.15. (a) A near monolayer of 51 Å diameter CdSe crystallites showing short-range hexagonal close packing. (b) Room temperature optical absorption spectra of CdSe nanocrystallites dispersed in hexane and ranging in size from ~12 to 115 Å. [C.B. Murray, D.J. Norris, and M.G. Bawendi, *J. Am. Chem. Soc.* **115**, 8706 (1993).]

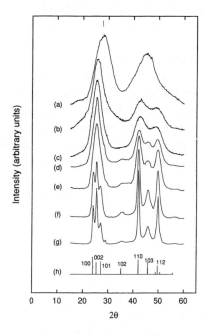

Fig. 3.16. Powder X-ray diffraction spectra of (a) 12, (b) 18, (c) 20, (d) 37, (e) 42, (f) 83 and (g) 115 Å diameter CdSe nanocrystallites compared with the bulk wurtzite peak positions (h). [C.B. Murray, D.J. Norris, and M.G. Bawendi, *J. Am. Chem. Soc.* **115**, 8706 (1993).]

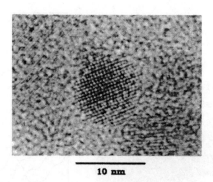

Fig. 3.17. An 80 Å diameter CdSe crystallite imaged in bright field with atom contrast shows the presence of stacking faults in the (002) direction. [C.B. Murray, D.J. Norris, and M.G. Bawendi, *J. Am. Chem. Soc.* **115**, 8706 (1993).]

diffraction peaks is evident, and excessive attenuation and broadening in (102) and (103) peaks are characteristic of stacking faults along the (002) axis.[49] Such defects are observed in high-resolution TEM image as shown in Fig. 3.17.[42]

Size-selective precipitation would further narrow the size distribution of the colloids prepared. For the fraction process to work well it is crucially important that the shape and surface derivation of the initial crystallites be uniform and that the initial polydispersity in size be relatively small.[42] It should be noted that although the subsequent growth of initial nuclei appears less important in the synthesis of monodispersed CdSe nanocrystallites as compared to that in the formation of monodispersed metal nanoparticles, due to the depletion of reagents as discussed above, the capping material provides an important steric barrier for diffusion and thus favors the diffusion controlled subsequent growth of existing nuclei.

Size-selective precipitation is a very useful method in the synthesis of monodispersed nanocrystals. For example, Guzelian *et al.*[50] prepared monodispersed InP nanocrystals of 2 to 5 nm in diameter via the reaction of $InCl_3$ and $P(Si(CH_3)_3)_3$ in trioctylphosphine oxide (TOPO) at elevated temperatures, and monodispersion is largely achieved by repeated size-selective precipitation. Since the synthesis is a slow process in which nucleation and growth occur simultaneously over long time scales, in contrast to temporally discrete nucleation and negligible subsequent growth in the synthesis of CdSe described above, InP nanoparticles have a broad size distribution. InP nanocrystals capped with dodecylamine are soluble in toluene and insoluble in methanol. Using stepwise addition of methanol to the reaction solution results in the incremental size-selective precipitation of the nanocrystals. From the same reaction mixture, isolated 2–5 nm nanocrystals are obtained, and if small enough volumes of methanol are used, a sufficiently careful precipitation series can resolve size distributions separated by as little as 0.15 nm.[50]

Thermal decomposition of complex precursor in a high-boiling solvent represents another method in the production of compound semiconductor nanoparticles with a narrow size distribution.[51,52] For example, when $GaCl_3$ is mixed with $P(SiMe_3)_3$ in a molar ratio of Ga : P of 1 : 1 in toluene at room temperature, a complex Ga and P precursor, $[Cl_2GaP(SiMe_3)_2]_2$ is formed.[53,54] Similar reactions may occur by mixing chloroindium oxalate and $P(SiMe_3)_3$ in a predetermined molar ratio in CH_3CN for the formation of InP complex precursor, or mixing chlorogallium oxalate, chloroindium oxalate and $P(SiMe_3)_3$ in a desired molar ratio in toluene at room temperature.[51] InP, GaP and $GaInP_2$ high-quality nanocrystallites are formed by heating the complex precursors dissolved in high-boiling solvent containing a mixture of TOP and TOPO as a colloidal stabilizer at elevated temperatures for several days. The typical thermal decomposition of InP precursor solution in TOP/TOPO at elevated temperatures produces InP nanocrystals capped with TOPO[52]:

$$\text{InP precursor} + (C_8H_{17})_3PO \rightarrow \text{InP-}(C_8H_{17})_3PO + \text{byproducts} \quad (3.37)$$

Such prepared nanoparticles of InP, GaP and $GaInP_2$ are well crystallized with bulk zinc blende structure. An increase in heating duration was found to improve the crystallinity of the nanoparticles. Different particle sizes ranging from 2.0 to 6.5 nm are obtained by changing the precursor concentration or by changing the temperature. The narrow size distribution is achieved due to (i) the slow process rate of the decomposition reaction of the complex precursors and possibly (ii) the steric diffusion barrier of the TOP and TOPO stabilizer monolayer on the growing surface of nanoparticles.[51] The addition of methanol into the colloidal solution results in the precipitation of nanoparticles.

Thermal decomposition of complex precursors is also applied in the synthesis of GaAs nanoparticles.[55,56] For example, when an appropriate amount of $Li(THF)_2As(SiMe_3)_2$ (THF = tetrahydrofuran) is added to a pentane solution of $[(C_5Me_5)_2GaCl]_2$, followed by filtration, evaporation of the solvent, and recrystallization, pure arsinogallane complex precursor, $(C_5Me_5)GaAs(SiMe_3)_2$ is produced. This complex precursor, when dissolved in organic solvents such as alcohol, undergoes thermal decomposition to form GaAs nanoparticles when heated above 60°C or exposed to air.[55] When tris(trimethylsilyl)arsine reacts with gallium chloride, complex GaAs precursors can be prepared.[57] GaAs nanocrystals can be prepared by heating the above complex precursor dissolved in polar organic solvents, such as in quinoline at 240°C for 3 days.[56]

Colloidal CdS and PbS dispersions with particle sizes ≤8 nm were prepared by mixing $Cd(OOCCH_3)_2 \cdot 2H_2O$ or $Pb(OOCCH_3)_2 \cdot 3H_2O$ with surfactants and thioacetamide (CH_3CSNH_2) in methanol solution.[58] Surfactants used in the preparation of CdS and PbS nanoparticles include: acetylacetone, 3-aminopropyltriethoxysilane, 3-aminopropyltrimethoxysilane and 3-mercaptopropyltrimethoxysilane (MPTMS). Among these surfactants, MPTMS was found to be the most effective surfactant in the preparation of nanoparticles of CdS and PbS.[58,59]

CdS nanoparticles can be synthesized by mixing $Cd(ClO_4)_2$ and $(NaPO_3)_6$ solutions with pH adjusted with NaOH and bubbled with argon gas. Desired amount of H_2S was injected into the gas phase and the solution was vigorously shaken.[45] The starting pH value was found to have a significant influence on the average size of the particles synthesized. The particle size increases with a decreasing starting pH value, and Fig. 3.18 shows the absorption and fluorescence spectra of the three CdS colloidal dispersions with different starting pH values.[45] For the smallest particle, i.e. for example, the onset of absorption is already shifted to a wavelength clearly shorter than 500 nm.

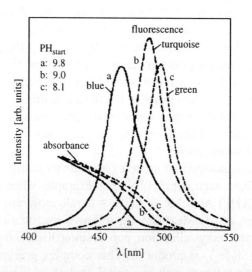

Fig. 3.18. Absorption and fluorescence spectra of the three CdS colloidal dispersions with different starting pH values. The particle size increases with a decreasing starting pH value. [L. Spanhel, M. Haase, H. Weller, and A. Henglein, *J. Am. Chem. Soc.* **109**, 5649 (1987).]

Synthesis of GaN nanocrystallites poses a different challenge. Typically GaN would be formed at temperatures higher than 600°C.[60,61] Even thermal pyrolysis of complex precursors such as [H_2GaNH_2]$_3$ and Ga(C_2H_5)$_3NH_3$ which already have Ga–N bond requires a post heat treatment at temperatures above 500°C.[62,63] The reaction of Li_3N with $GaCl_3$ in benzene at 280°C under pressure in an autoclave produces nanocrystallite GaN through a liquid–solid reaction[64]:

$$GaCl_3 + Li_3N \rightarrow GaN + 3LiCl \tag{3.38}$$

Such GaN nanocrystallites formed are of ~30 nm in diameter with mainly hexagonal structure with a small fraction of rock salt-phase with lattice constants close to that of bulk materials.[64]

Solution synthesis of colloidal GaN has also been developed.[65,66] For example, Mićić et al.[65] synthesized colloidal GaN nanoparticles of 3.0 nm in diameter with spherical shape and zinc blende crystal structure. First a GaN complex precursor, polymeric gallium imide, {Ga(NH)$_{3/2}$}$_n$, was prepared by the reaction of dimeric amidogallium, Ga$_2$[N(CH$_3$)$_2$]$_6$, with gaseous ammonia, NH_3, at room temperature.[67,68] The precursor was then heated in trioctylamine (TOA) at 360°C for 24 h to produce GaN nanocrystals under flowing ammonia at ambient pressure. The solution was cooled to 220°C and a mixture of TOA and hexadecylamine (HAD) was added and stirred at 220°C for 10 h. The GaN nanocrystals were capped with a mixture of TOA and HAD.

3.2.5. *Synthesis of oxide nanoparticles*

Compared to the synthesis of metallic and non-oxide nanoparticles, the approaches used in the fabrication of oxide nanoparticles are less elaborated and there are less defined general strategies for the achievement of monosized distribution. Although all the fundamental considerations, including a burst of homogeneous nucleation and diffusion controlled subsequent growth, are applicable to the oxide systems, the practical approaches vary noticeably from system to system. Reaction and growth in the formation of oxide nanoparticles are more difficult to manipulate, since oxides are generally more stable thermally and chemically than most semiconductors and metals. For example, Ostwald ripening is applied in the synthesis of oxide nanoparticles to reduce size distribution; the results may be less effective than in other materials. The most studied and best-established example of oxide colloidal is silica colloids[69] though various oxide nanoparticles have been studied.[70,71] Commonly oxide particles in colloidal dispersions are synthesized by sol-gel processing. Sol-gel processing is also commonly used in the fabrication of various core-shell nanostructures[72] and surface engineering of nanostructures.[73] Before discussing the general approaches for the synthesis of oxide nanoparticles, let us briefly discuss the sol-gel processing first.

3.2.5.1. Introduction to sol-gel processing

Sol-gel processing is a wet chemical route for the synthesis of colloidal dispersions of inorganic and organic–inorganic hybrid materials, particularly oxides and oxide-based hybrids. From such colloidal dispersions, powders, fibers, thin films and monoliths can be readily prepared. Although the fabrication of different forms of final products requires some specific considerations, the fundamentals and general approaches in the synthesis of colloidal dispersions are the same. Sol-gel processing offers many advantages, including low processing temperature and molecular level homogeneity. Sol-gel processing is particularly useful in making complex metal oxides, temperature sensitive organic–inorganic hybrid materials, and thermodynamically unfavorable or metastable materials. For more details, readers may wish to consult the abundant literature in this field. For instance, *Sol-Gel Science* by Brinker and Scherer,[74] *Introduction to Sol-Gel Processing* by Pierre,[75] and *Sol-Gel Materials* by Wright and Sommerdijk[76] provide an excellent and comprehensive coverage on sol-gel processing and materials. Typical sol-gel processing consists of hydrolysis and condensation of precursors. Precursors can be either metal alkoxides or inorganic and organic salts. Organic or aqueous solvents may be used to dissolve precursors, and catalysts are often added to promote hydrolysis and condensation reactions:

Hydrolysis:

$$M(OEt)_4 + xH_2O \Leftrightarrow M(OEt)_{4-x}(OH)_x + xEtOH \quad (3.39)$$

Condensation:

$$M(OEt)_{4-x}(OH)_x + M(OEt)_{4-x}(OH)_x$$
$$\Leftrightarrow (OEt)_{4-x}(OH)_{x-1} MOM(OEt)_{4-x}(OH)_{x-1} + H_2O \quad (3.40)$$

Hydrolysis and condensation reactions are both multiple-step processes, occurring sequentially and in parallel. Each sequential reaction may be reversible. Condensation results in the formation of nanoscale clusters of metal oxides or hydroxides, often with organic groups embedded or attached to them. These organic groups may be due to incomplete hydrolysis, or introduced as non-hydrolysable organic ligands. The size of the nanoscale clusters, along with the morphology and microstructure of the final product, can be tailored by controlling the hydrolysis and condensation reactions.

For the synthesis of colloidal dispersions of multiple-component materials, the challenges are to ensure hetero-condensation reactions between different constituent precursors, which typically have different chemical reactivities. The reactivity of a metal atom is dependent largely on the extent of charge transfer and the ability to increase its coordination number. As a rule of thumb, the electronegativity of a metal atom decreases and the ability to increase its coordination number increases with their ionic radius as shown in Table 3.3.[77] Accordingly the chemical reactivity of the corresponding alkoxides increases with their ionic radius. There are several ways to ensure hetero-condensation, and achieve a homogeneous mixture of multiple components at the molecular/atomic level.

First, the precursors can be modified by attaching different organic ligands. For a given metal atom or ion, large organic ligand or more complex organic ligand would result in a less reactive precursor.[74] For example, $Si(OC_2H_5)_4$ is less reactive than $Si(OCH_3)_4$, and $Ti(OPr^x)_4$ is less reactive than $Ti(OPr^i)_4$. Another way to control the reactivity of the alkoxides is to chemically modify the coordination state of the alkoxides with a chelating agent such as acetylacetone. Multiple step sol-gel processing is yet another way to overcome this problem. The less reactive precursor is first partially hydrolyzed, and more reactive precursor is hydrolyzed later.[78] In more extreme cases, one precursor can be fully hydrolyzed first and all water is depleted, if hydrolyzed precursor has a very low condensation rate, then the second precursor is introduced and forced to condensate with the hydrolyzed precursor by the reaction:

$$M(OEt)_4 + 4H_2O \Leftrightarrow M(OH)_4 + 4HOEt \qquad (3.41)$$

Table 3.3. Electronegativity, χ, partial charge, δM, ionic radius, r, and coordination number, n, of some tetravalent metals.[77]

Alkoxide	χ	δM	$r(\text{Å})$	n
$Si(OPr^i)_4$	1.74	+0.32	0.40	4
$Ti(OPr^i)_4$	1.32	+0.60	0.64	6
$Zr(OPr^i)_4$	1.29	+0.64	0.87	7
$Ce(OPr^i)_4$	1.17	+0.75	1.02	8

where OPr^i is $OCH_2CH_2CH_3$

Condensation reactions are only limited between hydrolyzed less reactive precursor with more reactive precursor:

$$M(OH)_4 + M'(OEt)_4 \Leftrightarrow (HO)_3\text{-MOM}'(OEt)_3 \qquad (3.42)$$

Incorporating organic components into an oxide system by sol-gel processing makes it easy to form organic–inorganic hybrids. One approach is to co-polymerize or co-condense both the inorganic precursor(s), which lead to the formation of the inorganic component, and the organic precursor(s), which consist of non-hydrolysable organic groups. Such organic–inorganic hybrids are a single-phase material, in which the organic and inorganic components are linked through chemical bonds. Another approach is to trap the desired organic components physically inside the inorganic or oxide network, by either homogeneously dispersing the organic components in the sol, or infiltrating the organic molecules into the gel network. Similar approaches can be applied for the incorporation of bio-components into oxide systems. Another method to incorporate bio-components into the oxide structure is to use functional organic groups to bridge inorganic and biological species. Organic–inorganic hybrid materials form a new family of materials, which promise a lot of important potential applications and will be discussed further in Chapter 6.

Another challenge in making complex oxide sols is that the constituent precursors may exert a catalytic effect on one another. As a result, the hydrolysis and condensation reaction rates when two precursors are mixed together may be significantly different from those when the precursors are processed separately.[79] In the sol preparation, not much attention has been paid to the control of crystallization or formation of crystal structure, although the formation of crystalline structure of complex oxides without high-temperature firing is desired for some applications. Matsuda and co-workers have demonstrated that it is possible to form the crystalline phase of $BaTiO_3$ without high temperature sintering by carefully controlling processing conditions, including concentrations and temperature.[80] However, there is still a lack of general understanding on the control of crystallization of complex oxides during sol preparation.

By a careful control of sol preparation and processing, monodispersed nanoparticles of various oxides, including complex oxides, organic–inorganic hybrids, and biomaterials, can be synthesized. The key issue is to promote temporal nucleation followed by diffusion-controlled subsequent growth.[81–83] The particle size can be varied by changing the concentration and aging time.[74] In a typical sol, nanoclusters formed by hydrolysis and condensation reactions commonly have a size ranging from 1 to 100 nm.

It should also be noted that in the formation of monodispersed oxide nanoparticles, the stabilization of colloids is generally achieved by electrostatic double layer mechanism. Therefore, polymer steric diffusion barrier existing in the formation of metal and non-oxide semiconductor colloids, is generally not present in the formation of metal oxides. So the diffusion controlled growth is achieved through other mechanisms, such as controlled release and a low concentration of growth species in the sol.

3.2.5.2. Forced hydrolysis

The simplest method for the generation of uniformly sized colloidal metal oxides is based on forced hydrolysis of metal salt solutions. It is well known that most polyvalent cations readily hydrolyze, and that deprotonation of coordinated water molecules is greatly accelerated with increasing temperature. Since hydrolysis products are intermediates to precipitation of metal oxides, increasing temperature results in an increasing amount of deprotonated molecules. When the concentration far exceeds the solubility, nucleation of metal oxides occurs. In principle, to produce such metal oxide colloids, one just needs to age hydrolyzed metal solutions at elevated temperatures. It becomes obvious that hydrolysis reaction should proceed rapidly and produce an abrupt supersaturation to ensure a burst of nucleation, resulting in the formation of a large number of small nuclei, eventually leading to the formation of small particles. This principle was demonstrated in the pioneer work on the formation of silica spheres by Stöber and co-workers.[83]

The procedures for the preparation of silica spheres were simple and straightforward. Various silicon alkoxides with different alkyl ligand sizes were used as precursors, ammonia was used as a catalyst, and various alcohols were used as solvents. First alcohol solvent, ammonia, and a desired amount of water were mixed, and then silicon alkoxide precursor was added under vigorous stirring. The formation of colloids or the change of optical appearance of the solution became noticeable just in a few minutes after the addition of precursors. Depending on the precursors, solvents and the amounts of water and ammonia used, spherical silica particles with mean sizes ranging from 50 nm to 2 μm were obtained. Figure 3.19 shows the first example of such prepared silica spheres.[83]

It was found that the reaction rate and particle size were strongly dependent on solvents, precursors, amount of water and ammonia. For the different alcoholic solvents, reaction rates were fastest with methanol, slowest with n-butanol. Likewise, final particle sizes obtained under

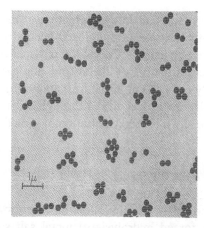

Fig. 3.19. SEM micrograph of silica spheres prepared in the ethanol-ethyl ester system. [W. Stober, A. Fink, and E. Bohn, *J. Colloid Interf. Sci.* **26**, 62 (1968).]

comparable conditions were smallest in methanol and biggest in *n*-butanol. However, there was a tendency toward wide size distributions with the higher alcohols. Similar relationship with regard to reaction rates and particle sizes was found when comparing results with different ligand sizes in the precursors. Smaller ligand resulted in faster reaction rate and smaller particle size, whereas larger ligands led to slower reaction rate and large particle size. Ammonia was found necessary for the formation of spherical silica particles, since condensation reaction under a basic condition yields three-dimensional structure instead of a linear polymeric chain which occurs under an acidic condition.[74]

Both hydrolysis and condensation reactions, as any other chemical reactions, are strongly dependent on reaction temperatures. An elevated temperature would result in a drastic increase of reaction rate. Preparation of spherical colloidal α-Fe_2O_3 nanoparticles of 100 nm in size can be used as another example to illustrate the typical procedure of forced hydrolysis.[84] First $FeCl_3$ solution is mixed with HCl, and diluted. The mixture is then added into preheated H_2O at 95–99°C with constant stirring. The solution is kept in a sealed preheated bottle at 100°C for 24 hr before being quenched in cold water. The high temperature favors a fast hydrolysis reaction and results in the high supersaturation, which in turn leads to the formation of a large number of small nuclei. Dilution before heating to high temperatures is very important to ensure a controlled nucleation and subsequent diffusion-limited growth. A long aging period would permit the occurrence of Ostwald ripening to further narrow the size distribution.

3.2.5.3. Controlled release of ions

Controlled release of constituent anions and/or cations has a significant influence on the kinetics of nucleation and subsequent growth of oxide nanoparticles, and is achieved by the spontaneous release of anions from organic molecules. For example, it is well known that solutions of urea, $CO(NH_2)_2$, when heated liberate hydroxide ions, which can cause precipitation of metal oxide or hydroxide.[85-87] For example, the decomposition of urea is used to control the nucleation process in the synthesis of Y_2O_3:Eu nanoparticles.[86] Yttrium and europium chlorides were dissolved in water and the pH was adjusted to ~1 with hydrochloride acid or potassium hydroxide. An excess of urea, typically 15x, was dissolved into the solution. The solution was then raised to > 80°C for 2 hours. The urea decomposed slowly and there was a burst of nucleation when a certain pH value of ~4–5 was reached.

In general, certain types of anions are commonly introduced into the system as a catalyst. In addition to the catalytic effect, anions commonly exert other influences on the processing and the morphology of the nanoparticles.[88] Figure 3.20 shows the TEM images of particles obtained from solutions of $FeCl_3$ and HCl under various conditions listed in

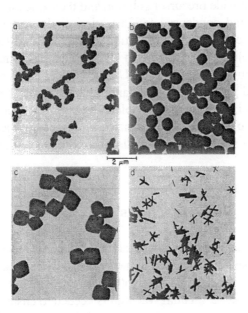

Fig. 3.20. TEM images of various iron oxide and iron hydroxide nanoparticles obtained from solutions of $FeCl_3$ and HCl under various conditions listed in Table 3.4. [E. Matijević, *J. Colloid Interf. Sci.* **58**, 374 (1977).]

Table 3.4. A summary of synthesis parameters including temperature and time of aging are used to obtain α-Fe$_2$O$_3$ (A, B, and C) or β-FeO(OH) nanoparticles shown in Fig. 3.19.[88,89]

	Fe^{3+} (M)	Cl^- (M)	Initial pH	Final pH	Temp (°C)	Time
A	0.018	0.104	1.3	1.1	100	24 hr
B	0.315	0.995	2.0	1.0	100	9 days
C	0.09	0.28	1.65	0.88	100	24 hr
D	0.09	0.28	1.65	0.70	100	6 hr

Table 3.4.[88,89] Systems a, b, and c represent hematite (α-Fe$_2$O$_3$) dispersions, where system d is rod-like akageneite, β-FeO(OH). Presence of anions may result in a change of the surface properties and interface energy of nanoparticles, and subsequently influence the growth behavior of the particle. Anions may be incorporated into the structure of nanoparticles, or adsorbed onto the surface of nanoparticles. Anions may also have significant influences on the stability of the colloidal dispersion, when nanoparticles are stabilized by electrostatic stabilization mechanism.

The preparation of crystalline ZnO nanoparticles is another example of controlled release of anions. First zinc acetate is dissolved into methanol to form zinc alkoxide precursor solution and then zinc alkoxide precursor is hydrolyzed and condensed to form zinc oxide colloid with lithium hydroxide as a catalyst with sonication at 0°C or room temperature.[90,91] Sonication accelerates the release of OH$^-$ groups, resulting in immediate reaction to form a stable ZnO sol. Use of NaOH, KOH or Mg(OH)$_2$ all produces turbid precipitates. ZnO nanoparticles are ~3.5 nm in diameter in fresh sols and ~5.5 nm in 5 day old ones. Aging of alcoholic ZnO colloids is known to produce larger particles.[92-94] Acetate groups are believed to attach to the surface of ZnO colloids and thus stabilize the colloidal dispersion.[90,94]

3.2.6. Vapor phase reactions

Nanoparticles can also be synthesized by vapor phase reactions, following the same mechanisms discussed in the synthesis of nanoparticles in liquid medium. In general, reaction and synthesis are carried out at elevated temperatures and under a vacuum. Vacuum is needed to ensure a low concentration of growth species so as to promote diffusion-controlled subsequent growth. Grown nanoparticles are normally collected on a non-sticking substrate placed down stream at a relatively low temperature. Obviously only a small fraction of nanoparticles do settle on the substrate surface. Furthermore, the nanoparticles that settled on the substrate surface may not represent the true particle size distribution. It is also difficult to introduce stabilization mechanism during synthesis to prevent the formation of agglomerates. Despite the aforementioned challenges, it has been demonstrated that various nanoparticles can be synthesized by vapor phase reactions. For example, the gas aggregation technique has been applied in the synthesis of silver nanoparticles of 2–3 nm in diameter.[95] Another example is the production of highly dispersed silica particles less than 100 nm in diameter by combustion of silicon tetrachloride in a hydrogen torch.[96]

It is noted that nanoparticles, formed through homogeneous nucleation and then deposited on substrates, may migrate and agglomerate.[97] Two types of agglomerates were found. One is the large size spherical particle and another is the needle-like particle. The formation of prolate particles commonly along step edges were found in the systems of Au on (100) NaCl[98] and (111) CaF[99] substrates, and Ag on (100) NaCl substrates.[95] However, the step edges are not always required for the formation of needle-like crystals. For example, crystal CdS nanorods with a length up to several hundred micrometers were formed.[100] Au particles with diameters in a few nanometers have been grown on various oxide substrates including iron oxide,[101] γ-alumina,[102] and titania.[103]

GaAs nanoparticles can be synthesized by homogeneous vapor phase nucleation from organometallic precursors.[104] Trimethyl gallium and AsH_3 are used as precursors and hydrogen is used as a carrier gas as well as a reduction reagent. Reaction and nucleation occur at a temperature of 700°C at atmospheric pressure. GaAs nanoparticles are collected thermophoretically on a holey carbon film downstream at a temperature of 350°C. The nanoparticles are found to be composed of highly faceted single crystal GaAs with diameters ranging from 10 to 20 nm. In addition, an increase in reaction and nucleation temperature results in an increased particle size. An increased concentration of precursors has a similar influence on the particle size. However, change in temperature and precursor concentrations is found to have negligible influence on the morphology of nanoparticles.

3.2.7. *Solid state phase segregation*

Nanoparticles of metals and semiconductors in glass matrix are commonly formed by homogeneous nucleation in solid state.[105,106] First the desired metal or semiconductor precursors were introduced to and homogeneously distributed in the liquid glass melt at high temperatures during glass making, before quenching to room temperature. Then the glass was annealed by heating to a temperature about the glass transition point and held for a pre-designed period of time. During the annealing, metal or semiconductor precursors were converted to metals and semiconductors. As a result, supersaturated metals or semiconductors formed nanoparticles through nucleation and subsequent growth via solid-state diffusion.

Homogeneous glasses are made by dissolving metals, in the form of ions, in the glass melts and then rapidly cooled to room temperature. In such glasses metals remain as ions.[107] Upon reheating to an intermediate temperature region, metallic ions are reduced to metallic atoms by certain reduction agents such as antimony oxide that is also added into the glasses. Metallic nanoparticles can also be nucleated by ultraviolet, X-ray, or γ-ray radiation if a radiation-sensitive ion such as cerium is present.[107] The subsequent growth of the nuclei takes place by solid-state diffusion.[108] For example, glasses with nanoparticles of gold,[107] silver,[109] and copper[110] can all be prepared with such an approach. Although metallic ions may be highly soluble in the glass melts or glasses, metallic atoms are not soluble in glasses. When heated to elevated temperatures, metallic atoms acquire needed diffusivity to migrate through the glasses and subsequently form nuclei. These nuclei would grow further to form nanoparticles of various sizes. Since solid-state diffusion is relatively slow, it is relatively easy to have a diffusion-controlled growth for the formation of monosized particles. Figure 3.21 shows the TEM micrographs of Cu and Ag nanoparticles in glass matrices.[111]

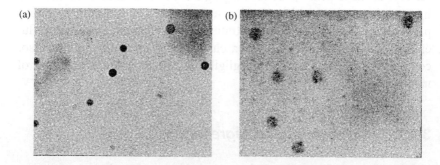

Fig. 3.21. TEM micrograph of Cu and Ag nanoparticles in BaO-P_2O_5 glass: (a) 50P_2O_5-50BaO-6SnO-6Cu_2O, and (b) 50P_2O_5-50BaO-4SnO-4Ag_2O. [K. Uchida, S. Kaneko, S. Omi, C. Hata, H. Tanji, Y. Asahara, and A.J. Ikushima, *J. Opt. Soc. Am.* **B11**, 1236 (1994).]

Nanoparticles dispersed in glass matrix can be synthesized through sol-gel processing as well. There are two approaches: (i) mixing pre-synthesized colloidal dispersion with matrix sol before gelation, and (ii) making a homogeneous sol containing desired ions for the formation of nanoparticles first and annealing the solid product at elevated temperatures.

For example, silica glasses doped with $Cd_xZn_{1-x}S$ were prepared by hydrolysis and polymerization of tetraethoxylsilane, $Si(OC_2H_5)_4$, TEOS, cadmium acetate, $Cd(CH_3COO)_2 \cdot 2H_2O$, zinc acetate, $Zn(CH_3COO)_2 \cdot 2H_2O$ in dimethylsulfoxide (DMSO), which serves as both solvent and sulfur precursor.[112] First cadmium and zinc precursors were dissolved into DMSO. When a homogenous solution was attained, TEOS and water were then added. The mixture was refluxed at 80°C for 2 days. The dry gels were first heat treated at 350°C in air to eliminate the residual organics and then heated again at 500 and 700°C in nitrogen for 30 min at each temperature. The gels before firing at elevated temperatures were colorless and transparent, indicating a homogeneous glass phase with absence of $Cd_xZn_{1-x}S$ nanoparticles. Glasses become yellow, when fired at 500°C in nitrogen, indicating the formation of $Cd_xZn_{1-x}S$ nanoparticles.

Nanoparticles of metals in polymer matrix can be synthesized through the reduction of metal ions by growing polymer chain radicals.[113–116] Typical preparative procedure can be illustrated by taking the synthesis of Ag nanoparticles in poly(methylmethacrylate) (PMMA), as an example. Silver trifluoroacetate ($AgCF_3CO_2$, AgTfa) and radical polymerization initiators, either 2,2'-azobisisobuyronitrile (AIBN) or benzoyl peroxide (BPO), were dissolved into methylmethacrylate (MMA). The solution was then heated at 60°C for over 20 hr to complete the polymerization of MMA; the resulting Ag-PMMA samples were further heat-treated at 120°C (which is slightly above the glass transition temperature of PMMA) for another 20 hr. In such a process, the metal ions were reduced to metal atoms by the growing polymer chain radicals, and consequently metal atoms nucleated to form nanoparticles. The post heating at higher temperatures was considered to promote further growth of already formed metallic nuclei. However, it is not clear how much is the enlargement of the nanoparticle size and the evolution of particle size distribution during such a post heat-treatment.

The type and concentration of polymerization initiators were found to have significant effects on size and size distribution of the grown metallic nanoparticles as shown in Fig. 3.22.[114] Although all the other experimental conditions were kept the same, the variation of the concentration and the type of the polymer radicals demonstrated distinct influences on the Ag particle sizes. Under a steady-state condition as applied in the above

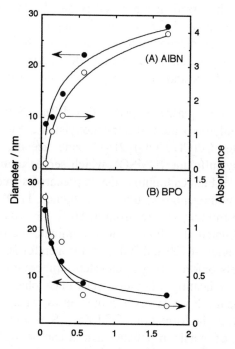

Fig. 3.22. Relationships between the average diameter of Ag particles (closed circle), and peak intensities of surface plasmon absorption of Ag clusters at ca. 420 nm (open circle), and the initiator concentration. [N. Yanagihara, K. Uchida, M. Wakabayashi, Y. Uetake, and T. Hara, *Langmuir* **15**, 3038 (1999).]

synthesis of Ag-PMMA composite, the concentration of the polymer radicals is proportional to the initial concentration of initiators.[117] Therefore, an increased concentration of polymer initiators are expected to result in an increased amount of polymer chain radicals, which would promote the reduction of metal ions and thus produce more metal atoms for nucleation (a higher concentration or supersaturation). Early discussion [Eqs. (3.5) and (3.9)] indicates that a higher supersaturation permits a smaller size but generates a larger number of nuclei. This explains the results presented in Fig. 3.22(b), which shows that the Ag nanoparticle size decreases with an increased concentration of BPO initiator. However, Fig. 3.22(a) shows an opposite relationship that the nanoparticle size increases with an increased concentration of AIBN initiator. A possible explanation for the results is that benzoyloxy radicals have an oxidation power against metal ions, whereas the isobutyronitrile radicals do not.[117,118] Furthermore, it was found that a high concentration of metal atoms would favor the surface process limited growth, leading to a wide size distribution.

Metallic nanoparticles were also prepared through precipitation or crystallization by annealing amorphous metal alloys at elevated temperatures.[119,120] Superparamagnetic nanocrystalline $Fe_{63.5}Cr_{10}Si_{13.5}B_9Cu_1Nb_3$ in the form of a ribbon of ~10 mm wide and ~25 μm thick was made by a melt spinning technique, followed by an annealing at elevated temperatures in argon.[121] The average grain size was found to range from ~5 nm to ~10 nm and to increase with the annealing temperature ranging from 775 K to 850 K.[121]

3.3. Nanoparticles through Heterogeneous Nucleation

3.3.1. *Fundamentals of heterogeneous nucleation*

When a new phase forms on a surface of another material, the process is called heterogeneous nucleation. Let us consider a heterogeneous nucleation process on a planar solid substrate. Assuming growth species in the vapor phase impinge on the substrate surface, these growth species diffuse and aggregate to form a nucleus with a cap shape as illustrated in Fig. 3.23. Similar to homogeneous nucleation, there is a decrease in the Gibbs free energy and an increase in surface or interface energy. The total change of the chemical energy, ΔG, associated with the formation of this nucleus is given by:

$$\Delta G = a_3 r^3 \Delta \mu_v + a_1 r^2 \gamma_{vf} + a_2 r^2 \gamma_{fs} - a_2 r^2 \gamma_{sv} \qquad (3.43)$$

where r is the mean dimension of the nucleus, $\Delta \mu_v$ is the change of Gibbs free energy per unit volume, γ_{vf}, γ_{fs}, and γ_{sv} are the surface or interface energy of vapor-nucleus, nucleus-substrate, and substrate-vapor interfaces,

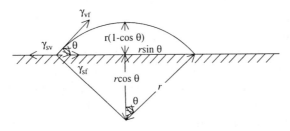

Fig. 3.23. Schematic illustrating heterogeneous nucleation process with all related surface energy in equilibrium.

respectively. Respective geometric constants are given by:

$$a_1 = 2\pi(1 - \cos\theta) \tag{3.44}$$

$$a_2 = \pi \sin^2\theta \tag{3.45}$$

$$a_3 = 3\pi(2 - 3\cos\theta + \cos^2\theta) \tag{3.46}$$

where θ is the contact angle, which is dependent only on the surface properties of the surfaces or interfaces involved, and defined by Young's equation:

$$\gamma_{sv} = \gamma_{fs} + \gamma_{vf}\cos\theta \tag{3.47}$$

Similar to homogeneous nucleation, the formation of new phase results in a reduction of the Gibbs free energy, but an increase in the total surface energy. The nucleus is stable only when its size is larger than the critical size, r^*:

$$r^* = \frac{-2(a_1\gamma_{vf} + a_2\gamma_{fs} - a_2\gamma_{sv})}{3a_3\Delta G_v} \tag{3.48}$$

and the critical energy barrier, ΔG^*, is given by:

$$\Delta G^* = \frac{4(a_1\gamma_{vf} + a_2\gamma_{fs} - a_2\gamma_{sv})^3}{27a_3^2 \Delta G_v} \tag{3.49}$$

Substituting all the geometric constants, we get:

$$r^* = \frac{2\pi\gamma_{vf}}{\Delta G_v}\left\{\frac{\sin^2\theta\cdot\cos\theta + 2\cos\theta - 2}{2 - 3\cos\theta + \cos^3\theta}\right\} \tag{3.50}$$

$$\Delta G^* = \left\{\frac{16\pi\gamma_{vf}}{3(\Delta G_v)^2}\right\}\left\{\frac{2 - 3\cos\theta + \cos^3\theta}{4}\right\} \tag{3.51}$$

Comparing this equation with Eq. (3.6), one can see that the first term is the value of the critical energy barrier for homogeneous nucleation, whereas the second term is a wetting factor. When the contact angle is 180°, i.e. the new phase does not wet on substrate at all, the wetting factor equals to 1 and the critical energy barrier becomes the same as that of homogeneous nucleation. In the case of the contact angle less than 180°, the energy barrier for heterogeneous nucleation is always smaller than that of homogeneous nucleation, which explains the fact that heterogeneous nucleation is easier than homogeneous nucleation in most cases. When the contact angle is 0°, the wetting factor will be zero and there is no energy barrier for the formation of new phase. One example of such cases is that the deposit is the same material as the substrate.

Zero-Dimensional Nanostructures: Nanoparticles

For the synthesis of nanoparticles or quantum dots on substrates, $\theta > 0$ is required and the Young's equation becomes:

$$\gamma_{sv} < \gamma_{fs} + \gamma_{vf} \tag{3.52}$$

Such heterogeneous nucleation is generally referred to as island (or Volmer–Weber) growth in the thin films community.[97] Other two nucleation-modes are layer (or Frank–van der Merwe) and island-layer (or Stranski–Krastanov) growth. Detailed discussion will be presented in Chapter 5.

3.3.2. Synthesis of nanoparticles

Various methods have been proposed to generate homogeneous surface defects to act as nucleation centers, including thermal oxidation,[122] sputtering and thermal oxidation,[123] and Ar plasma and ulterior thermal oxidation.[124] Evaporated metals such as silver and gold tend to form small particles on highly oriented pyrolitic graphite (HOPG) substrate.[125] Such metal nanoparticles formed were found closely associated with surface defects.[123,125,126] When edges are the only defects on substrate surfaces, the particles are concentrated only around these edges. For example, metal atoms on the substrates would diffuse and form particles concentrated at the step edges, since step edges on a substrate are preferred nucleation sites due to its high-energy state. However, for other defects such as pit holes, the nanoparticles were found to be distributed all over the substrate surfaces as demonstrated in Fig. 3.24.[126]

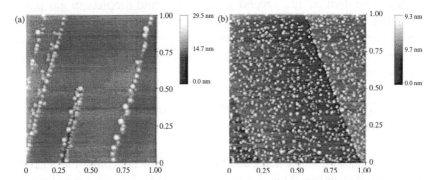

Fig. 3.24. Scanning force microscopy images of silver nanoparticles on HOPG-298 graphite substrates: (a) growth occurs only at the edge defects in the original substrate and (b) growth occurs wherever surface defects are present. [A. Stabel, K. Eichhorst-Gerner, J.P. Rabe, and A.R. González-Elipe, *Langmuir* **14**, 7324 (1998).].

Nickel nanoparticles of diameters ranging from 20 to 600 nm with a narrow size distribution on HOPG substrate were synthesized using a hydrogen co-evolution electrochemical deposition.[127] The chemicals used in the synthesis were $Ni(NO_3)_2 \cdot 6H_2O$, NH_4Cl, $NaCl$ and NH_4OH and the aqueous solution was kept at a pH of 8.3 during the synthesis.

GaAs nanoparticles in size range from 2.5 to 60 nm is grown on high surface area amorphous silica spheres of ~100 nm by molecular beam epitaxy (MBE).[128] The synthesis of GaAs nanoparticles takes place at ~580°C under conditions typically used for the growth of high quality epitaxial films. GaAs nanoparticles larger than 3.5 nm have a good crystalline order with a lattice constant equal to that of bulk material. Such prepared GaAs nanoparticles are covered with a shell of native oxides, Ga_2O_3 and As_2O_3, of 1.0 to 1.5 nm in thickness.

It should be noted that the formation of nanoparticles through heterogeneous nucleation is different from the synthesis by vapor phase reaction (Sec. 3.2.6). For homogeneous nucleation in vapor phase, particles are first formed in the vapor phase and then deposited onto substrate surfaces, whereas for heterogeneous nucleation, growth species impinge onto and form nuclei on substrate surfaces.

3.4. Kinetically Confined Synthesis of Nanoparticles

Kinetically controlled growth is to spatially confine the growth so that the growth stops when the limited amount of source materials is consumed or the available space is filled up. Many spatial confinements have been established for the synthesis of nanoparticles. In general, spatial confinement can be divided into several groups: (i) liquid droplets in gas phase including aerosol synthesis and spray pyrolysis, (ii) liquid droplets in liquid, such as micelle and micro emulsion synthesis, (iii) template-based synthesis, and (iv) self-terminating synthesis. All these methods will be briefly discussed in this section.

3.4.1. *Synthesis inside micelles or using microemulsions*

The synthesis of nanoparticles can be achieved by confining the reaction in a restricted space. This method is exemplified by the synthesis of nanoparticles inside micelles or in microemulsion. In micelle synthesis, reactions proceed among the reactants that are available only inside the micelle and the particle stops growing when the reactants are consumed. The formation of micelles will be discussed in detail later in Chapter 6; however, a brief description of the formation of micelles is given below. When surfactants or block polymers, typically consisting of two parts: one hydrophilic and another hydrophobic, are dissolved into a solvent, they preferentially self-assemble at air/aqueous solution or hydrocarbon/ aqueous solution interfaces. The hydrophilic part is turned towards the aqueous solution. When the concentration of the surfactants or block polymers exceeds a critical level, they self-assemble in such a way to form micelles. Surfactants or block polymers will reside at the interface separating hydrocarbon and aqueous solutions. A microemulsion is a dispersion of fine liquid droplets of an organic solution in an aqueous solution. Such a microemulsion system can be used for the synthesis of nanoparticles. The chemical reactions can take place either at the interfaces between the organic droplets and aqueous solution, when reactants are introduced separately into two non-mixable solutions, or inside the organic droplets when all the reactants are dissolved into the organic droplets.

In the following, we will use the work by Steigerwald et al.[43] on the synthesis of CdSe nanoparticles using organometallic reagents in inverse micellar solution as an example to illustrate the synthesis process. The surfactant bis(2-ethylhexyl) sulfosuccinate (aerosol-OT; AOT) of 33.3 g is dissolved in heptane (1300 mL), and then deoxygenated water (4.3 mL) is added. The mixture is stirred magnetically until the mixture becomes homogeneous, which gives a microemulsion with the ratio $W = [H_2O]/[AOT] = 3.2$. 1.12 mL of 1.0 M Cd^{2+} solution, prepared from $Cd(ClO_4)_2 \cdot 6H_2O$ and deoxygenated water, is added to the above microemulsion. Stirring gives an optically homogeneous microemulsion with $W = 4.0$. A solution of bis(trimethylsily)selenium, $Se(TMS)_2$ (210 μL) in heptane (50 mL) is added quickly to the microemulsion via syringe. A color develops throughout the homogeneous microemulsion as the semiconductor particles form. Under otherwise similar processing conditions, the ratio of $W = [H_2O]/[AOT]$ controls the size of CdSe crystallites. The same results were reported in the formation of colloidal crystallites from ionic reagents.[129,130]

The surface of the semiconductor nanoparticles prepared in inverse micellar solution can be further modified, and in general, surface modification is achieved by introducing silylorganoselenides, which react quickly with metal salts to form metal selenium covalent bonds.[43,131] For example, the surfactant stabilized CdSe is first coated with Cd^{2+} by the addition of 0.5 mL of 1.0 M Cd^{2+} solution and then with 350 μL of phenyl(trimethylsilyl)selenium, PhSeTMS in 50 mL of heptane. The mixture becomes cloudy and the colored precipitate is collected either by centrifugation or filtration. In this process, a Cd-rich surface is first generated on the CdSe nanocrystallites, and then reacts with PhSeTMS to form a layer of phenyl ligands which form covalent bonds with and cover the CdSe nanoparticle surface.[43]

Various monodispersed polymer particles can be prepared by carefully controlled emulsion polymerizations.[132–134] Typically a water-soluble polymerization initiator and a surfactant are added into a mixture of water and monomer. The hydrophobic monomer molecules form large droplets, typically 0.5 to 10 μm in diameter, which are stabilized by the surfactant molecules whose hydrophilic ends point outward and whose hydrophobic ends point inward toward the monomer droplet. The concentration of micelles, typically 10^{18} per mL, is far larger than that of the monomer droplets, 10^{10}–10^{11} per mL. Polymerization-initiators enter both monomer droplets and micelles. Polymerization proceeds in both monomer droplets and micelles with monomers transferred from monomer droplets. The resulting polymer particles are typically between 50 nm and 0.2 μm in diameter.[132] Such prepared polymer colloids were found to have exceedingly narrow size distribution and spherical shape.[135,136]

3.4.2. Aerosol synthesis

The formation of nanoparticles by aerosol method differs from other methods in several aspects. First of all, aerosol method can be considered as a top-down approach as compared with other methods, which have a bottom-up approach. Secondly, nanoparticles can be polycrystalline as compared with either single crystalline or amorphous structure of nanoparticles prepared by other methods. Thirdly, the nanoparticles prepared need to be collected and redispersed for many applications. In this method, a liquid precursor is first prepared. The precursor can be a simple mixture solution of desired constituent elements or a colloidal dispersion. Such a liquid precursor is then mistified to make a liquid aerosol, i.e. a dispersion of uniform droplets of liquid in a gas, which may simply solidify through evaporation of solvent or further react with the chemicals that are present in the gas. The resulting particles are spherical and their size is determined by the size of the initial liquid droplets and concentration of the solid. Aerosols can be relatively easily produced by sonication or spinning.[137] For example, TiO_2 particles can be produced from $TiCl_4$ or titanium alkoxide aerosols.[138] First amorphous spherical titania particles are formed, and then converted to anatase crystalline when calcined at elevated temperatures. Rutile phase is obtained when the powders are heated at 900°C. Following the same procedure with Al-2'-butoxide droplets, spherical alumina particles can be produced.[139]

The aerosol technique has also been used in the preparation of polymer colloids. The starting materials are droplets of organic monomers that can be either polymerized in contact with an initiator in gaseous state,[140] or copolymerized with another organic reactant.[141] For example, colloidal particles of poly(p-tertiarybutylstyrene) were prepared by polymerizing monomer droplets dispersed in helium gas with trifluoromethanesulfonic acid vapor, which acted as the polymerization initiator.[140] Polymer particles of styrene and divinylbenzene were synthesized through copolymerization between two monomers: styrene and divinylbenzene.[141] It should be noted that the polymer particles formed using aerosol synthesis are large particles, with diameters ranging from ~1 to 20 micron meters.

3.4.3. Growth termination

In the synthesis of nanoparticles, the size can be controlled by so-called growth termination. The approach is conceptually straightforward. When organic components or alien ions are attached to the growth surface strongly so that all the available growth sites are occupied, growth process stops. Herron and co-workers[142] synthesized colloidal particles of CdS based on the competitive growth and termination of CdS species in the presence of thiophenol surface capping agents. Cadmium acetate, thiophenol and anhydrous sodium sulfide were used for the synthesis and all the synthetic procedures and manipulations were carried out in a dry-box filled with nitrogen. Three stock solutions were prepared: (A) cadmium acetate dissolved into methanol, [Cd] = 0.1 M, (B) sodium sulfide in a mixture of water and methanol in 1 : 1 volume ratio, $[S^{2-}] = 0.1$ M, and (C) thiophenol in methanol, [PhSH] = 0.2 M. Stock solutions B and C were first thoroughly mixed and then stock solution A was added under stirring in an overall volume ratio of A : B : C = 2 : 1 : 1. The solution was stirred for 15 min, filtered, and suction dried through a filter by nitrogen. Such prepared CdS particles were crystalline and XRD spectra matched with that of bulk sphalerite CdS. The surface of the CdS particles was capped with thiophenol molecules as schematically illustrated in Fig. 3.25.[142] CdS particle sizes varied with the relative ratio of sulfide to thiophenol and ranged from less than 1.5 nm to ~3.5 nm. It was clearly demonstrated that an increasing amount of capping molecules relative to sulfide precursor resulted in a reduced particle size. Therefore, the size of these nanoparticles could be conveniently controlled by adjusting the relative

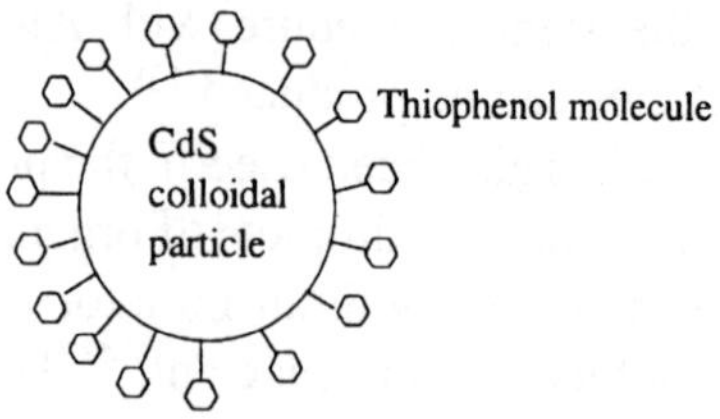

Fig. 3.25. Termination growth for the synthesis of nanoparticles. When organic components occupy all the surface growth sites, growth of nanoparticle stops. The final size of grown nanoparticles can be controlled by the concentration of organic ligands introduced to the system. [N. Herron, Y. Wang, and H. Eckert, *J. Am. Chem. Soc.* **112**, 1322 (1990).]

concentrations of capping molecules and precursors. Similar synthetic approach is applicable to the formation of metal oxide nanoparticles. For example, crystalline tetragonal ZrO_2 nanoparticles of 2 nm in diameter is formed by hydrolysis of acac-modified zirconium propoxide in the presence of para-toluene sulfonic acid and aging at 60–80°C.[143]

3.4.4. Spray pyrolysis

Spray pyrolysis is basically a solution process and has been widely used in the preparation of metal and metal oxide powders.[144,145] The process can be simply described as converting microsized liquid droplets of precursor or precursor mixture into solid particles through heating. In practice, spray pyrolysis involves several steps: (i) generating microsized droplets of liquid precursor or precursor solution, (ii) evaporation of solvent, (iii) condensation of solute, (iv) decomposition and reaction of solute, and (v) sintering of solid particles.

Kieda and Messing[146] reported the production of silver particles using precursor solutions of Ag_2CO_3, Ag_2O and $AgNO_3$ with NH_4HCO_3 at temperatures of 400°C or less. It was recognized that the ability of silver ions to form the ammine complexes plays a very important role in the production of nanoparticles in this low temperature spray pyrolysis. It was postulated that such a process would be applicable for most transition metals such as Cu, Ni, Zn, ions of which complexes can be formed with ammines.

Brennan et al.[147] prepared nanometer-sized particles of CdSe starting from either $Cd(SePh)_2$ or $[Cd(SePh)_2]_2[Et_2PCH_2CH_2PEt_2]$ through a mild solid state pyrolysis *in vacuo* at temperatures ranging from 320 to 400°C for 24 hr. Analogue process was used to produce nanoparticles of ZnS and CdS,[148] and CdTe and HgTe.[149]

Oxide nanoparticles can also be prepared by spray pyrolysis. Kang et al.[150] made Y_2O_3 nanoparticles doped with europium by a combination of sol-gel processing and spray pyrolysis. Colloidal solution was prepared using urea as a reduction reagent and spray pyrolysis was carried at about 1300°C. Nanoparticles were found to exhibit smooth surface, spherical shape and hollow structure.

3.4.5. Template-based synthesis

Iron oxide, Fe_3O_4 nanoparticles dispersed in a solid polymer matrix can be synthesized by infiltration of iron chloride solution.[151] The polymer matrices are cation exchange resins, which are formed by beads of 100–300 μm in diameter and contain micropores. The iron oxide nanoparticle synthesis is performed in nitrogen by dispersing the resin in an iron chloride solution. Matrix cations, Na^+ or H^+, are exchanged with Fe^{2+} and Fe^{3+}. The exchange is followed by hydrolysis and polymerization in an alkaline medium at 65°C with the formation of Fe_3O_4 nanoparticles within the resin macropores. The process is repeated to increase the load of Fe_3O_4 and thus the size of nanoparticles. Regularly shaped spheres of Fe_3O_4 with diameters ranging from 3 to 15 nm are prepared. CdSe nanoparticles have also been synthesized using zeolites as templates[152] and ZnS nanoparticles in silicate glasses.[153] Template can also be used as a shadow mask for the synthesis of nanoparticles by gas deposition. For example, ordered arrays of multiple metallic nanoparticles on silicon substrates were deposited by evaporation using anodic porous alumina membranes as masks.[154]

3.5. Epitaxial Core-Shell Nanoparticles

Nanoparticles have been subjected to a variety of surface engineering for various applications including self-assembly of organic components and bioactive species, and dielectric-metal core-shell nanostructures. This topic deserves special attention and will be discussed in detail in Chapter 6. However, the semiconductor–semiconductor core-shell structures will be discussed below, since such core-shell structures grow epitaxially and the shell can be considered as an extension of core structure with different chemical compositions. In addition, the growth of core and shell in these systems are very closely related.

Semiconductor nanoparticles can have quantum effects and have high emission yields across the visible and near infrared (NIR) spectrum. The surface of such nanoparticles or quantum dots largely determines the quantum yield and emission life-time of the band gap luminescence. High luminescence yields are achieved by the use of surface passivation to reduce the non-radiative surface recombination of charge carriers. Two methods of passivation are commonly employed. One is through so-called band gap engineering, whereby a larger band gap semiconductor with good lattice mismatch is epitaxially deposited onto the core surface.[155] Another method is to adsorb Lewis bases onto the surface.[156,157] One example of the latter is otylamine used to passivate the surface of CdSe and CdSe/ZnS quantum dots.[158]

For the growth of a layer of larger band gap semiconductor on the surface of a nanoparticle, the growth condition must be controlled such that no homogeneous nucleation would occur, but only a growth proceeds on the surface of the nanoparticles. Therefore, the concentration of the growth species needs to be controlled such that the supersaturation is not high enough for nucleation, but high enough for growth. There are two approaches applied to control the supersaturation of growth species. One is by the drop wise addition of growth precursor solution into the reaction mixture, which consists of grown nanoparticles (cores). Another method is to vary the growth temperatures. For example, in the synthesis of CdSe/ZnS core/shell nanostructures, the temperatures at which each individual size of nanoparticles was overcoated are as follows: 140°C for 2.3 and 3.0 nm diameters, 160°C for 3.5 nm, 180°C for 4.0 nm, 200°C for 4.8 nm, and 220°C for 5.5 nm.[159] Lower temperature is required for the growth on smaller nanoparticles, since the solubility and the supersaturation depends on the surface curvature as discussed in the previous chapter. Furthermore, the association between the surface atoms or ions of the nanoparticles (cores) and the capping materials should not be too strong, so that the growth species can displace the capping molecules or insert between the surface atoms and the capping molecules.

In the following, a few examples will be used to illustrate the general approach in fabricating core-shell nanostructures. First, let us look at the preparation of ZnS-capped CdSe nanocrystals.[155] The CdSe nanocrystallites are prepared by a method described earlier in Sec. 3.2.4.[42] The Zn and S stock solution was prepared with 0.52 mL of bis-trimethylsilyl sulfide, $(TMS)_2S$ (0.0025 mol) in 4.5 mL of TOP, adding 3.5 mL of dimethylzinc, Me_2Zn solution (0.0035 mol), and diluting with 16 mL of TOP in a nitrogen filled dry-box. When the TOP capped CdSe colloidal dispersion was prepared and cooled to ~300°C, the Zn/S/TOP solution was injected into CdSe colloidal dispersion five times at approximately 20 s intervals. A total molar ratio of injected reagents Cd/Se : Zn/S was 1 : 4. Upon cooling the reaction mixture was stirred at 100°C for 1 h. A layer of ZnS of

Zero-Dimensional Nanostructures: Nanoparticles 103

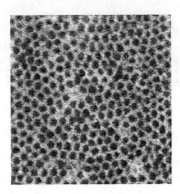

Fig. 3.26. TEM image of ZnS-capped CdSe nanocrystals. This picture is 95 × 95 nm. [M.A. Hines and P. Guyot-Sionnest, *J. Phys. Chem.* **100**, 468 (1996).]

~0.6 nm was coated onto the surface of CdSe nanoparticles as supported by X-ray photoelectron spectroscopy and transmission electron microscopy. Figure 3.26 shows the TEM picture of ZnS-capped CdSe nanocrystals.[155]

The epitaxial growth of shell material on the core nanocrystallites can eliminate both the anionic and cationic surface dangling bonds, and also generate a new nanocrystal system, as demonstrated by Peng *et al.*[160] The wurtzite CdSe/CdS structure is ideal in many aspects. The lattice mismatch of 3.9% is small enough to allow heteroepitaxial growth, while still large enough to prevent alloying, and the difference in bandgaps is large enough for shell growth to increase the quantum yield and the stability of the cores. The synthesis procedure of CdSe/CdS core/shell nanostructure is described below.[160] First stock solution for CdSe nanocrystal synthesis was prepared. CdSe stock solution was made by adding the desired amount of $Cd(CH_3)_2$ to a solution of Se powder dissolved in tributylphosphine (TBP) in a dry-box under nitrogen, with the Cd : Se molar ratio kept as 1 : 0.7 or 1 : 0.9. TOPO, used as a high-boiling point solvent as well as stabilizer, was heated to 360°C under argon before a stock solution was quickly injected. The reaction was either stopped immediately by quick removal of the heating or allowed to continue after lowering the temperature to 300°C. Nanocrystals were precipitated by the addition of methanol to the cooled, room temperature reaction mixture. After centrifugation and drying under nitrogen, CdSe nanocrystals capped with TOPO and of 3.5 nm in diameter were obtained. For the shell growth, the above CdSe nanocrystals were dissolved into anhydrous pyridine and refluxed overnight under argon. CdS stock solution, made by adding $(TMS)_2S$ to a solution of $Cd(CH_3)_2$ dissolved in TBP under nitrogen with a Cd : S molar ratio of 1 : 2.1, was added drop wise (1 drop per second) to the reaction

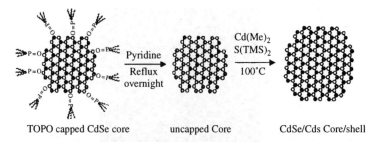

Fig. 3.27. Schematic synthesis of CdSe/CdS core/shell nanocrystals [X. Peng, M.C. Schlamp, A.V. Kadavanich, and A.P. Alivisatos, *J. Am. Chem. Soc.* **119**, 7019 (1997).]

solution at 100°C. Stopping the addition of CdS stock solution and removing of the heating source would terminate the shell growth. Dodecylamine was added to the reaction solution at room temperature until the nanocrystals precipitated. When CdSe nanoparticle is refluxed in pyridine overnight, TOPO could be almost completely removed from CdSe nanocrystals without affecting the nanocrystal structure. Pyridine displaces TOPO and forms a weak bond to a surface Cd atom, providing simultaneous chemical stability and access to the surface, permitting the growth of CdS shell to CdSe core. This reaction is schematically shown in Fig. 3.27.[160]

3.6. Summary

The preparation of monodispersed nanoparticles can be achieved through many different approaches, either homogeneous or heterogeneous nucleation, in gaseous, liquid or solid medium. There are some common fundamentals for the synthesis of nanoparticles with monodispersion. (i) Temporal nucleation, i.e. nucleation occurs in a very short time. Such a temporal nucleation is achieved through generating an abrupt supersaturation. Introduction of monosized seeds for heterogeneous nucleation and growth is another approach. (ii) Subsequent growth needs to be diffusion controlled. This is achieved through introducing a diffusion barrier, such as a polymer monolayer on the growth surface, using a low concentration of growth species, or slowly generating growth species. (iii) Ostwald ripening is often used to narrow the size distribution. (iv) Size-selective precipitation is applied to further separate large particles from small ones, though it is done after the synthesis. In contrast to spontaneous growth of monodispersed nanoparticles, spatial confinement is also applied to the synthesis of nanoparticles. The technical approach here is very straightforward: only a certain amount of growth species or a limited space is available for the formation of individual nanoparticles.

References

1. E.H.C. Parker (ed.) *The Technology and Physics of Molecular Beam Epitaxy*, Plenum, New York, 1985.
2. J.J. Jewell, J.P. Harbison, A. Scherer, Y.H. Lee, and L.T. Florez, *IEEE J. Quant. Electron.* **27**, 1332 (1991).
3. M. Haruta and B. Delmon, *J. Chim. Phys.* **83**, 859 (1986).
4. A.E. Nielsen, *Kinetic of Precipitation*, MacMillan, New York, 1964.
5. R. Williams, P.M. Yocom, and F.S. Stofko, *J. Colloid Interf. Sci.* **106**, 388 (1985).
6. A. Henglein, *Chem. Rev.* **89**, 1861 (1989).
7. Z.L. Wang, *Adv. Mater.* **10**, 13 (1998).
8. G. Schmid, *Chem. Rev.* **92**, 1709 (1992).
9. G. Schmid (ed.), *Clusters and Colloids*, VCH, New York, 1994.
10. G. Schon and U. Simon, *Colloid Polym Sci.* **273**, 101 (1995).
11. M. Faraday, *Phil. Trans.* **147**, 145 (1857).
12. J. Turkevich, J. Hillier, and P.C. Stevenson, *Discuss. Faraday Soc.* **11**, 55 (1951).
13. J. Turkevich, *Gold Bull.* **18**, 86 (1985).
14. H. Hirai, Y. Nakao, N. Toshima, and K. Adachi, *Chem. Lett.* 905 (1976).
15. H. Hirai, Y. Nakao, and N. Toshima, *J. Macromol. Sci.-Chem.* **A12**, 1117 (1978).
16. A. Henglein, B.G. Ershov, and M. Malow, *J. Phys. Chem.* **99**, 14129 (1995).
17. J. Turkevich and G. Kim, *Science* **169**, 873 (1970).
18. J. Turkevich, K. Aika, L.L. Ban, I. Okura, and S. Namba, *J. Res. Inst. Catal. Hokkakaida Univ.* **24**, 54 (1976).
19. L.D. Rampino and F.F. Nord, *J. Am. Chem. Soc.* **63**, 2745 (1941).
20. F.A. Cotton and G. Wilkison, *Advanced Inorganic Chemistry*, 5th edition. John Wiley, New York, 1988.
21. R.A. Salkar, P. Jeevanandam, S.T. Aruna, Y. Koltypin and A. Gedanken, *J. Mater. Chem.* **9**, 1333 (1999).
22. M. Gutierrez and A. Henglein, *J. Phys. Chem.* **91**, 6687 (1987).
23. M.T. Reetz and W. Helbig, *J. Am. Chem. Soc.* **116**, 7401 (1994).
24. J.A. Becker, R. Schafer, R. Festag, W. Ruland, J.H. Wendorff, J. Pebler, S.A. Quaiser, W. Helbig, and M.T. Reetz, *J. Chem. Phys.* **103**, 2520 (1995).
25. K.H. Lieser, *Angew. Chem. Int. Ed. Engl.* **8**, 188 (1969).
26. V.K. La Mer, *Ind. Eng. Chem. Res.* **44**, 1270 (1952).
27. W.O. Miligan and R.H. Morriss, *J. Am. Chem. Soc.* **86**, 3461 (1964).
28. M.T. Reetz and M. Maase, *Adv. Mater.* **11**, 773 (1999).
29. D.G. Duff, P.P. Edwards, and B.F.G. Johnson, *J. Phys. Chem.* **99**, 15934 (1995).
30. K. Chou and C. Ren, *Mater. Chem. Phys.* **64**, 241 (2000).
31. A. Henglein, *Chem. Mater.* **10**, 444 (1998).
32. T.S. Ahmadi, Z.L. Wang, T.C. Green, A. Henglein, and M.A. El-Sayed, *Science* **272**, 1924 (1996).

33. Y. Yin, Z. Li, Z. Zhong, B. Gates, Y. Xia, and S. Venkateswaran, *J. Mater. Chem.* **12**, 522 (2002).
34. A.G. Ingalls, *Amateur Telescope Making (Book One)*, Scientific American Inc., New York, p. 101, 1981.
35. P. Toneguzzo, G. Viau, O. Acher, F. Fiévet-Vincent, and F. Fiévet, *Adv. Mater.* **13**, 1032 (1998).
36. M.L. Steigerwald and L.E. Brus, *Acc. Chem. Res.* **23**, 183 (1990).
37. A.P. Alivisatos, *Science* **271**, 933 (1996).
38. M.G. Bawendi, M.L. Steigerwald, and L.E. Brus, *Annu. Rev. Phys. Chem.* **41**, 477 (1990).
39. Y. Wang, *Acc. Chem. Res.* **24**, 133 (1991).
40. C.B. Murray, C.R. Kagan, and M.G. Bawendi, *Ann. Rev. Mater. Sci.* **30**, 545 (2000).
41. S.A. Majetich and A.C. Carter, *J. Phys. Chem.* **97**, 8727 (1993).
42. C.B. Murray, D.J. Norris, and M.G. Bawendi, *J. Am. Chem. Soc.* **115**, 8706 (1993).
43. M.L. Steigerwald, A.P. Alivisatos, J.M. Gibson, T.D. Harris, R. Kortan, A.J. Muller, A.M. Thayer, T.M. Duncan, D.C. Douglas, and L.E. Brus, *J. Am. Chem. Soc.* **110**, 3046 (1988).
44. S.M. Stuczynski, J.G. Brennan, and M.L. Steigerwald, *Inorg. Chem.* **28**, 4431 (1989).
45. L. Spanhel, M. Haase, H. Weller, and A. Henglein, *J. Am. Chem. Soc.* **109**, 5649 (1987).
46. M.G. Bawendi, A. Kortan, M.L. Steigerwald, and L.E. Brus, *J. Chem. Phys.* **91**, 7282 (1989).
47. A.L. Smith, *Particle Growth in Suspensions*, Academic Press, New York, 1983.
48. H. Reiss, *J. Chem. Phys.* **19**, 482 (1951).
49. A. Guinier, *X-Ray Diffraction*, W.H. Freeman, San Francisco, CA, 1963.
50. A.A. Guzelian, J.E.B. Katari, A.V. Kadavanich, U. Banin, K. Hamad, E. Juban, A.P. Alivisatos, R.H. Wolters, C.C. Arnold, and J.R. Heath, *J. Phys. Chem.* **100**, 7212 (1996).
51. O.I. Mićić, J.R. Sprague, C.J. Curtis, K.M. Jones, J.L. Machol, A.J. Nozik, H. Giessen, B. Fluegel, G. Mohs, and N. Peyghambarian, *J. Phys. Chem.'* **99**, 7754 (1995).
52. O.I. Mićić, C.J. Curtis, K.M. Jones, J.R. Sprague, and A.J. Nozik, *J. Phys. Chem.* **98**, 4966 (1994).
53. R.L. Wells, M.F. Self, A.T. MaPhail, S.R. Auuchon, R.C. Wandenberg, and J.P. Jasinski, *Organometallics* **12**, 2832 (1993).
54. S.R. Aubuchon, A.T. McPhail, R.L. Wells, J.A. Giambra, and J.R. Bowser, *Chem. Mater.* **6**, 82 (1994).
55. E.K. Byrne, L. Parkanyi, and K.H. Theopold, *Science* **241**, 332 (1988).
56. M.A. Olshavsky, A.N. Goldstein, and A.P. Alivisatos, *J. Am. Chem. Soc.* **112**, 9438 (1990).
57. R.L. Wells, C.G. Pitt, A.T. McPhail, A.P. Purdy, S. Shafieezad, and R.B. Hallock, *Chem. Mater.* **1**, 4 (1989).
58. M. Guglielmi, A. Martucci, E. Menegazzo, G.C. Righini, S. Pelli, J. Fick, and G. Vitrant, *J. Sol-Gel Sci. Technol.* **8**, 1017 (1997).
59. L. Spanhel, E. Arpac, and H. Schmidt, *J. Non-Cryst. Solids* **147&148**, 657 (1992).
60. W.C. Johnson, J.B. Parsons, and M.C. Crew, *J. Phys. Chem.* **36**, 2561 (1932).
61. A. Addaniano, *J. Electrochem. Soc.* **108**, 1072 (1961).
62. J.W. Hwang, S.A. Hanson, D. Britton, J.F. Evans, K.F. Jensen, and W.L. Gladfelter, *Chem. Mater.* **7**, 517 (1995).

63. J.E. Andrews and M.A. Littlejohn, *J. Electrochem. Soc.* **122**, 1273 (1975).
64. Y. Xie, Y. Qian, W. Wang, S. Zhang, and Y. Zhang, *Science* **272**, 1926 (1996).
65. O.I. Mićić, S.P. Ahrenkiel, D. Bertram, and A.J. Nozik, *Appl. Phys. Lett.* **75**, 478 (1999).
66. A. Manz, A. Birkner, M. Kolbe, and R.A. Fischer, *Adv. Mater.* **12**, 569 (2000).
67. J.F. Janik and R.L. Wells, *Chem. Mater.* **8**, 2708 (1996).
68. J.L. Coffer, M.A. Johnson, L. Zhang, and R.L. Wells, *Chem. Mater.* **9**, 2671 (1997).
69. R.K. Iler, *The Chemistry of Silica: Solubility, Polymerization, Colloid and Surface Properties, and Biochemistry*, Wiley, New York, 1979.
70. E. Matijević, *Chem. Mater.* **5**, 412 (1993).
71. E. Matijević, *Langmuir* **10**, 8 (1994).
72. S.T. Selvan, C. Bullen, M. Ashokkumar, and P. Mulvaney, *Adv. Mater.* **13**, 985 (2000).
73. F. Caruso, *Adv. Mater.* **13**, 11 (2001).
74. C.J. Brinker and G.W. Scherer, *Sol-Gel Science: the Physics and Chemistry of Sol-Gel Processing*, Academic Press, San Diego, CA, 1990.
75. Alain C. Pierre, *Introduction to Sol-Gel Processing*, Kluwer, Boston, MA, 1998.
76. J.D. Wright and N.A.J.M. Sommerdijk, *Sol-Gel Materials: Chemistry and Applications*, Gordon and Breach Science Publishers, Amsterdam, 2001.
77. J. Livage, F. Babonneau, and C. Sanchez, in *Sol-Gel Optics: Processing and Applications*, ed. L.C. Klein, Kluwer, Boston, MA, p. 39, 1994.
78. B.E. Yoldas, *J. Non-Cryst. Solids* **38–39**, 81 (1980).
79. C.M. Chan, G.Z. Cao, H. Fong, M. Sarikaya, T. Robinson, and L. Nelson, *J. Mater. Res.* **15**, 148 (2000).
80. H. Matsuda, N. Kobayashi, T. Kobayashi, K. Miyazawa, and M. Kuwabara, *J. Non-Cryst. Solids*, **271**, 162 (2000).
81. E. Matijević, *Acc. Chem. Res.* **14**, 22 (1981).
82. E. Matijević, *Prog. Colloid Polym. Sci.* **57**, 95 (1976).
83. W. Stöber, A. Finx, and E. Bohn, *J. Colloid Interf. Sci.* **26**, 62 (1968).
84. E. Matijević and P. Scherner, *J. Colloid Interf. Sci.* **63**, 509 (1978).
85. E. Matijević and W.P. Hsu, *J. Colloid Interf. Sci.* **118**, 506 (1987).
86. D. Sordelet and M. Akinc, *J. Colloid Interf. Sci.* **122**, 47 (1988).
87. G. Wakefield, E. Holland, P.J. Dobson, and J.L. Hutchison, *Adv. Mater.* **13**, 1557 (2001).
88. E. Matijević, *Ann. Rev. Mater. Sci.* **15**, 483 (1985).
89. E. Matijević, *J. Colloid Interf. Sci.* **58**, 374 (1977).
90. L. Spanhel and M.A. Anderson, *J. Am. Chem. Soc.* **113**, 2826 (1991).
91. S. Sakohara, M. Ishida, and M.A. Anderson, *J. Phys. Chem.* **B102**, 10169 (1998).
92. U. Koch, A. Fojtik, H. Weller, and A. Henglein, *Chem. Phys. Lett.* **122**, 507 (1985).
93. M. Haase, H. Weller, and A. Henglein, *J. Phys. Chem.* **92**, 482 (1988).
94. D.W. Bahnemann, C. Karmann, and M.R. Hoffmann, *J. Phys. Chem.* **91**, 3789 (1987).
95. S.A. Nepijko, D.N. Levlev, W. Schulze, J. Urban, and G. Ertl, *Chem. Phys. Chem* **3**, 140 (2000).
96. E. Wagner and H. Brünner, *Angew. Chem.* **72**, 744 (1960).
97. M. Ohring, *The Material Science of Thin Films*, Academic Press, San Diego, CA, 1992.
98. S.A. Nepijko, H. Hofmeister, H. Sack-Kongehl, and R. Schlögl, *J. Cryst. Growth* **213**, 129 (2000).
99. J. Viereck, W. Hoheisel, and F. Trager, *Surf. Sci.* **340**, L988 (1995).

100. A.E. Romanov, I.A. Polonsky, V.G. Gryaznov, S.A. Nepijko, T. Junghannes, and N.I. Vitryhovski, *J. Cryst. Growth* **129**, 691 (1993).
101. M. Haruta, *Catal. Today* **36**, 153 (1997).
102. R.J.H. Grisel and B.E. Nieuwenhuys, *J. Catal.* **199**, 48 (2001).
103. M. Valden, X. Lai, and D.W. Goodman, *Science* **281**, 1647 (1998).
104. P.C. Sercel, W.A. Saunders, H.A. Atwater, and K.J. Vahala, *Appl. Phys. Lett.* **61**, 696 (1992).
105. M. Yamane and Y. Asahara, *Glasses for Photonics*, Cambridge Univ. Press, Cambridge, 2000.
106. R.H. Doremus, *Glass Science*, 2nd edition, John Wiley & Sons, New York, 1994.
107. S.D. Stookey, *J. Am. Ceram. Soc.* **32**, 246 (1949).
108. R.H. Doremus, in *Nucleation and Crystallization in Glasses and Melts*, the American Ceramic Society, Columbus, OH, p.117, 1967.
109. R.H. Doremus, *J. Chem. Phys.* **41**, 414 (1965).
110. R.H. Doremus, S.-C. Kao, and R. Garcia, *Appl. Opt.* **31**, 5773 (1992).
111. K. Uchida, S. Kaneko, S. Omi, C. Hata, H. Tanji, Y. Asahara, and A.J. Ikushima, *J. Opt. Soc. Am.* **B11**, 1236 (1994).
112. E. Cordoncillo, J.B. Carda, M.A. Tena, G. Monros, and P. Escribano, *J. Sol-Gel Sci. Technol.* **8**, 1043 (1997).
113. N. Yanagihara, *Chem. Lett.* 305 (1998)
114. N. Yanagihara, K. Uchida, M. Wakabayashi, Y. Uetake, and T. Hara, *Langmuir* **15**, 3038 (1999).
115. Y. Nakao, *J. Chem. Soc., Chem. Commun.* 826 (1993).
116. Y. Nakao, *J. Colloid Interf. Sci.* **171**, 386 (1995).
117. F.W. Billmeyer, *Textbook of Polymer Science*, 3rd edition, John Wiley & Sons, New York, 1984.
118. H.G. Elias, *Macromolecules*, 2nd edition, Plenum, New York, 1984.
119. J.J. Becker, *Trans. Am. Inst. Mining Met. Petrol. Engr.* **209**, 59 (1957).
120. A.E. Berkowitz and P.J. Flanders, *J. Appl. Phys.* **30**, 111S (1959).
121. V. Franco, C.F. Conde, A. Conde, L.F. Kiss, D. Kaptás, T. Kemény, and I. Vincze, *J. Appl. Phys.* **90**, 1558 (2001).
122. H. Change and A. Bard, *J. Am. Chem. Soc.* **113**, 5588 (1991).
123. H. Hövel, Th. Becker, A. Bettac, B. Reihl, M. Tschudy, and E.J. Williams, *J. Appl. Phys.* **81**, 154 (1997).
124. X.Q. Zhong, D. Luniss, and V. Elings, *Surf. Sci.* **290**, 688 (1993).
125. Y.O. Ahn and M. Seidl, *J. Appl. Phys.* **77**, 5558 (1995).
126. A. Stabel, K. Eichhorst-Gerner, J.P. Rabe, and A.R. González-Elipe, *Langmuir* **14**, 7324 (1998).
127. M.P. Zach and R.M. Penner, *Adv. Mater.* **12**, 878 (2000).
128. C.J. Sandroff, J.P. Harbison, R. Ramesh, M.J. Andrejco, M.S. Hegde, D.M. Hwang, C.C. Change, and E.M. Vogel, *Science* **245**, 391 (1989).
129. M. Meyer, C. Wallberg, K. Kurihara, and J.H. Fendler, *J. Chem. Soc. Chem. Commun.* 90 (1984).
130. J.H. Fendler, *Chem. Rev.* **87**, 877 (1987).
131. J.W. Anderson, G.K. Banker, J.E. Drake, and M. Rodgers, *J. Chem. Soc., Dalton Trans.* 1716 (1973).
132. J.R. Fried, *Polymer Science and Technology*, Prentice Hall, Upper Saddle River, NJ, p. 51, 1995.

133. I Piirma (ed.), *Emulsion Polymerization*, Academic Press, New York, 1982.
134. G.W. Poehlein, R.H. Ottewill, and J.W. Goodwin, (eds.), *Science and Technology of Polymer Colloids*, Vol. II, Martinus Nijhoff, Boston, MA, 1983.
135. R.C. Backus and R.C. Williams, *J. Appl. Phys.* **20**, 224 (1949).
136. E. Bradford and J. Vanderhoff, *J. Appl. Phys.* **26**, 864 (1955).
137. N.A. Fuchs and A.G. Sutugin, in *Aerosol Science*, ed. C.N. Davies, Academic Press, New York, p. 1, 1966.
138. M. Visca and E. Matijević, *J. Colloid Interf. Sci.* **68**, 308 (1978).
139. B.J. Ingebrethsen and E. Matijević, *J. Aerosol Sci.* **11**, 271 (1980).
140. R. Partch, E. Matijević, A.W. Hodgson, and B.E. Aiken, *J. Polymer Sci. Polymer Chem. Ed.* **21**, 961 (1983).
141. K. Nakamura, R.E. Partch, and E. Matijević, *J. Colloid Interf. Sci.* **99**, 118 (1984).
142. N. Herron, Y. Wang, and H. Eckert, *J. Am. Chem. Soc.* **112**, 1322 (1990).
143. M. Chatry, M. In, M. Henry, C. Sanchez, and J. Livage, *J. Sol-Gel Sci. Technol.* **1**, 233 (1994).
144. G.L. Messing, S.C. Zhang, and G.V. Jayanthi, *J. Am. Ceram. Soc.* **76**, 2707 (1993).
145. A. Gurav, T. Kodas, T. Pluym, and Y. Xiong, *Aerosol Sci. Technol.* **19**, 411 (1993).
146. N. Kieda and G.L. Messing, *J. Mater. Res.* **13**, 1660 (1998).
147. J.G. Brennan, T. Siegrist, P.J. Carroll, S.M. Stuczynski, L.E. Brus, M.L. Steigerwald, *J. Am. Chem. Soc.* **111**, 4141 (1989).
148. K. Osakada and T. Yamamoto, *J. Chem. Soc., Chem. Commun.* 1117 (1987).
149. M.L. Steigerwald and C.R. Sprinkle, *J. Am. Chem. Soc.* **109**, 7200 (1987).
150. Y.C. Kang, H.S. Roh, and S.B. Park, *Adv. Mater.* **12**, 451 (2000).
151. A.M. Testa, S. Foglia, L. Suber, D. Fiorani, LI. Casas, A. Roig, E. Molins, J.M. Grenéche, and J. Tejada, *J. Appl. Phys.* **90**, 1534 (2001).
152. Y. Wang and N. Herron, *J. Phys. Chem.* **91**, 257 (1987).
153. M.G. Bawendi, M.L. Steigerwald, and L.E. Brus, *Annu. Rev. Phys. Chem.* **41**, 477 (1990).
154. H. Masuda, K. Yasui, and K. Nishio, *Adv. Mater.* **12**, 1031 (2000).
155. M.A. Hines and P. Guyot-Sionnest, *J. Phys. Chem.* **100**, 468 (1996).
156. S.A. Majetich and C. Carter, *J. Phys. Chem.* **97**, 8727 (1993).
157. F. Seker, K. Meeker, T.F. Kuech, and A.B. Ellis, *Chem. Rev.* **100**, 2505 (2000).
158. S.T. Selvan, C. Bullen, M. Ashokkumar, and P. Mulvaney, *Adv. Mater.* **13**, 985 (2001).
159. B.O. Dabbousi, J. Rodriguez-Viejo, F.V. Mikulec, J.R. Heine, H. Mattoussi, R. Ober, K.F. Jensen, and M.G. Bawendi, *J. Phys. Chem.* **B101**, 9463 (1997).
160. X. Peng, M.C. Schlamp, A.V. Kadavanich, and A.P. Alivisatos, *J. Am. Chem. Soc.* **119**, 7019 (1997).

Chapter 4

One-Dimensional Nanostructures: Nanowires and Nanorods

4.1. Introduction

One-dimensional nanostructures have been called by a variety of names including: whiskers, fibers or fibrils, nanowires and nanorods. In many cases, nanotubules and nanocables are also considered one-dimensional structures. Although whiskers and nanorods are in general considered to be shorter than fibers and nanowires, the definition is often a little arbitrary. In addition, one-dimensional structures with diameters ranging from several nanometers to several hundred microns were referred to as whiskers and fibers in the early literature, whereas nanowires and nanorods with diameters not exceeding a few hundred nanometers are used predominantly in the recent literature. One will find, from reading this chapter, the fact that many fundamental understandings and techniques of growth of one-dimensional nanostructures are based on the early work on the growth of whiskers and fibers, *albeit* with less emphasis on nanometer scale. In this chapter, various terms of one-dimensional structures will be used interchangeably, though nanowires in general have a high aspect ratio than that of nanorods.

Many techniques have been developed in the synthesis and formation of one-dimensional nanostructured materials, though some techniques have been explored extensively, while others have attracted far less attention.

One-Dimensional Nanostructures: Nanowires and Nanorods 111

These techniques can be generally grouped into four categories:

(1) Spontaneous growth:
   (a) Evaporation (or dissolution)–condensation
   (b) Vapor (or solution)–liquid–solid (VLS or SLS) growth
   (c) Stress–induced recrystallization
(2) Template-based synthesis:
   (a) Electroplating and electrophoretic deposition
   (b) Colloid dispersion, melt, or solution filling
   (c) Conversion with chemical reaction
(3) Electrospinning
(4) Lithography

Spontaneous growth, template-based synthesis and electrospinning are considered as a bottom-up approach, whereas lithography is a top-down technique. Spontaneous growth commonly results in the formation of single crystal nanowires or nanorods along a preferential crystal growth direction depending on the crystal structures and surface properties of the nanowire materials. Template-based synthesis mostly produces polycrystalline or even amorphous products. All the above techniques for the preparation of one-dimensional nanostructured materials will be discussed in this chapter, following the above order just for the sake of clarity. Lithography will be discussed briefly only; more detail and general discussion on lithography will be presented in Chapter 7. Similar to the previous chapter, this chapter will cover all the nanowires, nanorods and nanotubules of various materials including metals, semiconductors, polymers and insulating oxides. This chapter focuses on the fundamentals and principles of the major synthesis methods. For detailed information on specific materials, the readers are referred to a comprehensive review on the synthesis, characterization and applications of one-dimensional nanostructures.[1] Carbon nanotubes, as a special family of nanomaterials, deserve special attention and, therefore, will be discussed separately later on in Chapter 6.

4.2. Spontaneous Growth

Spontaneous growth is a process driven by the reduction of Gibbs free energy or chemical potential. The reduction of Gibbs free energy is commonly realized by phase transformation or chemical reaction or the release of stress. For the formation of nanowires or nanorods, anisotropic growth is required, i.e. the crystal grows along a certain orientation faster than other directions. Uniformly sized nanowires, i.e. the same diameter along the longitudinal direction of a given nanowire, can be obtained when crystal growth proceeds along one direction, whereas no growth along other directions. In spontaneous growth, for given material and growth conditions, defects and impurities on the growth surfaces can play a significant role in determining the morphology of the final products.

4.2.1. Evaporation (dissolution)–condensation growth

4.2.1.1. Fundamentals of evaporation (dissolution)–condensation growth

Evaporation–condensation process is also referred to as a vapor–solid (VS) process; however, the discussion in this section will not only be limited to simple evaporation–condensation process. Chemical reactions among various precursors may be involved to produce desired materials. Of course, the growth of nanorods from solution is also included. The driving force for the synthesis of nanorods and nanowires by spontaneous growth is a decrease in Gibbs free energy, which arises from either recrystallization or a decrease in supersaturation. Nanowires and nanorods grown by evaporation–condensation methods are commonly single crystals with fewer imperfections. The formation of nanowires, nanorods or nanotubules through evaporation (or dissolution)–condensation is due to the anisotropic growth. Several mechanisms are known to result in anisotropic growth, for example:

(1) Different facets in a crystal have different growth rate. For example, in silicon with a diamond structure, the growth rate of {111} facets is smaller than that of {110}.
(2) Presence of imperfections in specific crystal directions such as screw dislocation.
(3) Preferential accumulation of or poisoning by impurities on specific facets.

Before discussing the growth of various nanowires by evaporation–condensation method in detail, let us first review the fundamentals of crystal growth. Crystal growth can be generally considered as a heterogeneous reaction, and a typical crystal growth proceeds following the sequences, as sketched in Fig. 4.1:

(1) Diffusion of growth species from the bulk (such as vapor or liquid phase) to the growing surface, which, in general, is considered to proceed rapid enough and, thus, not at a rate limiting process.

(2) Adsorption and desorption of growth species onto and from the growing surface. This process can be rate limiting, if the supersaturation or concentration of growth species is low.
(3) Surface diffusion of adsorbed growth species. During surface diffusion, an adsorbed species may either be incorporated into a growth site, which contributes to crystal growth, or escape from the surface.
(4) Surface growth by irreversibly incorporating the adsorbed growth species into the crystal structure. When a sufficient supersaturation or a high concentration of growth species is present, this step will be the rate-limiting process and determines the growth rate.
(5) If by-product chemicals were generated on the surface during the growth, by-products would desorb from the growth surface, so that growth species can adsorb onto the surface and the process can continue.
(6) By-product chemicals diffuse away from the surface so as to vacate the growth sites for continuing growth.

For most crystal growth, rate-limiting step is either adsorption–desorption of growth species on the growth surface (step 2) or surface growth (step 4). When step 2 is rate limiting, the growth rate is determined by condensation rate, J (atoms/cm^2 sec), which is dependent on the number of growth species adsorbed onto the growth surface, which is directly proportional to the vapor pressure or concentration, P, of the growth species in the vapor as given by:

$$J = \frac{\alpha \sigma P_0}{\sqrt{2\pi mkT}} \qquad (4.1)$$

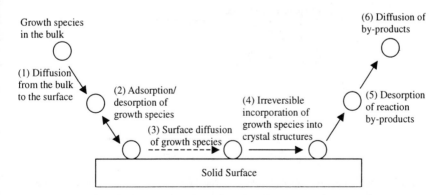

Fig. 4.1. Schematic illustrating six steps in crystal growth, which can be generally considered as a heterogeneous reaction, and a typical crystal growth proceeds following the sequences.

where α is the accommodation coefficient, $\sigma = (P-P_0)/P_0$ is the supersaturation of the growth species in the vapor in which P_0 is the equilibrium vapor pressure of the crystal at temperature T, m is the atomic weight of the growth species and k is Boltzmann constant. α is the fraction of impinging growth species that becomes accommodated on the growing surface, and is a surface specific property. A surface with a high accommodation coefficient will have a high growth rate as compared with low α surfaces. A significant difference in accommodation coefficients in different facets would result in anisotropic growth. When the concentration of the growth species is very low, the adsorption is more likely a rate-limiting step. For a given system, the growth rate increases linearly with the increase in the concentration of growth species. Further increase in the concentration of growth species would result in a change from an adsorption limited to surface growth limited process. When the surface growth becomes a limiting step, the growth rate becomes independent of the concentration of growth species as schematically shown in Fig. 4.2. A high concentration or vapor pressure of growth species in the vapor phase would increase the probability of defect formation, such as impurity inclusion and stack faults. Further, a high concentration may result in a secondary nucleation on the growth surface or even homogeneous nucleation, which would effectively terminate the epitaxial or single crystal growth.

An impinging growth species onto the growth surface can be described in terms of the residence time and/or diffusion distance before escaping

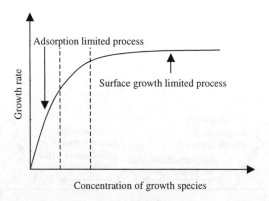

Fig. 4.2. Relation between growth rate and reactant concentration. At low concentration, growth is diffusion limited and thus increases linearly with increasing reactant concentration. At high concentration, surface reaction is the limit step and thus the growth rate becomes independent of reactant concentration.

back to the vapor phase. The residence time, τ_s, for a growth species on the surface is described by:

$$\tau_s = \frac{1}{\nu}\exp\left(\frac{E_{des}}{kT}\right) \quad (4.2)$$

where ν is the vibrational frequency of the adatom, i.e. adsorbed growth species, on the surface (typically $10^{12}\,\text{sec}^{-1}$), and E_{des} is the desorption energy required for the growth species escaping back to the vapor. While residing on the growth surface, a growth species will diffuse along the surface with the surface diffusion coefficient, D_s, given by:

$$D_s = \frac{1}{2}a_0\nu\exp\left(\frac{-E_s}{kT}\right) \quad (4.3)$$

where E_s is the activation energy for surface diffusion and a_0 is the size of the growth species. So the mean diffusion distance, X, for a growth species from the site of incidence:

$$X = \sqrt{2D_s\tau_s} = a_0\exp\left(\frac{E_{des}-E_s}{kT}\right) \quad (4.4)$$

It is clear that in a crystal surface, if the mean diffusion distance is far longer than the distance between two growth sites such as kinks or ledges, all adsorbed growth species will be incorporated into the crystal structure and the accommodation coefficient would be unity. If the mean diffusion distance is far shorter than the distance between growth sites, all adatoms will escape back to the vapor and the accommodation coefficient will be zero. The accommodation coefficient is dependent on desorption energy, activation energy of surface diffusion and the density of growth sites.

When step 2 proceeds sufficiently rapid, surface growth, i.e. step 4, becomes a rate-limiting process. In Chapter 2, we have discussed the fact that in a crystal, different facets have different surface energy. Different facets in a given crystal have different atomic density and atoms on different facets have a different number of unsatisfied bonds (also referred to as broken or dangling bonds), leading to different surface energy. Such a difference in surface energy or the number of broken chemical bonds leads to different growth mechanisms and varied growth rates. According to Periodic Bond Chain (PBC) theory developed by Hartman and Perdok,[2] all crystal facets can be categorized into three groups based on the number of broken periodic bond chains on a given facet: flat surface, stepped surface and kinked surface. The number of broken periodic bond chains can be understood as the number of broken bonds per atom on a given facet in a simplified manner. Let us first review the growth mechanisms on a flat surface.

For a flat surface, the classic step-growth theory was developed by Kossel, Stranski and Volmer, which is also called the KSV theory.[3] They recognized that the crystal surface, on the atomic scale, is not smooth, flat or continuous, and such discontinuities are responsible for the crystal growth. To illustrate the step growth mechanism, we consider a {100} surface of a simple cubic crystal as an example and each atom as a cube with a coordination number of six (six chemical bonds), as schematically sketched in Fig. 4.3. When an atom adsorbs onto the surface, it diffuses randomly on the surface. When it diffuses to an energetically favorable site, it will be irreversibly incorporated into the crystal structure, resulting in the growth of the surface. However, it may escape from the surface back to the vapor. On a flat surface, an adsorbed atom may find different sites with different energy levels. An atom adsorbed on a terrace would form one chemical bond between the atom and the surface; such an atom is called an adatom, which is in a thermodynamically unfavorable state. If an adatom diffuses to a ledge site, it would form two chemical bonds and become stable. If an atom were incorporated to a ledge-kink site, three chemical bonds would be formed. An atom incorporated into a kink site would form four chemical bonds. Ledge, ledge-kink and kink sites are all considered as growth sites; incorporation of atoms into these sites is irreversible and results in growth of the surface. The growth of a flat surface is due to the advancement of the steps (or ledges). For a given crystal facets and a given growth condition, the growth rate will be dependent on the step density. A misorientation would result in an increased density of steps and consequently lead to a high growth rate. An increased step density would favor the irreversible incorporation of adatoms by reducing the

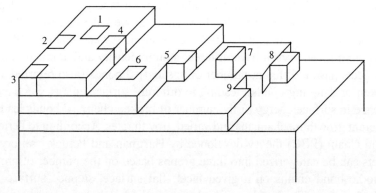

Fig. 4.3. Schematic illustrating the step growth mechanism, considering a {100} surface of a simple cubic crystal as an example and each atom as a cube with a coordination number of six (six chemical bonds) in bulk crystal.

surface diffusion distance between the impinging site and the growth site, before adatoms escape back to the vapor phase.

The obvious limitation of this growth mechanism is the regeneration of growth sites, when all available steps are consumed. Burton, Cabrera and Frank[4] proposed screw dislocation serves as a continuous source to generate growth sites so that the stepped growth would continue (as shown in Fig. 4.4). The crystal growth proceeds in a spiral growth, and this crystal growth mechanism is now known as BCF theory. The presence of screw dislocation will not only ensure the continuing advancement of the growth surface, but also enhance the growth rate. The growth rate of a given crystal facet under a given experimental condition would increase with an increased density of screw dislocations parallel to the growth direction. It is also known that different facets can have a significantly different ability to accommodate dislocations. The presence of dislocations on a certain facet can result in anisotropic growth, leading to the formation of nanowires or nanorods.

The PBC theory offers a different perspective in understanding the different growth rate and behavior in different facets.[2,5,6] Let us take a simple cubic crystal as an example to illustrate the PBC theory as shown in Fig. 4.5.[2] According to the PBC theory, {100} faces are flat surfaces (denoted as F-face) with one PBC running through one such surface, {110} are stepped surfaces (S-face) that have two PBCs, and {111} are kinked surfaces (K-face) that have three PBCs. For {110} surfaces, each surface site is a step or ledge site, and thus any impinging atom would be incorporated wherever it adsorbs. For {111} facets, each surface site is a kink site and would irreversibly incorporate any incoming atom adsorbed onto the surface. For both {110} and {111} surfaces, the above growth is referred to as a random addition mechanism and no adsorbed atoms would escape back to the vapor

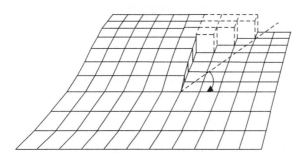

Fig. 4.4. The crystal growth proceeds in a spiral growth, known as BCF theory, in which screw dislocation serves as a continuous source to generate growth sites so that the stepped growth would continue.

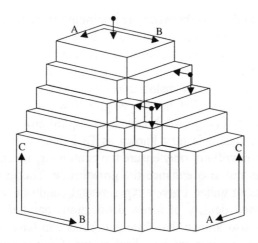

Fig. 4.5. Schematic illustrating the PBC theory. In a simple cubic crystal, {100} faces are flat surfaces (denoted as F-face) with one PBC running through one such surface, {110} are stepped surfaces (S-face) that have two PBCs, and {111} are kinked surfaces (K-face) that have three PBCs. [P. Hartman and W.G. Perdok, *Acta Crystal.* **8**, 49 (1955).]

phase. It is obvious that both {110} and {111} faces have faster growth rate than that of {100} surface in a simple cubic crystal. In a general term, S-faces and K-faces have a higher growth rate than F-faces. For both S- and K-faces, the growth process is always adsorption limited, since the accommodation coefficients on these two type surfaces are unity, all impinging atoms are captured and incorporated into the growth surface. For F-faces, the accommodation coefficient varies between zero (no growth at all) and unity (adsorption limited), depending on the availability of kink and ledge sites.

The above theories enable us to understand better why some facets in a given crystal grow much faster than others. However, facets with fast growth rate tend to disappear, i.e. surfaces with high surface energy will disappear. In a thermodynamically equilibrium crystal, only those surfaces with the lowest total surface energy will survive as determined by the Wulff plot.[7,8] Therefore, the formation of high aspect ratio nanorods or nanowires entirely based on different growth rates of various facets is limited to materials with special crystal structures. In general, other mechanisms are required for the continued growth along the axis of nanorods or nanowires, such as defect-induced growth and impurity-inhibited growth.

It should be noted that for an anisotropic growth, a low supersaturation is required. Ideally, the concentration is higher than the equilibrium concentration (saturation) of the growth surface, but equal or lower than that of other non-growth surfaces. A low supersaturation is required for anisotropic growth, whereas a medium supersaturation supports bulk crystal growth, and a high supersaturation results in secondary or homogeneous nucleation leading to the formation of polycrystalline or powder.

4.2.1.2. Evaporation–condensation growth

Sears[9] was the first to explain the growth of mercury whiskers (or nanowires, with a diameter of ~200 nm and a length of 1–2 mm) by axial screw-dislocation induced anisotropic growth in 1955. The mercury whiskers or nanowires were grown by a simple evaporation–condensation method, with a condensation temperature of $-50°C$ under vacuum, and the estimated axial growth rate was of approximately 1.5 μm/sec under a supersaturation of 100, which is defined as a ratio of pressure over equilibrium pressure. However, it was found that the whiskers or nanowires remained at constant radius throughout the axial growth, and thus implied that there was no or negligible lateral growth. In a subsequent article, Sears[10] also demonstrated that fine whiskers of other materials including zinc, cadmium, silver and cadmium sulfide could be grown by evaporation–condensation method. The experimental conditions varied with the material in question. The growth temperature varied from 250°C for cadmium to 850°C for silver whiskers, with a supersaturation ranging from ~2 for cadmium sulfide to ~20 for cadmium.

Subsequently, a lot of research has been devoted to confirm the presence of axial screw dislocation for the growth of nanowires; however, in most cases, various techniques including electron microscopy and etching, all failed to reveal the presence of axial screw dislocation.[11] Micro-twins and stacking faults are observed in many nanowires or nanorods grown by evaporation–condensation method and are suggested to be responsible for the anisotropic growth. However, many other researches revealed no axial defects at all in the grown nanorods and nanowires. It is obvious that the growth of nanorods or nanowires is not necessarily controlled by the presence of microtwins, though formation of twins is very important in determining the final crystal morphology.[12] Such an anisotropic growth is also not possible to explain by means of anisotropic crystal structures. Obviously, more work is needed to understand the growth of nanowires and nanorods by evaporation–condensation method.

Another related issue is the fact that the observed growth rate of the nanowires exceeds the condensation rate calculated using the equation for a flat surface [Eq. (4.1)], assuming the accommodation coefficient is unity. That means the growth rate of nanowires is faster than all the growth species arrived at the growth surface. To explain such a significantly enhanced growth rate of a whisker or nanowire, a dislocation–diffusion theory was proposed.[13] In this model, the fast growth rate was explained as follows: the depositing materials at the tip are originated from two resources: direct condensation of growth species from the vapor and the migration of adsorbed growth species on side surfaces to the growth tip. However, an adatom migrating over an edge from side surfaces to the growth surface on the tip is unlikely, since the edge serves as an energy barrier for such a migration.[14–16]

Wang and his co-workers[17] reported the growth of single crystal nanobelts of various semiconducting oxides simply by evaporating the desired commercially available metal oxides at high temperatures under a vacuum of 300 torr and condensing on an alumina substrate, placed inside the same alumina tube furnace, at relatively lower temperatures. The oxides include zinc oxide (ZnO) of wurtzite hexagonal crystal structure, tin oxide (SnO_2) of rutile structure, indium oxide (In_2O_3) with C-rare-earth crystal structure, and cadmium oxide (CdO) with NaCl cubic structure. We will just focus on the growth of ZnO nanobelts to illustrate their findings, since similar phenomena were found in all four oxides. Figure 4.6 shows

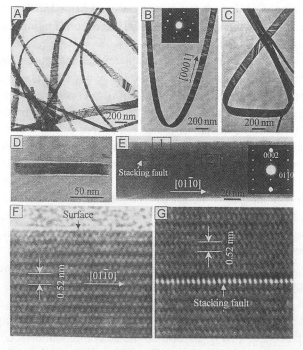

Fig. 4.6. SEM and TEM pictures of ZnO nanobelts [Z.W. Pan, Z.R. Dai, and Z.L. Wang, *Science* **291**, 1947 (2001).]

the SEM and TEM pictures of ZnO nanobelts.[17] The typical thickness and width-to-thickness ratios of the ZnO nanobelts are in the range of 10 to 30 nm and ~5 to 10, respectively. Two growth directions were observed: [0001] and [0110]. No screw dislocation was found throughout the entire length of the nanobelt, except a single stacking fault parallel to the growth axis in the nanobelts grown along [0110] direction. The surfaces of the nanobelts are clean, atomically sharp and free of any sheathed amorphous phase. Their further TEM analysis also revealed the absence of amorphous globules on the tip of nanobelts. The above observations imply that the growth of nanobelts is not due to the VLS mechanism, which will be discussed later in this chapter. The growth of nanobelts cannot be attributed to either screw dislocation induced anisotropic growth, nor impurity inhibited growth. Furthermore, since four oxides in question all have different crystal structures, it is not likely that the growth of nanobelts is directly related to their crystal structures. Nanobelts of other oxides such as Ga_2O_3 with a crystal structure of monoclinic and PbO_2 (rutile) were also synthesized by the same technique.[18] It seems worthwhile to note that the shape of nanowires and nanobelts may also depend on growth temperature. Early work showed that single crystal mercury grown at different temperatures would have either a platelet shape or a whisker form.[9,19] CdS ribbons were also grown by evaporation–condensation method.[10]

Kong and Wang[20] further demonstrated that by controlling growth kinetics, left-handed helical nanostructures and nano-rings can be formed by rolling up single crystal ZnO nanobelts. This phenomenon is attributed to a consequence of minimizing the total energy attributed by spontaneous polarization and elasticity. The spontaneous polarization results from the noncentrosymmetric ZnO crystal structure. In (0001) facet-dominated single crystal nanobelts, positive and negative ionic charges are spontaneously established on the zinc- and oxygen-terminated ±(0001) surfaces, respectively. Figure 4.7 shows SEM images of the synthesized ZnO nanobelt helical nanostructures.[20]

Liu et al.[21] synthesized SnO_2 nanorods by converting nanoparticles at elevated temperatures. The nanoparticles were chemically synthesized from $SnCl_4$ by inverse microemulsion using non-ionic surfactant, and have an average size of 10 nm and are highly agglomerated. SnO_2 nanoparticles are likely to be amorphous. When heated to temperatures ranging from 780°C to 820°C in air, single crystal SnO_2 nanorods with rutile structure were formed. Nanorods are straight and have uniform diameters ranging from 20 to 90 nm and lengths from 5 to 10 μm, depending on annealing temperature and time. Various oxide nanowires, such as ZnO, Ga_2O_3 and MgO, and CuO were synthesized by such evaporation–condensation

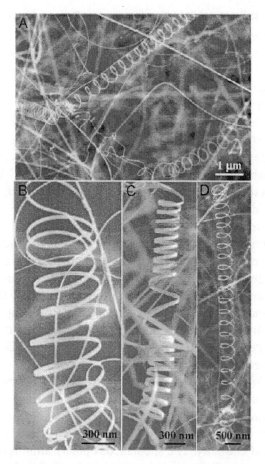

Fig. 4.7. SEM images of the synthesized single crystal ZnO nanobelt helical nanostructures. The typical width of the nanobelts is ~30 nm, and pitch distance is rather uniform. The helixes are left-handed. [X.Y. Kong and Z.L. Wang, *Nano Lett.* **3**, 1625 (2003).]

method.[22] Figure 4.8 shows such grown CuO nanowires by heating copper wire in air to a temperature of 500°C for 4 hrs.[23] In addition, Si_3N_4 and SiC nanowires were also synthesized by simply heating the commercial powders of these materials to elevated temperatures.[24]

Chemical reactions and the formation of intermediate compounds play important roles in the synthesis of various nanowires by evaporation–condensation methods. Reduction reactions are often used to generate volatile deposition precursors and hydrogen, water and carbon are commonly used as reduction agent. For example, hydrogen and water were used in the growth of binary oxide nanowires, such as Al_2O_3, ZnO and SnO_2 through a two-step process (reduction and oxidation).[25,26] Silicon

One-Dimensional Nanostructures: Nanowires and Nanorods

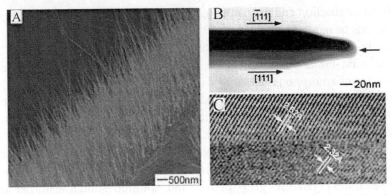

Fig. 4.8. (A) SEM and (B) TEM micrographs of CuO nanowires synthesized by heating a copper wire (0.1 mm in diameter) in air to a temperature of 500°C for 4 hr. Each CuO nanowire was a bicrystal as shown by its electron diffraction pattern and high-resolution TEM characterization (C). [X. Jiang, T. Herricks, and Y. Xia, *Nano Lett.* **2**, 1333 (2002).]

nanowires could be synthesized by thermal evaporation of silicon monoxide under a reducing environment.[27] SiO powder was simply heated to a temperature of 1300°C, and the vapor of silicon monoxide was carried by a mixture of argon and 5% hydrogen. The (100) silicon substrate was maintained at 930°C for the growth. It was found the as-grown nanowires of 30 nm in diameter consist of a silicon core of 20 nm in diameter and a shell of silicon dioxide of 5 nm in thickness. The silicon core is believed to form through the reduction of silicon monoxide by hydrogen. The silicon dioxide shell may serve as a stopper for the side growth, resulting in a uniform diameter throughout a nanowire. Carbon was used in the synthesis of MgO nanowires.[28]

Although it is known that the impurities have differential adsorption on various crystal facets in a given crystal and the adsorption of impurity would retard the growth process, no nanorods have been grown by vapor condensation methods based on impurity poisoning by design. However, impurity poisoning has often been cited as one of the reasons, which resulted in anisotropic growth during the synthesis of nanowires and nanorods.

4.2.1.3. *Dissolution–condensation growth*

Dissolution–condensation process differs from evaporation–condensation in growth media. In dissolution–condensation process, the growth species first dissolve into a solvent or a solution, and then diffuse through the solvent or solution and deposit onto the surface resulting in the growth of nanorods or nanowires.

Gates et al.[29] prepared uniform single crystal nanowires of selenium by dissolution–condensation methods. In the first step, spherical colloidal particles of amorphous selenium with sizes of ~300 nm in aqueous solution were prepared through the reduction of selenious acid with excess hydrazine at 100°C. When the solution was cooled to room temperature, some nanocrystalline selenium with trigonal structure was precipitated. In the second step, when the solution aged at room temperature in dark, amorphous selenium colloid particles dissolved into the solution, whereas the selenium crystallites grew. In this solid–solution–solid transformation, the morphology of the crystalline selenium products was determined by the anisotropic growth, which is attributed to the one-dimensional characteristics of the infinite, helical chains of selenium in the trigonal structure. Trigonal Se crystals were found to grow predominantly along the [001] direction.[30] Se nanowires grown by this method were found free of defects, such as kinks and dislocations.

The chemical solution method was also explored for the synthesis of nanorods of crystalline Se_xTe_y compound.[31] In an aqueous medium (refluxed at ~100°C), a mixture of selenious acid and orthotelluric acid was reduced by excess hydrazine[32]:

$$xH_2SeO_3 + yH_6TeO_6 + \left(x + \frac{3y}{2}\right)N_2H_4$$
$$\rightarrow Se_xTe_{y(s)} + \left(x + \frac{3y}{2}\right)N_{2(g)} + \left(x + \frac{3y}{2}\right)H_2O \qquad (4.5)$$

Tellurium could easily precipitate out as crystalline hexagonal nanoplatelets under the experimental conditions through a homogeneous nucleation process.[33] It is postulated that the selenium and tellurium atoms subsequently produced by the above reduction reaction would grow into nanorods on the nanoplatelet tellurium seeds with a growth direction along [001]. The as-synthesized nanorods typically have a mean length of <500 nm and a mean diameter of ~60 nm, with a stoichiometric chemical composition of SeTe, and a trigonal crystal structure, similar to that of Se and Te. Hydrazine can also promote the growth of nanorods directly from metal powders in solution; for example, single crystal ZnTe nanorods with diameters of 30–100 nm and lengths of 500–1200 nm were synthesized using Zn and Te metal powders as reactants and hydrazine hydrate as solvent by a solvothermal process.[34] It was postulated that hydrazine could promote anisotropic growth in addition to its role as a reduction agent.

Wang et al.[35] have grown single crystal Mn_3O_4 nanowires of 40–80 nm diameter and lengths up to 150 μm in a molten NaCl flux. $MnCl_2$ and

Na$_2$CO$_3$ were mixed with NaCl and nonylphenyl ether (NP-9) and heated to 850°C. After cooling, the NaCl was removed by washing in distilled water. NP-9 was used to prevent the formation of small particles at the expense of nanowires. The nanowires are believed to grow through Ostwald ripening, with NP-9 acting to reduce the eutectic temperature of the system, as well as stabilizing the smaller precursor particles.

Nanowires can grow on alien crystal nanoparticles, which serve as seeds for heteroepitaxial growth, by solution processing. Sun et al.[36] synthesized crystalline silver nanowires of 30–40 nm in diameter and ~50 μm in length using platinum nanoparticles as growth seeds. The growth species of Ag is generated by the reduction of AgNO$_3$ with ethylene glycol, whereas the anisotropic growth was achieved by introducing surfactants such as polyvinyl pyrrolidone (PVP) in the solution. Polymer surfactants adsorbed on some growth surfaces, so that kinetically blocked (or poisoning) the growth, resulting in the formation of uniform crystalline silver nanowires. TEM analyses further revealed that the growth directions of face-center-cubic silver nanowires were [2$\bar{1}\bar{1}$] and [01$\bar{1}$]. Figure 4.9 shows the silver nanowires grown in solution using Pt nanoparticles as growth seeds.[36] Dissolution–condensation can also grow nanowires on a substrate. Govender et al.[37] formed ZnO nanorods on glass substrates from a solution of zinc acetate or zinc formate and hexamethylenetetramine at room temperature. These faceted nanorods were preferentially oriented in

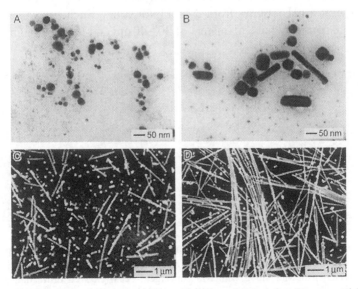

Fig. 4.9. SEM images of silver nanowires grown in solution using Pt nanoparticles as growth seeds. [Y. Sun, B. Gates, B. Mayers, and Y. Xia, *Nano Lett.* **2**, 165 (2002).]

the [0001] direction (that is, along the c-axis), with diameters of about 266 nm and a length of ~3 μm.

Nanowires can also be grown using the same methods commonly used for the synthesis of nanocrystals, i.e. by decomposing of organometallic compounds in the presence of coordinating organics. For example, Urban and co-workers[38,39] synthesized single crystal $BaTiO_3$ nanowires with diameters ranging from 5 to 70 nm and lengths up to >10 μm, by solution-phase decomposition of barium titanium isopropoxide, $BaTi[OCH(CH_3)_2]_6$. In a typical reaction, an excess of 30% H_2O_2 was added at 100°C to a heptadecane solution containing a 10:1 molar ratio of $BaTi[OCH(CH_3)_2]_6$ to oleic acid. The reaction mixture was then heated to 280°C for 6 hrs, resulting in a white precipitate composed of nanowire aggregates. Well-isolated nanowires were obtained by sonication and fractionation between water and hexane. Figure 4.10 shows the TEM images and convergent-beam electron diffraction patterns of $BaTiO_3$ nanowires.[39] Elemental analysis, X-ray diffraction, and electron diffraction all indicate that the grown nanowires are single crystalline perovskite $BaTiO_3$ with the [001] direction aligned along the wire axis. It should be noted that the diameters and lengths of grown nanowires vary substantially and no strategy is available for a controlled growth of uniformly sized nanowires.

Hydrothermal growth is another method having been explored in the formation of nanorods from inorganic salts. $CdWO_4$, with a monoclinic crystal structure, nanorods were synthesized directly by the reaction of cadmium chloride ($CdCl_2$) and sodium tungstate ($NaWO_4$) at 130°C under pressure at a pH ranging from 3 to 11 for 5 hrs. Such grown cadmium tungstate nanorods have diameters of 20–40 nm and lengths ranging from

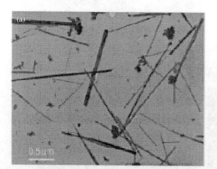 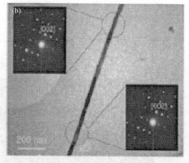

Fig. 4.10. (a) TEM images of $BaTiO_3$ nanowires, showing that the reaction produces mainly nanowires and small quantities (~10%) of nanoparticle aggregates. (b) TEM image of a $BaTiO_3$ nanowire along with two convergent beam electron diffraction patterns. [J.J. Urban, J.E. Spanier, L. Ouyang, W.S. Yun, and H. Park, *Adv. Mater.* **15**, 423 (2003).]

80 to 280 nm.[40] The growth of nanorods is attributed to anisotropic growth, though no specific growth direction was identified. Nanotubes of $H_2Ti_3O_7$ have been hydrothermally synthesized from TiO_2 powder dissolved in a NaOH aqueous solution at 130°C by Chen et al.[41] The synthesized products were hollow tubes about 9 nm in diameter, with lengths from 100 to several hundred nanometers.

Nanowires or nanorods by the evaporation (dissolution)–condensation deposition most likely have faceted morphology and are generally short in length with relatively small aspect ratios, particularly when grown in liquid medium. However, anisotropic growth induced by axial imperfections, such as screw dislocation, microtwins and stacking faults, or by impurity poisoning, can result in the growth of nanowires with very large aspect ratios.

4.2.2. Vapor (or solution)–liquid–solid (VLS or SLS) growth

4.2.2.1. Fundamental aspects of VLS and SLS growth

In the VLS growth, a second phase material, commonly referred to as either impurity or catalyst, is purposely introduced to direct and confine the crystal growth on to a specific orientation and within a confined area. A catalyst forms a liquid droplet by itself or by alloying with growth material during the growth, which acts as a trap of growth species. Enriched growth species in the catalyst droplets subsequently precipitates at the growth surface resulting in the one-directional growth. Wagner et al.[42,43] first proposed the VLS theory over 40 years ago to explain the experimental results and observations in the growth of silicon nanowires or whiskers that could not be explained by the evaporation–condensation theory. These phenomena include:

(1) There are no screw dislocations or other imperfections along the growth direction.
(2) The growth direction $\langle 111 \rangle$ is the slowest as compared with other low index directions such as $\langle 110 \rangle$ in silicon.
(3) Impurities are always required and
(4) A liquid-like globule is always found in the tip of nanowires.

Wagner[44] summarized the experimental details, results, and the VLS theory in a truly elegant way in a classical paper, and Givargizov[11] further elaborated the experimental observations and models and theories developed regarding the VLS process. The readers who want to learn more about this subject are strongly recommended to read those literatures. Although an extensive research in this field has been carried out in the recent years, fundamentals of the VLS method have not been changed significantly. Wagner[44] summarized the requirements for the VLS growth 30 years ago, which are still valid in today's understanding:

(1) The catalyst or impurity must form a liquid solution with the crystalline material to be grown at the deposition temperature,
(2) The distribution coefficient of the catalyst or impurity must be less than unity at the deposition temperature.
(3) The equilibrium vapor pressure of the catalyst or impurity over the liquid droplet must be very small. Although the evaporation of the catalyst does not change the composition of the saturated liquid composition, it does reduce the total volume of the liquid droplet. Unless more catalyst is supplied, the volume of the liquid droplet reduces. Consequently, the diameter of the nanowire will reduce and the growth will eventually stop, when all the catalyst is evaporated.
(4) The catalyst or impurity must be inert chemically. It must not react with the chemical species such as by-products presented in the growth chamber.
(5) The interfacial energy plays a very important role. The wetting characteristics influence the diameter of the grown nanowire. For a given volume of liquid droplet, a small wetting angle results in a large growth area, leading to a large diameter of nanowires.
(6) For a compound nanowire growth, one of the constituents can serve as the catalyst.
(7) For controlled unidirectional growth, the solid–liquid interface must be well defined crystallographically. One of the simplest methods is to choose a single crystal substrate with desired crystal orientation.

In a VLS growth, the process can be simply described as following as sketched in Fig. 4.11. The growth species is evaporated first, and then diffuses and dissolves into a liquid droplet. The surface of the liquid has a large accommodation coefficient, and is therefore a preferred site for deposition. Saturated growth species in the liquid droplet will diffuse to and precipitate at the interface between the substrate and the liquid. The precipitation will first follow nucleation and then crystal growth. Continued precipitation or growth will separate the substrate and the liquid droplet, resulting in the growth of nanowires.

Let us take the growth of silicon nanowires with gold as a catalyst as an example to illustrate the experimental process of the VLS growth. A thin layer of gold is sputtered on a silicon substrate and annealed at an elevated

One-Dimensional Nanostructures: Nanowires and Nanorods

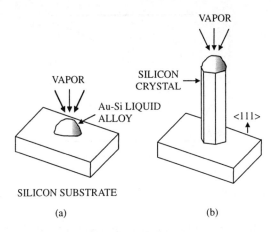

Fig. 4.11. Schematic showing the principal steps of the vapor–liquid–solid growth technique: (a) initial nucleation and (b) continued growth.

temperature (above the eutectic point of 385°C of the silicon–gold system), which is typically the same as the growth temperature. During the annealing, silicon and gold react and form a liquid mixture, which forms a droplet on the silicon substrate surface. During the growth, an equilibrium composition is reached at the growth temperature as determined by the binary phase diagram as shown in Fig. 4.12. When silicon species is evaporated from the source and preferentially condensed at the surface of the liquid droplet, the liquid droplet will become supersaturated with silicon. Subsequently, the supersaturated silicon will diffuse from the liquid–vapor interface and precipitate at the solid–liquid interface resulting in the growth of silicon. The growth will proceed unidirectionally perpendicular to the solid–liquid interface. Once the growth species is adsorbed onto the liquid surface, it will dissolve into the liquid. The material transport in the liquid is diffusion-controlled and occurs under essentially isothermal conditions. At the interface between the liquid droplet and growth surface, the crystal growth proceeds essentially the same as that in the Czochraski crystal growth.

Crystalline defects, such as screw dislocations, are not essential for VLS growth. However, defects present at the interface may promote the growth and lower the required supersaturation. From the above discussion, it is clear that the growth of nanowires by the VLS method is not restricted by the type of substrate materials and the type of catalysts. The nanowires can be single crystal, polycrystalline or amorphous depending on the substrates and growth conditions.

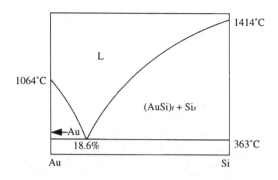

Fig. 4.12. Phase diagram of the gold–silicon binary system.

The preferential adsorption of growth species onto the liquid droplet surface can be understood. For a perfect or an imperfect crystal surface, an impinging growth species diffuse along the surface. During the diffusion, the growth species may be irreversibly incorporated into the growth site (ledge, ledge-kink, or kink). If the growth species did not find a preferential site in a given period of time (the residence time), the growth species will escape back to the vapor phase. A liquid surface is distinctly different from a perfect or imperfect crystal surface, and can be considered as a "rough" surface. Rough surface is composed of only ledge, ledge-kink, or kink sites. That is every site over the entire surface is to trap the impinging growth species. The accommodation coefficient is unit. Consequently, the growth rate of the nanowires or nanorods by VLS method is much higher than that without liquid catalyst. Wagner and Ellis[45] reported that the growth rate of silicon nanowires using a liquid Pt–Si alloy is about 60 times higher than directly on the silicon substrate at 900°C. It is likely that in addition to acting as a sink for the growth species in the vapor phase, the catalyst or impurity forming the liquid with the growth material can act as a catalyst for the heterogeneous reaction or deposition.

As discussed in Chapter 2, the equilibrium vapor pressure or solubility is dependent on the surface energy and a radius (or curvature of a surface) at a given condition defined by the Kelvin equation.

$$\ln\left(\frac{P}{P_0}\right) = -\frac{2\gamma\Omega}{kTr} \qquad (4.6)$$

Where P is the vapor pressure of a curved surface, P_0 is the vapor pressure of a flat surface, γ is the surface energy, Ω is the atomic volume, r is the surface radius, and k is the Boltzmann constant. For the growth of nanowires, if facets are developed during the growth, longitudinal and lateral growth rates of nanowires or nanorods will be solely determined by the growth behavior of individual facets. However, if nanowires were cylindrical in shape, the lateral growth rate would be significantly smaller than the longitudinal one, assuming all surfaces have the same surface energy. Convex surface (the side surface) with very small radius (<100 nm) would have a significantly higher vapor pressure as compared with that of a flat growing surface. A supersaturated vapor pressure or concentration of the growth species for the growing surface may be well below the equilibrium vapor pressure of the convex surface of thin nanowires. For the growth of uniform high quality crystalline nanowires or nanorods, in general supersaturation should be kept relatively low, so that there would be no growth on the side surface. A high supersaturation would result in growth of other facets, just as in the vapor–solid growth discussed before. Further high supersaturation would lead to secondary nucleation on the growth surface or homogeneous nucleation, resulting in termination of epitaxial growth.

Figure 4.13 compared the axial (V_{II}) and lateral ($V_{\perp}$) growth rates for Si and Ge nanowires as well as the substrate (i.e. film) growth rate of these materials, with SiH_4 and GeH_4 as precursors and numerous metals (Au, Ag, Cu, Ni, Pd) as catalysts.[46] This figure indicates that the lateral and the substrate growth rates are essentially the same, whereas the axial rate by the VLS process for both Si and Ge nanowires, are approximately two orders of magnitude higher than the VS growth rates under the same conditions.

The enhanced growth rate can also be partly due to the fact that the condensation surface area for the growth species in the VLS growth is larger than the surface area of the crystal growth. While the growth surface is the interface between the liquid droplet and the solid surface, the condensation surface is the interface between the liquid droplet and vapor phase. Depending on the contact angle, the liquid surface area can be several times of the growth surface.

4.2.2.2. VLS growth of various nanowires

Growth of elementary Si and Ge nanowires has been well established.[47–49] Figure 4.14 shows typical Si nanowires grown by the VLS method.[48] Although gold was initially used for the growth of silicon nanowires with the VLS method, other catalysts have been found to be effective in the formation of nanowires of various materials. For example, Si nanowires can be synthesized using Fe as a catalyst[50,51] at a relatively high growth

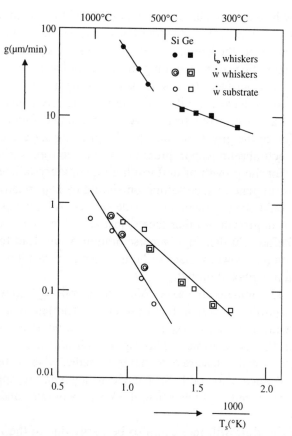

Fig. 4.13. Comparison of the axial ($V_{\parallel}$) and lateral ($V_{\perp}$) growth rates for Si and Ge nanowires as well as the substrate (i.e. film) growth rate of these materials, with SiH_4 and GeH_4 as precursors and numerous metals (Au, Ag, Cu, Ni, Pd) as catalysts. [G.A. Bootsma and H.J. Gassen, *J. Crystal Growth* **10**, 223 (1971).]

temperature of 1200°C. A mixture of silicon powder with 5 wt% Fe was ablated by either laser or simple heating to 1200°C, which was also the growth temperature. The nanowires have nominal diameters of ~15 nm and a length varying from a few tens to several hundreds micrometers. An amorphous layer of silicon oxide of ~2 nm in thickness over coated the outside of silicon nanowires. The amorphous oxide layer was likely to be formed during the growth at high temperature, when a small amount of oxygen leaked into the deposition chamber.

Nanowires of compound materials can also grow using VLS method. For example, Duan and Lieber[52] grew semiconductor nanowires of the III-V materials GaAs, GaP, GaAsP, InAs, InP, InAsP, the II-VI materials ZnS,

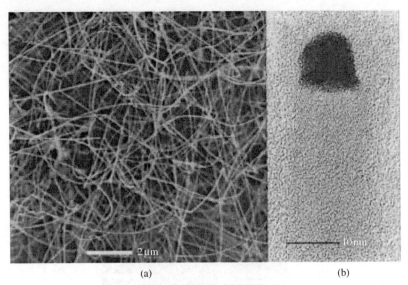

Fig. 4.14. (a) Field-emission scanning electron micrograph of silicon nanowires produced by VLS method and (b) High-resolution TEM image of the end of a 15 nm diameter wire exhibiting Si lattice planes and a gold nanocluster. [J. Hu, T.W. Odom, and C.M. Lieber, *Acc. Chem. Res.* **32**, 435 (1999).]

ZnSe, CdS, CdSe and IV-IV alloys of SiGe. Table 4.1 summarizes the single crystal compound nanowires synthesized. GaAs nanowires were grown using gold, silver and copper as catalysts, whereas all other nanowires were grown using gold as the catalyst. From the table, it is clear that most crystals have a preferential growth direction of [111], except CdS, which grew preferably along [100] or [002], and CdSe, which has a growth direction of [110]. It is further noticed that all the nanowires grew with desired stoichiometric compositions. For nanowires of Si_xGe_{1-x} alloys, the composition varied with deposition temperature, with x varying from 0.95 to 0.13 from the same $(Si_{0.7}Ge_{0.3})_{0.95}Au_{0.05}$ target in one experiment. Such a variation of chemical composition in the grown nanowires was explained by the fact that the optimal growth temperatures of the two individual nanowire materials were quite different. Figure 4.15 shows the SEM images of compound semiconductor nanowires grown by VLS method.[52] The authors further pointed out that catalysts for the VLS growth can be chosen in the absence of detailed phase diagrams by identifying metals in which the nanowire component elements are soluble in the liquid phase but that do not form solid compounds more stable than the desired nanowire phase; i.e. the ideal metal catalyst should be physically active but chemically stable or inert.

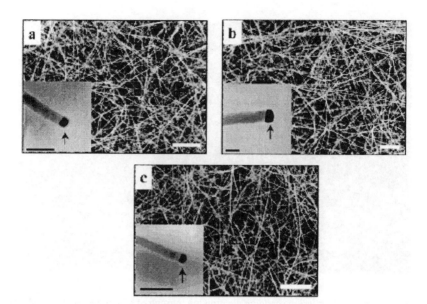

Fig. 4.15. Field-emission SEM images of compound semiconductor nanowires grown by VLS method: (a) GaAs, (b) GaP, and (c) GaAs$_{0.6}$P$_{0.4}$. The scale bars are 2 μm. [X. Duan and C.M. Lieber, *Adv. Mater.* **12**, 298 (2000).]

4.2.2.3. Control of the size of nanowires

The size of nanowires grown by VLS method is solely determined by the size of the liquid catalyst droplets. To grow thinner nanowires, one can simply reduce the size of the liquid droplets. Typical method used to form small liquid catalyst droplets is to coat a thin layer of catalyst on the growth substrate and to anneal at elevated temperatures.[53] During annealing, catalyst reacts with the substrate to form a eutectic liquid and further lead to reduction in the overall surface energy. Au as a catalyst and silicon as a substrate is a typical example. The size of the liquid catalyst droplets can be controlled by the thickness of the catalyst film on the substrate. In general, a thinner film forms smaller droplets, giving smaller diameters of nanowires subsequently grown. For example, 10 nm Au film yields single crystal germanium nanowires of 150 nm in diameter, while 5 nm Au film results in the growth of 80 nm sized germanium nanowires.[54] However, further reduction in the catalyst film thickness did not result in a decreased diameter of germanium nanowires.[54] No further reduction in diameter of nanowires indicated that there is a minimum size of liquid droplets achievable by applying thin films.

Further reduction of diameters of nanowires could be achieved by dispersing monosized catalyst colloids on the substrate surface, instead of a thin film of catalyst.[55,56] GaP nanowires were grown by laser catalytic growth synthetic method[49] using gold colloids.[56] Gold colloids or nanoclusters were supported on a silica substrate and the reactants Ga and P were generated from a solid target of GaP by laser ablation. Single crystal GaP nanowires show a growth direction of [111] and have a stoichiometric composition of 1:0.94 confirmed by EDAX. The diameters of GaP nanowires were determined by the size of the catalyst gold nanoclusters. GaP nanowires grown from 8.4, 18.5 and 28.2 nm diameter gold colloids were found to be 11.4, 20 and 30.2 nm, respectively. Similar technique was applied to the growth of InP nanowires.[55] The growth substrate temperature was controlled to be approximately 500–600°C, and a constant flow of Ar at 100 standard cubic centimeter per minute under a pressure of 200 torr was maintained during the growth. The laser for ablation used was an ArF excimer laser with a wavelength of 193 nm. InP nanowires were found to be single crystal and grew along the [111] direction. Figure 4.16 shows the general concepts of control of the diameters and length of nanowires grown by growth time and the size of catalyst colloids.[55] Detailed analysis further revealed an amorphous oxide layer of 2–4 nm in thickness presented on all nanowires. The existence of an amorphous oxide layer was explained by the overgrowth of an amorphous InP on the side faces and subsequent oxidation after the samples were exposed to air. The overgrowth on side faces is not catalyst activated and implies supersaturated vapor concentrations of growth constituents in the system.

Since the diameters of nanowires grown by VLS method is solely controlled by the size of the liquid catalyst droplets, thinner wires can be grown by using smaller liquid droplets. However, this approach has its limit. From Eq. (4.6), we already know that the equilibrium vapor pressure of a solid surface is dependent on the surface curvature. The same dependence is found for a solubility of a solute in a solvent. As the size of the droplets was reduced, the solubility would increase. For the growth of very thin nanowires, a very small droplet is required. However, a convex surface with a very small radius would have a very high solubility. As a result, a high supersaturation in the vapor phase has to be generated. A high supersaturation in the vapor phase may promote the lateral growth on the side surface of nanowires with the vapor–solid mechanism. Therefore, a conical structure may be developed instead of uniformly sized nanowires. Further, a high supersaturation may initiate homogeneous nucleation in the gas phase or secondary nucleation at the surface of nanowires.

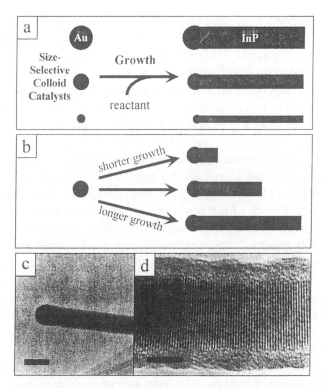

Fig. 4.16. Schematic illustrating the general concepts of control of the diameters and length of nanowires grown by growth time and the size of catalyst colloids. [M.S. Gudiksen, J. Wang, and C.M. Lieber, *J. Phys. Chem.* **B105**, 4062 (2001).]

Another characteristic in the VLS method should be noted. According to the Kelvin equation, an equilibrium solubility and supersaturation of growth species in larger liquid catalyst droplets can be obtained easier than that in smaller droplets. The growth of nanowires will proceed only when the concentration of growth species is above the equilibrium solubility. When the concentration or supersaturation in the vapor phase is appropriately controlled, vapor pressure could be kept below the equilibrium solubility in small liquid droplets, and the growth of nanowires of thinnest nanowires would terminate. When the growth proceeds at high temperatures and the grown nanowires are very thin, radial size instability is often observed as shown in Fig. 4.17.[11] Such instability is explained by the oscillation of the size of the liquid droplet on the growth tip and the concentration of the growth species in the liquid droplet.[11] Such instability could be another barrier for the synthesis of very thin nanowires, which may require high deposition temperatures.

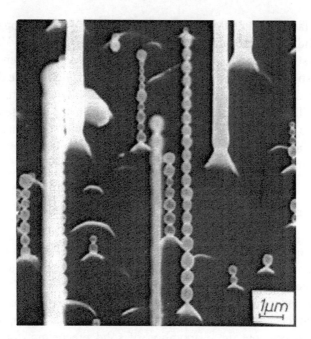

Fig. 4.17. Radial size instability in nanowires grown by VLS. [E.I. Givargizov, *Highly Anisotropic Crystals*, D. Reidel, Dordrecht, 1986.]

The diameter of nanowires grown by VLS method is determined by the minimum size of the liquid catalyst droplet under the equilibrium conditions.[44] Methods to achieve small sizes of liquid catalyst droplets are straightforward. For example, laser ablation can be used to deposit catalytic material on a heated substrate to form nanometer diameter clusters with controlled pressure and temperature.[57] In a similar manner, many other evaporation techniques could be used to deposit nanometer catalyst clusters on substrates for the growth of nanowires.

Nanowires or nanorods grown by VLS method in general have a cylindrical morphology, i.e. without facets on the side surface and having a uniform diameter. The physical conditions of both Czochraski and VLS methods are very similar; growth proceeds very close to the melting points or liquid–solid equilibrium temperature. Surfaces may undergo a transition from faceted (smooth) to "rough" surface, known as the roughening transition.[58] Below the roughening temperature, a surface is faceted, and above this temperature, thermal motion of the surface atoms overcomes the interfacial energy and causes a faceted crystal to roughen. From melt, only a restricted group of materials including silicon and bismuth can grow faceted single crystals.[59] However, facets may develop if there is vapor–solid (VS) deposition on the side surface. Although the VS deposition rate is much smaller than the VLS growth rate for a given temperature, it is still effective in controlling the morphology. Since the difference in the two deposition rates decreases with increasing temperature, the VS deposit will greatly influence the morphology in the high temperature range. It is noted that the diameter of the nanowire may change if the growth conditions vary or the catalyst evaporates or is incorporated into the nanowires.

4.2.2.4. Precursors and catalysts

A variety of precursors have been used for the VLS growth as that for evaporation–condensation methods. Gaseous precursors such as $SiCl_4$ for silicon nanowires[53] are convenient sources. Evaporation of solids by heating to elevated temperatures is another common practice.[60] Laser ablation of solid targets is yet another method used in generating vapor precursors.[49,61] To promote the evaporation of solid precursors, formation of intermediate compounds may be an appropriate approach. For example, Wu et al.[54] used a mixture of Ge and GeI_4 as precursors for the growth of Ge nanowires. The precursors evaporated through the formation of volatile compound via the following chemical reaction:

$$Ge_{(s)} + GeI_{4(g)} \rightarrow 2GeI_{2(g)} \tag{4.7}$$

The GeI_2 vapor was transported to the growth chamber, condensed into the liquid catalyst (here is Au/Si) droplets, and disproportionated according to

$$2GeI_{2(g)} \rightarrow Ge_{(l)} + GeI_{4(g)} \tag{4.8}$$

Other precursors have also been explored in the VLS growth of nanowires including ammonia and gallium acetylacetonate for GaN nanorods,[62] closo-1,2-dicarbadodecaborane ($C_2B_{10}H_{12}$) for B_4C nanorods,[63] and methyltrichlorosilane for SiC.[64]

ZnO nanowires have been grown on Au-coated (thickness ranging from 2 to 50 nm) silicon substrates by heating a 1 : 1 mixture of ZnO and graphite powder to 900–925°C under a constant flow of argon for 5–30 min.[65] The grown ZnO nanowires vary with the thickness of the initial Au coatings. For a 50 nm Au coating, the diameters of the nanowires normally range from 80 to 120 nm and their lengths are 10–20 μm. Thinner nanowires of 40–70 nm with lengths of 5–10 μm were grown on 3 nm Au-coated substrates. The grown ZnO nanowires are single crystals with a preferential growth direction of ⟨001⟩. The growth process of ZnO is believed to be different from that of elementary nanowires. The process involves the reduction of ZnO by graphite to form Zn and CO vapor at high temperatures (above 900°C). The Zn vapor is transported to and reacted with the Au catalyst, which would have already reacted with silicon to form a eutectic Au–Si liquid on silicon substrates, located downstream at a lower temperature to form Zu–Au–Si alloy droplets. As the droplets become supersaturated with Zn, crystalline ZnO nanowires are formed, possibly through the reaction between Zn and CO at a lower temperature. The above process could be easily understood by the fact that the reaction:

$$ZnO + C \leftrightarrow Zn + CO \tag{4.9}$$

is a reversible at temperatures around 900°C.[66] Although the presence of a small amount of CO is not expected to change the phase diagram significantly, no ZnO nanowires were grown on substrates in the absence of graphite.

A variety of materials can be used as catalysts for the VLS growth of nanowires. For example, silicon nanowires were grown using iron as a catalyst.[49] Any materials or mixtures can be used as catalyst as far as they meet the requirements described by Wagner.[44] For example, a mixture of Au and Si was used for the growth of germanium nanowires.[54]

Single crystal monoclinic gallium oxide (β-Ga_2O_3) nanowires were synthesized with a conventional DC arc-discharge method.[67] GaN powder mixed with 5 wt% of transition metal powders (Ni/Co = 1:1 and Ni/Co/Y = 4.5:4.5:1) was pressed into a small hole of the graphite anode. A total pressure of 500 torr of argon and oxygen gases in a ratio of 4:1 was maintained during the growth. The typical diameter of the nanowires is about 33 nm with a growth direction of [001], and no amorphous layer was founded on the surface. Possible chemical reaction for the formation of Ga_2O_3 is proposed to be:

$$2GaN + \left(\frac{3}{2} + x\right)O_{2(g)} \rightarrow Ga_2O_3 + 2NO_{x(g)} \tag{4.10}$$

Single crystal GeO_2 nanowires were grown by evaporation of a mixture of Ge powder and 8 wt% Fe at 820°C under a flow (130 sccm) of argon gas under a pressure of 200 torr.[68] The nanowires have diameters ranging from 15 to 80 nm. Although Fe was added as a catalyst to direct the growth of nanowires, no globules were found on the tips of grown nanowires. The authors argued that the GeO_2 nanowires were grown by mechanisms other than VLS method. It is also noticed that during the experiment, no oxygen was intentionally introduced into the system. Oxygen may leak into the reaction chamber and react with germanium to form germanium oxide.

The catalyst can be introduced *in situ* as well. In this case, the growth precursor is mixed with the catalyst and evaporated simultaneously at a higher temperature. Both the growth precursor or species and the catalyst condense at the substrate surface when a supersaturation is reached at a temperature lower than the evaporation temperature. The mixture of the growth species and catalyst react either in the vapor phase or on the substrate surface to form a liquid droplet. The subsequent nanowire growth would proceed as discussed before.

Yu *et al.*[69] reported the synthesis of amorphous silica nanowires by VLS method. A mixture of silicon with 20 wt% silica and 8 wt% Fe was ablated using an excimer laser of 246 nm wavelength under flowing argon at 100 torr. Fe was used as a catalyst and the growth temperature was 1200°C. The nanowires have a chemical composition of $Si:O = 1:2$, and a uniform size distribution with a diameter of 15 nm and a length up to hundreds micrometers.

GaN nanowires were prepared using elemental indium as a catalyst in the reaction of gallium and ammonia.[70] Nanowires have diameters ranging from 20 to 50 nm and lengths up to several micrometers, and they are high-purity crystalline with a preferred $\langle 100 \rangle$ growth direction. It should also be noted that GaN nanowires has to be grown with Fe as the catalyst.[71] However, no GaN nanowires were grown when gold was used as a catalyst.[70]

NiO and FeO have also been reported to act as catalysts for the growth of GaN nanowires.[72] Solid gallium was reacted with ammonia at temperatures of 920–940°C. The single crystal GaN nanowires have diameters of 10–40 nm and a maximum length of ~500 μm, with a preferential growth direction of [001]. It is assumed that under the growth conditions, NiO and FeO were first reduced to metals and the metals reacted with gallium to form liquid droplets permitting the growth of GaN nanowires via VLS method.

4.2.2.5. SLS growth

In general, a high temperature and a vacuum are required in the growth of nanowires by VLS method. An alternative method called solution–liquid–solid (SLS) growth method was developed by Buhro's research group,[73–75] and first applied for the synthesis of InP, InAs and GaAs nanowires with solution-phase reactions at relative lower temperatures (≤203°C). SLS method is very similar to VLS theory; Fig. 4.18 compares the similarities of differences between these two methods.[73] Nanowires were found to be polycrystalline or near-single-crystal with a diameter of 10–150 nm and a length of up to several micrometers. Let us take the growth of InP nanowires as an example to illustrate the SLS growth process. Precursors used were typical organometallic compounds: In(t-Bu)$_3$ and PH$_3$, which were dissolved into hydrocarbon solvent with protic catalyst such as MeOH, PhSH, Et$_2$NH$_2$ or PhCO$_2$H. In the solution, precursors reacted to form In and P species for the growth of InP nanowires with the following organometallic reaction, which is commonly used in chemical vapor deposition[76]:

$$In(t\text{-Bu})_3 + PH_3 \rightarrow InP + 3(t\text{-Bu})H \qquad (4.11)$$

Indium metal functions as the liquid phase or catalyst for the growth of InP nanowires. Indium melts at 157°C and forms liquid drops. It is postulated that both P and In dissolve into the In droplets and precipitate to

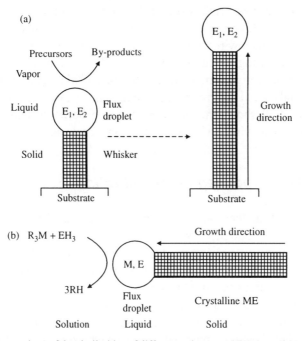

Fig. 4.18. Comparison of the similarities of differences between VLS (a) and SLS (b) growth techniques. [T.J. Trentler, K.M. Hickman, S.C. Goel, A.M. Viano, P.C. Gobbons, and W.E. Buhro, *Science* **270**, 1791 (1995).]

form nanowires of InP. The growth direction of InP nanowires was found to be predominated by ⟨111⟩, similar to that with VLS method.

Holmes et al.[77] used colloidal catalysts to control the diameters of silicon nanowires grown by solution–liquid–solid growth. Bulk quantities of defect-free silicon nanowires with nearly uniform diameter ranging from 4 to 5 nm were grown to a length of several micrometers. Alkanethiol-cated gold nanoclusters of 2.5 nm in diameter were used to direct the growth of silicon nanowires in a solution heated and pressurized above its critical point. The solution consisted of hexane and diphenysilane, as silicon precursor, and were heated to 500°C and pressurized at 200 or 270 bar. Under the above growth conditions, the diphenysilane decomposes to silicon atoms. Silicon atoms diffuse to and react with gold nanoclusters to a silicon–gold alloy droplet. When silicon concentration reaches a supersaturation, silicon precipitates out from the alloy droplets, resulting in the formation of silicon nanowires. The supercritical conditions are required to form the alloy droplets and to promote silicon crystallization. The growth directions of silicon nanowires were found to be pressure-dependent. The wires formed at 200 bar exhibited a preponderance of [100] orientation, whereas the samples synthesized at 270 bar oriented almost exclusively along the [111] direction. A thin coating of oxide or hydrocarbon was found on all the nanowires, though it was not possible to tell if the coatings were formed during or after the growth. The diameter and length of nanowires grown by SLS methods can be controlled by controlling the size of liquid catalyst and the growth time, in the same way as that in VLS methods. Figure 4.19 shows the linear relationship between diameters of the grown GaAs nanowires and In catalyst nanoparticle sizes.[75]

4.2.3. Stress-induced recrystallization

It is worth noting that nanowires can be synthesized by stress-induced recrystallization, though it has attracted little attention in the nanotechnology community. Application of pressure on solids at elevated temperatures is known to result in the growth of whiskers or nanowires with diameters as small as 50 nm.[78] It was demonstrated that the growth rate of tin whiskers increased proportionally with the applied pressure[78] and could be four orders of magnitude when a pressure of 7,500 psi was applied.[79] The growth of such nanowires or whiskers is based on a dislocation at the base of the whisker[80] and the growth proceeds from the base and not from the tip.[81] The formation of metallic nanorods is likely due to the confined growth at the surface between the metallic film and the

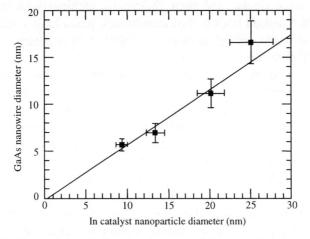

Fig. 4.19. The linear relationship between diameters of the grown GaAs nanowires and In catalyst nanoparticle sizes. [H. Yu and W.E. Buhro, *Adv. Mater.* **15**, 416 (2003).]

grown nanowires, whereas no growth is possible in other directions (side faces of nanowires). It should be noted that this technique is not widely explored in the recent studies on the growth of nanorods and nanowires.

4.3. Template-Based Synthesis

Template-based synthesis of nanostructured materials is a very general method and can be used in fabrication of nanorods, nanowires and nanotubules of polymers, metals, semiconductors and oxides. Various templates with nanosized channels have been explored for the template growth of nanorods and nanotubule. The most commonly used and commercially available templates are anodized alumina membrane,[82] radiation track-etched polymer membranes.[83] Other membranes have also been used as templates such as nanochannel array glass,[84] radiation track-etched mica,[85] and mesoporous materials,[86] porous silicon by electrochemical etching of silicon wafer,[87] zeolites[88] and carbon nanotubes.[89,90] Alumina membranes with uniform and parallel porous structure are made by anodic oxidation of aluminum sheet in solutions of sulfuric, oxalic, or phosphoric acids.[82,91] The pores are arranged in a regular hexagonal array, and densities as high as 10^{11} pores/cm^2 can be achieved.[92] Pore size ranging from 10 nm to 100 μm can be created.[92,93] The polycarbonate membranes are made by bombarding a nonporous polycarbonate sheet, with typical thickness ranging from 6–20 μm, with nuclear fission fragments to create damage tracks, and then chemically etching these tracks into pores.[83] In radiation track etched membranes, pores have a uniform size as small as 10 nm, though randomly distributed. Pore densities can be as high as 10^9 pores/cm^2.

In addition to the desired pore or channel size, morphology, size distribution and density of pores, template materials must meet certain requirements. First, the template materials must be compatible with the processing conditions. For example, an electrical insulator is required for a template to be used in electrochemical deposition. Except for the template directed synthesis, template materials should be chemically and thermally inert during the synthesis. Secondly, depositing materials or solution must wet the internal pore walls. Thirdly, for the synthesis of nanorods or nanowires, the deposition should start from the bottom or one end of the template channels and proceed from one side to another. However, for the growth of nanotubules, the deposition should start from the pore wall and proceed inwardly. Inward growth may result in the pore blockage, so that should be avoided in the growth of "solid" nanorods or nanowires. Kinetically, enough surface relaxation permits maximal packing density, so a diffusion-limited process is preferred. Other considerations include the easiness of release of nanowires or nanorods from the templates and of handling during the experiments.

4.3.1. Electrochemical deposition

Electrochemical deposition, also known as electrodeposition, can be understood as a special electrolysis resulting in the deposition of solid material on an electrode. This process involves (i) oriented diffusion of charged growth species (typically positively charged cations) through a solution when an external electric field is applied, and (ii) reduction of the charged growth species at the growth or deposition surface which also serves as an electrode. In general, electrochemical deposition is only applicable to electrical conductive materials such as metals, alloys, semiconductors and electrical conductive polymers, since after the initial deposition, the electrode is separated from the depositing solution by the deposit and the electrical current must go through the deposit to allow the deposition process to continue. Electrochemical deposition is widely used in making metallic coatings; the process is also known as electroplating.[94] When deposition is confined inside the pores of template membranes, nanocomposites are produced. If the template membrane is removed, nanorods or nanowires are prepared. Let us briefly review the fundamentals of electrochemistry, before we get into detailed discussion of the growth of nanorods by electrochemical deposition.

In Chapter 2, we have discussed the electrical properties of a solid surface. When a solid immerses in a polar solvent or an electrolyte solution, surface charge will be developed. At an interface between an electrode and an electrolyte solution, a surface oxidation or reduction reaction occurs, accompanied with charge transfer across the interface, until equilibrium is reached. For a given system, the electrode potential or surface charge density, E, is described by the Nernst equation:

$$E = E_0 + \frac{R_g T}{n_i F} \ln(a_i) \qquad (4.12)$$

Where E_0 is the standard electrode potential, or the potential difference between the electrode and the solution, when the activity, a_i, of the ions is unity, F, the Faraday's constant, R_g, the gas constant and T, temperature. When the electrode potential is more negative (higher) than the energy level of vacant molecular orbital in the electrolyte solution, electrons will transfer from the electrode to the solution, accompanied with dissolution or reduction of electrode as shown in Fig. 4.20(a).[95] If the electrode potential is more positive (lower) than the energy level of the occupied molecular orbital, the electrons will transfer from the electrolyte solution to the electrode, and the deposition or oxidation of electrolyte ions on the electrode will proceed simultaneously as illustrated in Fig. 4.20(b).[95] The reactions stop when equilibrium is achieved.

When two electrodes of different materials immerse into one electrolyte solution, each electrode will establish equilibrium with the electrolyte solution. Such equilibrium will be destroyed, if two electrodes are connected with an external circuit. Since different electrodes have different electrode potentials, this difference in electrode potential would drive electrons migration from the electrode with a higher electrode potential to the lower one. Let us take the Cu and Zn electrodes immersed in an aqueous solution as an example to illustrate the electrochemical process.[96] Assuming both activities of copper and zinc ions in the aqueous solution are unity in the beginning, the copper electrode has a more positive electrode potential (0.34 V) than that of the zinc electrode (-0.76 V). In the external circuit, electrons flow from the more negative electrode (Zn) to the more positive electrode (Cu). At the zinc-solution interface, the following electrochemical reactions take place:

$$Zn \rightarrow Zn^{2+} + 2e^- \qquad (4.13)$$

This reaction generates electrons at the interface, which then flow through the external circuit to another electrode (Cu). At the same time, Zn continues

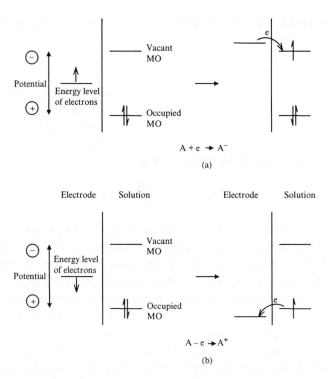

Fig. 4.20. Representation of (a) reduction and (b) oxidation process of a species A in solution. The molecular orbitals (MO) of species A shown are the highest occupied MO and the lowest vacant MO. As shown, these correspond in an approximate way to the E_0's of the A/A$^-$ and A$^+$/A couples, respectively. [A.J. Bard and L.R. Faulkner, *Electrochemical Methods*, John Wiley & Sons, New York, 1980.]

to dissolve from the electrode into the solution. At the copper-solution interface, a reduction reaction takes place resulting in deposition of Cu onto the electrode through the following reduction reaction:

$$Cu^{2+} + 2e^- \rightarrow Cu \qquad (4.14)$$

This spontaneous process ends only when new equilibrium is reached. From the Nernst equation, one can see that the copper electrode potential decreases due to a decrease in copper ion activity in the solution, whereas the zinc electrode potential increases due to an increased activity of zinc ions in the solution as both electrochemical reactions proceed. This system is a typical example of galvanic cell, in which chemical potential is converted into electricity. This process can be altered or even reversed, when an external electric field is introduced to the system.

When an external electric field is applied to two dissimilar electrodes, electrode potentials can be changed so that electrochemical reactions at both electrode-solution interfaces are reversed and the electrons flow from a more positive electrode to a more negative electrode. This process is called electrolysis, which converts electrical energy to chemical potential, and is a process widely used for applications of energy storage and materials processing. The system used for the electrolysis process is called electrolytic cell; in such a system the electrode connected to the positive side of the power supply is an anode, at which an oxidation reaction takes place, whereas the electrode connected to the negative side of the power supply is a cathode, at which a reduction reaction proceeds, accompanied with deposition. Sometimes, electrolytic deposition is therefore also called cathode deposition.

In an electrolytic cell, it is not necessary that anode dissolves into the electrolytic solution and the deposit is the same material as cathode. Which electrochemical reaction takes place at an electrode (either anode or cathode) is determined by the relative electrode potentials of the materials present in the system. Noble metals are often used as an inert electrode in electrolytic cells. A typical electrolytic process composes of a series of steps; each step could be a rate-limiting process:

(1) Mass transfer through the solution from one electrode to another.
(2) Chemical reactions at the interfaces between electrodes-solution.
(3) Electrons transfer at the electrode surfaces and through the external circuit.
(4) Other surface reactions such as adsorption, desorption or recrystallization.

Electrochemical deposition has been explored in the fabrication of nanowires of metals, semiconductors and conductive polymers without the use of porous templates and such growth of nanowires of conductive materials is a self-propagating process.[97] When little fluctuation yields the formation of small rods, the growth of rods or wires will continue, since the electric field and the density of current lines between the tips of nanowires and the opposing electrode are greater, due to a shorter distance, than that between two electrodes. The growth species will be more likely deposit onto the tip of nanowires, resulting in continued growth. However, this method is hardly used in practice for the synthesis of nanowires, since it is very difficult, if not impossible, to control the growth. Therefore, templates with desired channels are used for the growth of nanowires in electrochemical deposition. Figure 4.21 illustrates the common set-up for the template-based growth of nanowires using electrochemical deposition.[98] Template is attached onto the cathode, which is subsequently brought into contact with the deposition solution. The anode is placed in the deposition solution parallel to the cathode.

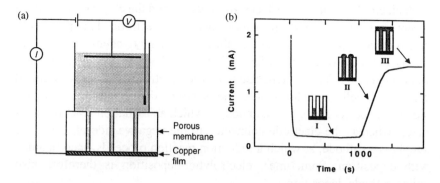

Fig. 4.21. Common experimental set-up for the template-based growth of nanowires using electrochemical deposition. (a) Schematic illustration of electrode arrangement for deposition of nanowires. (b) Current-time curve for electrodeposition of Ni into a polycarbonate membrane with 60 nm diameter pores at −1.0 V. Insets depict the different stages of the electrodeposition. [T.M. Whitney, J.S. Jiang, P.C. Searson, and C.L. Chien, *Science* **261**, 1316 (1993).]

When an electric field is applied, cations diffuse toward and reduce at the cathode, resulting in the growth of nanowires inside the pores of template. This figure also schematically shows the current density at different deposition stages when a constant electric field is applied. Possin[85] prepared various metallic nanowires by electrochemical deposition inside pores of radiation track-etched mica. Williams and Giordano[99] grew silver nanowires with diameters below 10 nm. The potentiostatic electrochemical template synthesis yielded different metal nanowires, including Ni, Co, Cu and Au with nominal pore diameters between 10 and 200 nm and the nanowires were found to be true replicas of the pores.[100] Whitney et al.[98] fabricated the arrays of nickel and cobalt nanowires by electrochemical deposition of the metals into track-etched templates. Single crystal antimony nanowires have been grown by Zhang et al.[101] in anodic alumina membranes using pulsed electrodeposition. Single crystal and polycrystalline superconducting lead nanowires were also prepared by pulse electrodeposition.[102] It is unexpected that the growth of single crystal lead nanowires required a greater departure from equilibrium conditions (greater overpotential) than the growth of polycrystalline ones. Semiconductor nanorods by electrodeposition include CdSe and CdTe synthesized by Klein et al.[103] in anodic alumina templates, and Schönenberger et al.[104] have made conducting polyporrole electrochemically in porous polycarbonate. Figure 4.22 shows some SEM images of metal nanowires grown by electrochemical deposition in templates.[101]

Hollow metal tubules can also be prepared using electrochemical deposition.[105,106] For growth of metal tubules, the pore walls of the template

One-Dimensional Nanostructures: Nanowires and Nanorods 149

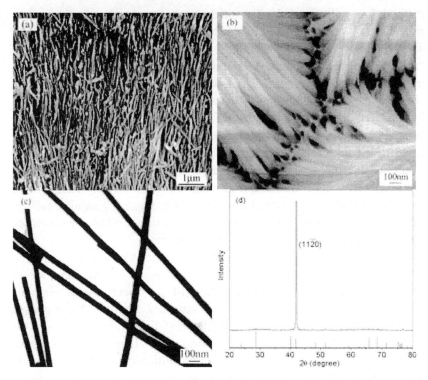

Fig. 4.22. (a) Field-emission SEM image of the general morphology of the antimony nanowire array. (b) Field emission SEM showing the filling degree of the template and height variation of the nanowires. (c) TEM image of antimony nanowires showing the morphology of individual nanowires. (d) XRD pattern of the antimony nanowire array; the sole diffraction peak indicates the same orientation of all nanowires. [Y. Zhang, G. Li, Y. Wu, B. Zhang, W. Song, and L. Zhang, Adv. Mater. 14, 1227 (2002).]

need to be chemically derivatized first so that the metal will preferentially deposit onto the pore walls instead of the bottom electrode. Such surface chemistry of the pore walls is achieved by anchoring silane molecules. For example, the pore surface of an anodic alumina template were covered with cyanosilanes, subsequent electrochemical deposition resulted in the growth of gold tubules.[107]

An electroless electrolysis process has also been applied in the fabrication of nanowires or nanorods.[105,108–110] Electroless deposition is actually a chemical deposition and involves the use of a chemical agent to plate a material from the surrounding phase onto a template surface.[111] The significant difference between electrochemical deposition and electroless deposition is that in the former, the deposition begins at the bottom electrode and the deposited materials must be electrically conductive, whereas the latter method does not require the deposited materials to be electrically conductive and the deposition starts from the pore wall and proceeds inwardly. Therefore, in general, electrochemical deposition results in the formation of "solid" nanorods or nanowires of conductive materials, whereas the electroless deposition often grows hollow fibrils or nanotubules. For electrochemical deposition, the length of nanowires or nanorods can be controlled by the deposition time, whereas the length of the nanotubules is solely dependent on the length of the deposition channels or pores, which often equal to the thickness of membranes. Variation of deposition time would result in a different wall thickness of nanotubules. An increase in deposition time leads to a thick wall and a prolonged deposition may form a solid nanorod. However, a prolonged deposition time does not guarantee the formation of solid nanorods. For example, the polyaniline tubules never closed up, even with prolonged polymerization time.[112]

It is noticed that in general polymer nanotubules are formed, even using electrochemical deposition, in contrast to "solid" metal nanorods or nanowires. Deposition or solidification of polymers insides template pores starts at the surface and proceeds inwardly. Martin[113] has proposed to explain this phenomenon by the electrostatic attraction between the growing polycationic polymer and anionic sites along the pore walls of the polycarbonate membrane. In addition, although the monomers are highly soluble, the polycationic form of the polymers is completely insoluble. Hence, there is a solvophobic component, leading the deposition at the surface of the pores.[114,115] Furthermore, the diffusion of monomers through the pores could become a limiting step and monomers inside the pores could be quickly depleted. The deposition of polymer inside pores stops and the entrance becomes corked Fig. 4.23 shows SEM images of such polymer nanotubes.[116]

Although many research groups have reported growth of uniformly sized nanorods and nanowires grown on polycarbonate template membranes, Schönenberger et al.[104] reported that the channels of carbonate membranes were not always uniform in diameter. They grew metal, including Ni, Co, Cu and Au and polyporrole nanowires using polycarbonate membranes with nominal pore diameters between 10 and 200 nm by electrolysis. From both potentiostatic study of growth process and SEM analysis of nanowire morphology, they concluded that the pores are in general not cylindrical with a constant cross-section, but are rather cigar-like. For the analyzed pores with a nominal diameter of 80 nm, the middle section of the pores is wider by up to a factor of 3. Figure 4.24 shows some such non-uniformly sized metal nanowires grown in polycarbonate membranes by electrochemical deposition.[104]

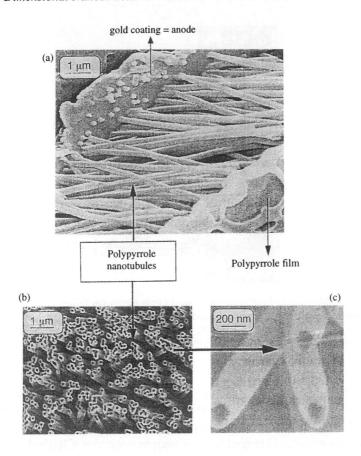

Fig. 4.23. SEM images of polymer nanotubes. [L. Piraux, S. Dubois, and S. Demoustier-Champagne, *Nucl. Instrum. Meth. Phys. Res.* **B131**, 357 (1997).]

4.3.2. *Electrophoretic deposition*

The electrophoretic deposition technique has been widely explored, particularly in film deposition of ceramic and organoceramic materials on cathode from colloidal dispersions.[117–119] Electrophoretic deposition differs from electrochemical deposition in several aspects. First, the deposit by electrophoretic deposition method need not be electrically conductive. Secondly, nanosized particles in colloidal dispersions are typically stabilized by electrostatic or electrosteric mechanisms. As discussed in the previous section, when dispersed in a polar solvent or an electrolyte solution,

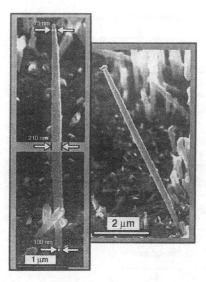

Fig. 4.24. SEM images of non-uniformly sized metal nanowires grown in polycarbonate membranes by electrochemical deposition. [C. Schönenberger, B.M.I. van der Zande, L.G.J. Fokkink, M. Henny, C. Schmid, M. Krüger, A. Bachtold, R. Huber, H. Birk, and U. Staufer, *J. Phys. Chem.* **B101**, 5497 (1997).]

the surface of nanoparticles develops an electrical charge via one or more of the following mechanisms: (i) preferential dissolution or (ii) deposition of charges or charged species, (iii) preferential reduction or (iv) oxidation, and (v) adsorption of charged species such as polymers. Charged surfaces will electrostatically attract oppositely charged species (typically called counter-ions) in the solvent or solution. A combination of electrostatic forces, Brownian motion and osmotic forces would result in the formation of a so-called double layer structure, as discussed in detail in Chapter 2 and schematically illustrated in Fig. 2.13. The figure depicts a positively charged particle surface, the concentration profiles of negative ions (counter ions) and positive ions (surface-charge determining ions) and the electric potential profile. The concentration of counter ions gradually decreases with distance from the particle surface, whereas that of charge determining ions increases. As a result, the electric potential decreases with distance. Near to the particle surface, the electric potential decreases linearly, in the region known as the Stern layer. Outside of the Stern layer, the decrease follows an exponential relationship, and the region between Stern layer and the point where the electric potential equals zero is called the diffusion layer. Together, the Stern layer and diffusion layer are called the double layer structure in the classic theory of electrostatic stabilization.

Upon application of an external electric field to a colloidal system or a sol, the constituent charged particles are set in motion in response to the electric field, as schematically illustrated in Fig. 4.25.[123] This type of motion is referred to as electrophoresis. When a charged particle is in motion, some of the solvent or solution surrounding the particle will move with it, since part of the solvent or solution is tightly bound to the particle. The plane that separates the tightly bound liquid layer from the rest of the liquid is called the slip plane. The electric potential at the slip plane is known as the zeta potential. Zeta potential is an important parameter in determining the stability of a colloidal dispersion or a sol; a zeta potential larger than about 25 mV is typically required to stabilize a system.[120] Zeta potential is determined by a number of factors, such as the particle surface charge density, the concentration of counter ions in the solution, solvent polarity and temperature. The zeta potential, ζ, around a spherical particle can be described as[121]:

$$\zeta = \frac{Q}{4\pi\varepsilon_r a(1 + \kappa a)} \quad (4.15)$$

with
$$\kappa = \sqrt{\frac{e^2 \Sigma n_i z_i^2}{\varepsilon_r \varepsilon_0 kT}}$$

where Q is the charge on the particle, a is the radius of the particle out to the shear plane, ε_r is the relative dielectric constant of the medium, and n_i and z_i are the bulk concentration and valence of the ith ion in the system, respectively. It is worthwhile to note that a positively charged surface

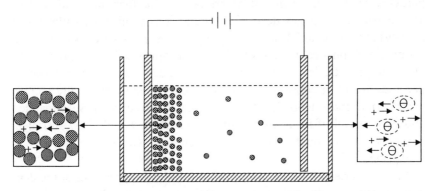

Fig. 4.25. Schematic showing the electrophoresis. Upon application of an external electric field to a colloidal system or a sol, the constituent charged nanoparticles or nanoclusters are set in motion in response to the electric field, whereas the counter-ions diffuse in the opposite direction.

results in a positive zeta potential in a dilute system. A high concentration of counter ions, however, can result in a zeta potential of the opposite sign.

The mobility of a nanoparticle in a colloidal dispersion or a sol μ is dependent on the dielectric constant of the liquid medium ε_r, the zeta potential of the nanoparticle ε, and the viscosity of the fluid η. Several forms for this relationship have been proposed, such as the Hückel equation[121]:

$$\mu = \frac{2\varepsilon_r\varepsilon_0\zeta}{3\pi\eta} \qquad (4.16)$$

Double layer stabilization and electrophoresis are extensively studied subjects. Readers may find additional detailed information in books on sol-gel processing[122–124] and colloidal dispersions.[121,125]

Electrophoretic deposition simply uses such an oriented motion of charged particles to grow films or monoliths by enriching the solid particles from a colloidal dispersion or a sol onto the surface of an electrode. If particles are positively charged (more precisely speaking, having a positive zeta potential), then the deposition of solid particles will occur at the cathode. Otherwise, deposition will be at the anode. At the electrodes, surface electrochemical reactions proceed to generate or receive electrons. The electrostatic double layers collapse upon deposition on the growth surface, and the particles coagulate. There is not much information on the deposition behavior of particles at the growth surface. Some surface diffusion and relaxation is expected. Relatively strong attractive forces, including the formation of chemical bonds between two particles, develop once the particles coagulate. The films or monoliths grown by electrophoretic deposition from colloidal dispersions or sols are essentially a compaction of nanosized particles. Such films or monoliths are porous, i.e. there are voids inside. Typical packing densities, defined as the fraction of solid (also called green density) are less than 74%, which is the highest packing density for uniformly sized spherical particles.[126] The green density of films or monoliths by electrophoretic deposition is strongly dependent on the concentration of particles in sols or colloidal dispersions, zeta potential, externally applied electric field and reaction kinetics between particle surfaces. Slow reaction and slow arrival of nanoparticles onto the surface would allow sufficient particle relaxation on the deposition surface, so that a high packing density is expected.

Many theories have been proposed to explain the processes at the deposition surface during electrophoretic deposition. Electrochemical process at the deposition surface or electrodes is complex and varies from system to system. However, in general, a current exists during electrophoretic deposition, indicating reduction and oxidation reactions occur at electrodes and/or deposition surface. In many cases, films or monoliths grown by electrophoretic deposition are electric insulators. However, the films or monoliths are porous and the surface of the pores would be electrically charged just like the nanoparticle surfaces, since surface charge is dependent on the solid material and the solution. Furthermore, the pores are filled with solvent or solution that contains counter ions and charge determining ions. The electrical conduction between the growth surface and the bottom electrode could proceed via either surface conduction or solution conduction.

Limmer et al.[127–129] combined sol-gel preparation and electrophoretic deposition in the growth of nanorods of various oxides including complex oxides such as lead zirconate titanate and barium titanate. In their approach, conventional sol-gel processing was applied for the synthesis of various sols. By appropriate control of the sol preparation, nanometer particles with desired stoichiometric composition were formed, electrostatically stabilized by adjusting to an appropriate pH and uniformly dispersed in the solvent. When an external electric field is applied, these electrostatically stabilized nanoparticles will respond and move towards and deposit on either cathode or anode, depending on the surface charge (more precisely speaking, the zeta potential) of the nanoparticles. Using radiation tracked-etched polycarbonate membranes with an electric field of ~1.5 V/cm, they grew nanowires with diameters ranging from 40 to 175 nm and a length of 10 μm corresponding to the thickness of the membrane. The materials include anatase TiO_2, amorphous SiO_2, perovskite structured $BaTiO_3$ and $Pb(Ti,Zr)O_3$, layered structured perovskite $Sr_2Nb_2O_7$. Nanorods grown by sol electrophoretic deposition are polycrystalline or amorphous. One of the advantages of this technique is the ability of synthesis of complex oxides and organic–inorganic hybrids with desired stoichiometric composition; Fig. 4.26 shows the nanorods and X-ray diffraction spectra of $Pb(Zr,Ti)O_3$ nanorods.[127] Another advantage is the applicability of a variety of materials; Fig. 4.27 shows the nanorods of SiO_2, TiO_2, $Sr_2Nb_2O_7$ and $BaTiO_3$.[128]

Wang et al.[130] used electrophoretic deposition to form nanorods of ZnO from colloidal sols. ZnO colloidal sol was prepared by hydrolyzing an alcoholic solution of zinc acetate with NaOH, with a small amount of zinc nitrate added to act as a binder. This solution was then deposited into the pores of anodic alumina membranes at voltages in the range of 10–400 V. It was found that lower voltages led to dense, solid nanorods, while higher voltages caused the formation of hollow tubules. The suggested mechanism is that the higher voltages cause dielectric breakdown of the anodic alumina, causing it to become charged similarly to the cathode.

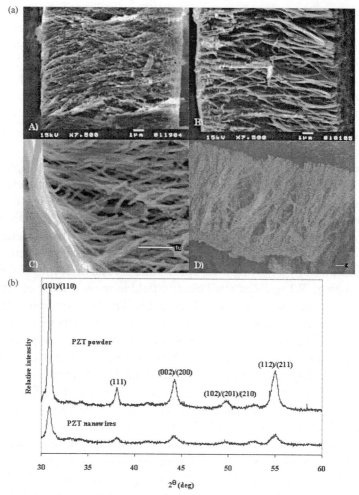

Fig. 4.26. (a) SEM micrograph of nanorods and (b) X-ray diffraction spectra of Pb(Zr,Ti)O$_3$ nanorods grown by template-based sol-gel electrophoretic deposition. [S.J. Limmer, S. Seraji, M.J. Forbess, Y. Wu, T.P. Chou, C. Nguyen, and G.Z. Cao, *Adv. Mater.* **13**, 1269 (2001).]

Electrostatic attraction between the ZnO nanoparticles and the pore walls then leads to tubule formation.

Miao et al.[131] prepared single crystalline TiO$_2$ nanowires by template-based electrochemically induced sol-gel deposition. Titania electrolyte solution was prepared using a method developed by Natarajan and Nogami,[132] in which Ti powder was dissolved into a H$_2$O$_2$ and NH$_4$OH aqueous solution and formed TiO^{2+} ionic clusters. When an external electric field was applied, TiO^{2+} ionic clusters diffused to cathode and underwent hydrolysis

One-Dimensional Nanostructures: Nanowires and Nanorods 157

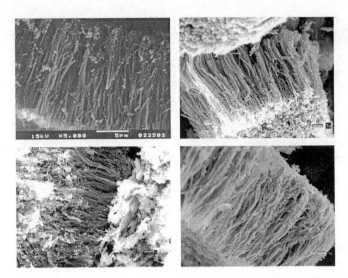

Fig. 4.27. SEM images of the nanorods of (A) SiO_2, (B) TiO_2, (C) $Sr_2Nb_2O_7$ and (D) BaTiO3 grown by template-based sol-gel electrophoretic deposition. [S.J. Limmer, S. Seraji, M.J. Forbess, Y. Wu, T.P. Chou, C. Nguyen, and G.Z. Cao, *Adv. Func. Mater.* **12**, 59 (2002).]

and condensation reactions, resulting in deposition of nanorods of amorphous TiO_2 gel. After heat treatment at 240°C for 24 hrs in air, nanowires of single crystal TiO_2 with anatase structure and with diameters of 10, 20, and 40 nm and lengths ranging from 2 to 10 μm were synthesized. However, no axis crystal orientation was identified. The formation of single crystal TiO_2 nanorods here is different from that reported by Martin's group.[133] Here the formation of single crystal TiO_2 is via crystallization of amorphous phase at an elevated temperature, whereas nanoscale crystalline TiO_2 particles are believed to assemble epitaxially to form a single crystal nanorod. Epitaxial agglomeration of two nanoscale crystalline particles has been reported,[134] though no large single crystals have been produced by assembling nanocrystalline particles. Figure 4.28 shows the micrograph of single crystal nanorods of TiO_2 grown by template-based electrochemically induced sol-gel deposition.[131]

4.3.3. Template filling

Direct template filling is the most straightforward and versatile method in preparation of nanorods and nanotubules. Most commonly, a liquid precursor or precursor mixture is used to fill the pores. There are several concerns

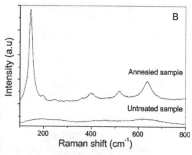

Fig. 4.28. SEM micrograph of single crystal nanorods of TiO$_2$ grown by template-based electrochemically induced sol-gel deposition. [Z. Miao, D. Xu, J. Ouyang, G. Guo, Z. Zhao, and Y. Tang, *Nano Lett.* **2**, 717 (2002).]

in the template filling. First of all, the wetability of the pore wall should be good enough to permit the penetration and complete filling of the liquid precursor or precursor mixture. For filling at low temperatures, the surface of pore walls can be easily modified to be either hydrophilic or hydrophobic by introducing a monolayer of organic molecules. Second, the template materials should be chemically inert. Thirdly, control of shrinkage during solidification is required. If adhesion between the pore walls and the filling material is weak or solidification starts at the center, or from one end of the pore, or uniformly, solid nanorods are most likely to form. However, if the adhesion is very strong, or the solidification starts at the interfaces and proceeds inwardly, it is most likely to form hollow nanotubules.

4.3.3.1. Colloidal dispersion filling

Martin and his co-workers[133,135] have studied the formation of various oxide nanorods and nanotubules by simply filling the templates with colloidal dispersions. Colloidal dispersions were prepared using appropriate sol-gel processing. The filling of the template was to place a template in a stable sol for various periods of time. The capillary force is believed to drive the sol into the pores, when the surface chemistry of the template pores were appropriately modified to have a good wetability for the sol. After the pores were filled with sol, the template was withdrawn from the sol and dried prior to firing at elevated temperatures. The firing at elevated temperatures served two purposes: removal of template so that free standing nanorods can be obtained and densification of the sol-gel derived green nanorods. Figure 4.29 show SEM micrographs of TiO$_2$ and ZnO nanorods made by filling the templates with sol-gel.[133]

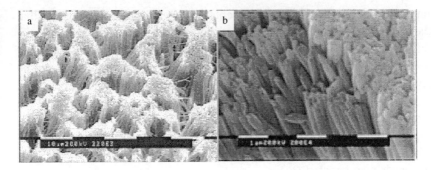

Fig. 4.29. SEM micrographs of oxide nanorods made by filling the templates with sol-gels: (a) ZnO and (b) TiO$_2$. [B.B. Lakshmi, P.K. Dorhout, and C.R. Martin, *Chem. Mater.* **9**, 857 (1997).]

Fig. 4.30. Hollow nanotubes formed by incomplete filling of the template. [B.B. Lakshmi, P.K. Dorhout, and C.R. Martin, *Chem. Mater.* **9**, 857 (1997).]

In the previous chapter, we discussed about sol-gel processing and knew that the typical sol consists of a large volume fraction of solvent up to 90% or higher.[122] Although the capillary force may ensure the complete filling of colloidal dispersion inside pores of the template, the amount of the solid filled inside the pores can be very small. Upon drying and subsequent firing processes, a significant amount of shrinkage would be expected. However, the results showed that most nanorods shrank a little only as compared with the size of the template pores and indicated that there are some unknown mechanisms, which enrich the concentration of solid inside pores. One possible mechanism could be the diffusion of solvent through the membrane, leading to the enrichment of solid along the internal surface of template pores, a process used in ceramic slip casting.[136] The formation of nanotubules (as shown in Fig. 4.30[133]) by such a sol filling process may imply such a process is indeed present. However, considering the fact that the templates typically were emerged into sol for just a few minutes, the diffusion through membrane and enrichment of solid inside the pores had to be a rather rapid process. This is a very versatile method and can be applied for any material, which can be made by sol-gel processing. However, the drawback is the difficult to ensure the complete filling of the template pores. It is also noticed that the nanorods made by template filling are commonly polycrystalline or amorphous. The exception was found, when the diameter of nanorods is smaller than 20 nm, single crystal TiO_2 nanorods were created.[133]

4.3.3.2. Melt and solution filling

Metallic nanowires can also be synthesized by filling a template with molten metals.[137,26] One example is the preparation of bismuth nanowires by pressure injection of molten bismuth into the nanochannels of an anodic alumina template.[138] The anodic alumina template was degassed and immersed in the liquid bismuth at 325°C (T_m = 271.5°C for Bi), and then high pressure Ar gas of ~300 bar was applied to inject liquid Bi into the nanochannels of the template for 5 hr. Bi nanowires with diameters of 13–110 nm and large aspect ratios of several hundred have been obtained. Individual nanowires are believed to be single crystal. When exposed to air, bismuth nanowires are readily to be oxidized. An amorphous oxide layer of ~4 nm in thickness was observed after 48 hr. After 4 weeks, bismuth nanowires of 65 nm in diameter were found to be totally oxidized. Nanowires of other metals, In, Sn and Al, and semiconductors, Se, Te, GaSb and Bi_2Te_3 were prepared by injection of melt liquid into anodic alumina templates.[25]

Polymeric fibrils have been made by filling a monomer solution, which contain the desired monomer and a polymerization reagent, into the template pores and then polymerizing the monomer solution.[139–141] The polymer preferentially nucleates and grows on the pore walls, resulting in tubules at short deposition times, as discussed previously in the growth of conductive polymer nanowires or nanotubules by electrochemical deposition and fibers at long times. Cai et al.[142] synthesized polymeric fibrils using this technique.

Similarly, metal and semiconductor nanowires have been synthesized through solution techniques. For example, Han et al.[143] synthesized Au, Ag and Pt nanowires in mesoporous silica templates. The mesoporous templates were filled with aqueous solutions of the appropriate metal salts (such as $HAuCl_4$), and after drying and treatment with CH_2Cl_2 the samples were reduced under H_2 flow to convert the salts to pure metal. Chen et al.[144] filled the pores of a mesoporous silica template with an aqueous solution of Cd and Mn salts, dried the sample, and reacted it with H_2S gas to convert to (Cd,Mn)S. $Ni(OH)_2$ nanorods have been grown in carbon-coated anodic alumina membranes by Matsui et al.[145] by filling the template with ethanolic $Ni(NO_3)_2$ solutions, drying, and hydrothermally treating the sample in NaOH solution at 150°C.

4.3.3.3. Chemical vapor deposition

Some researchers have used chemical vapor deposition (CVD) as a means to form nanowires. Ge nanowires were grown by Leon et al.[146] by diffusing Ge_2H_6 gas into mesoporous silica and heating. They believed that the precursor reacted with residual surface hydroxyl groups in the template, forming Ge and H_2. Lee et al.[147] used a platinum organometallic compound to fill the pores of mesoporous silica templates, followed by pyrolysis under H_2/N_2 flow to yield Pt nanowires.

4.3.3.4. Deposition by centrifugation

Template filling of nanoclusters assisted with centrifugation force is another inexpensive method for mass production of nanorod arrays. Figure 4.31 shows SEM images of lead zirconate titanate (PZT) nanorod arrays with uniform sizes and unidirectional alignment.[148] Such nanorod arrays were

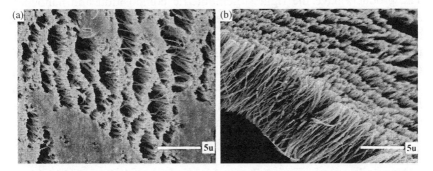

Fig. 4.31. SEM images of the top view (left) and side view (right) of lead zirconate titanate (PZT) nanorod arrays grown in polycarbonate membrane from PZT sol by centrifugation at 1500 rpm for 60 min. Samples were attached to silica glass and fired at 650°C in air for 60 min. [T.L. Wen, J. Zhang, T.P. Chou, and G.Z. Cao, unpublished (2003).]

grown in polycarbonate membrane from PZT sol by centrifugation at 1500 rpm for 60 min. The samples were attached to silica glass and fired at 650°C in air for 60 min. Nanorod arrays of other oxides including silica and titania have also been grown in this method. The advantages of centrifugation include its applicability to any colloidal dispersion systems including those consisting of electrolyte-sensitive nanoclusters or molecules. However, in order to grow nanowire arrays, the centrifugation force must be larger than the repulsion force between two nanoparticles or nanoclusters.

4.3.4. Converting through chemical reactions

Nanorods or nanowires can also be synthesized using consumable templates.[149] Nanowires of compounds can be synthesized or prepared using a template-directed reaction. First nanowires or nanorods of constituent element is prepared, and then reacted with chemicals containing desired element to form final products. Gates et al.[150] converted single crystalline trigonal selenium nanowires into single crystalline nanowires of Ag_2Se by reacting with aqueous $AgNO_3$ solutions at room temperature. The trigonal selenium nanowires were prepared first using solution synthesis method.[29] Selenium nanowires were either dispersed in water or supported on TEM grids during reaction with aqueous $AgNO_3$. The following chemical reaction was suggested:

$$3Se_{(s)} + 6Ag^+_{(aq)} + 3H_2O \rightarrow 2Ag_2Se_{(s)} + Ag_2SeO_{3(aq)} + 6H^+_{(aq)} \quad (4.17)$$

The products have the right stoichiometric composition and nanowires are single crystalline, with either tetragonal (low temperature phase) or orthorhombic structure (high temperature phase, the bulk phase transformation temperature is 133°C). It was further noticed that nanowires with diameters larger than 40 nm tended to have orthorhombic structure. Both crystallinity and morphology of the template were retained with high fidelity. Other compound nanowires can be synthesized by reacting selenium nanowires with desired chemicals using a similar approach. For example, Bi_2Se_3 nanowires may be produced by reacting Se nanowires with Bi vapor.[151]

Nanorods can also be synthesized by reacting volatile metal halide or oxide species with formerly obtained carbon nanotubes to form solid carbide nanorods with diameters between 2 and 30 nm and lengths up to 20 μm as shown schematically in Fig. 4.32.[152,153] Carbon nanotubes were

One-Dimensional Nanostructures: Nanowires and Nanorods

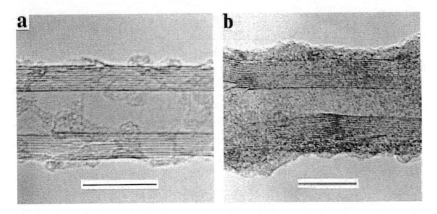

Fig. 4.32. TEM images of nanorods synthesized by reacting volatile metal halide or oxide species with formerly obtained carbon nanotubes to form solid titanium carbide nanorods with diameters between 2 and 30 nm and lengths up to 20 μm: (a) an unreacted carbon nanotube and (b) titanium carbide. The scale bars are 10 nm. [E.W. Wong, B.W. Maynor, L.D. Burns, and C.M. Lieber, *Chem. Mater.* **8**, 2041 (1996).]

used as removable template in the synthesis of silicon and boron nitride nanorods.[153] Silicon nitride nanorods of 4–40 nm in diameter were also prepared by reacting carbon nanotubes with a mixture of silicon monoxide vapor and flowing nitrogen at 1500°C[155]:

$$3SiO_{(g)} + 3C_{(s)} + 2N_{2(g)} \rightarrow Si_3N_{4(s)} + 3CO_{(g)} \quad (4.18)$$

Silicon monoxide is generated by heating a solid mixture of silicon and silica in an alumina crucible at 1500°C. The total transformation of carbon nanotubes into silicon nitride nanorods was observed.

ZnO nanowires were prepared by oxidizing metallic zinc nanowires.[156] In the first step, polycrystalline zinc nanowires without preferential crystal orientation were prepared by electrodeposition using anodic alumina membrane as a template, and in the second step, grown zinc nanowires were oxidized at 300°C for up to 35 hr in air, yielding polycrystalline ZnO nanowires with diameters ranging from 15 to 90 nm and length of ~50 μm. Although ZnO nanowires are embedded in the anodic alumina membranes, free standing nanowires may be obtained by selectively dissolving alumina templates.

Hollow nanotubules of MoS_2 of ~30 μm long and 50 nm in external diameter with wall thickness of 10 nm were prepared by filling a solution mixture of molecular precursors, $(NH_4)_2MoS_4$ and $(NH_4)_2Mo_3S_{13}$ into the pores of alumina membrane templates. Then template filled with the molecular precursors was heated to an elevated temperature and the molecular precursors thermally decomposed into MoS_2.[157]

Certain polymers and proteins were also reported to have been used to direct the growth of nanowires of metals or semiconductors. For example, Braun et al.[158] reported a two-step procedure to use DNA as a template for the vectorial growth of a silver nanorods of 12 μm in length and 100 nm in diameter. CdS nanowires were prepared by polymer-controlled growth.[159] For the synthesis of CdS nanowires, cadmium ions were well distributed in a polyacrylamide matrix. The Cd^{2+} containing polymer was treated with thiourea (NH_2CSNH_2) solvothermally in ethylenediamine at 170°C, resulting in degradation of polyacrylamide. Single crystal CdS nanowires of 40 nm in diameter and up to 100 μm in length with a preferential orientation of [001] were then simply filtered from the solvent.

4.4. Electrospinning

Electrospinning, also known as electrostatic fiber processing, technique has been originally developed for generating ultrathin polymer fibers.[160,161] Electrospinning uses electrical forces to produce polymer fibers with nanometer-scale diameters. Electrospinning occurs when the electrical forces at the surface of a polymer solution or melt overcome the surface tension and cause an electrically charged jet to be ejected. When the jet dries or solidifies, an electrically charged fiber remains. This charged fiber can be directed or accelerated by electrical forces and then collected in sheets or other useful geometrical forms. More than 30 polymer fibers with diameters ranging from 40 nm to 500 nm have been successfully produced by electrospinning.[162,163] The morphology of the fibers depends on the process parameters, including solution concentration, applied electric field strength, and the feeding rate of the precursor solution. Recently, electrospinning has also been explored for the synthesis of ultrathin organic–inorganic hybrid fibers.[164–167] For example, porous anatase titania nanofibers was made by ejecting an ethanol solution containing both poly(vinyl pyrrolidone) (PVP) and titanium tetraisopropoxide through a needle under a strong electric field, resulting in the formation of amorphous TiO_2/PVP composite nanofibers as shown in Fig. 4.33.[166] Upon pyrolysis of PVP at 500°C in air, porous TiO_2 fibers with diameter ranging from 20 to 200 nm, depending on the processing parameters are obtained.

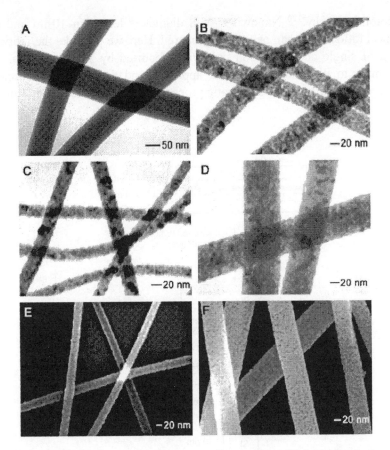

Fig. 4.33. (A) TEM image of TiO$_2$/PVP composite nanofibers fabricated by electrospinning an ethanol solution that contained 0.03 g/mL PVP and 0.1 g/mL Ti(OP$_r^i$)$_4$. (B) TEM image of the same sample after it had been calcined in air at 500°C for 3 hr. (C, D) TEM images of nanofibers made of anatase that were prepared under the same conditions except that the precursor solution contained (C) 0.025 g/mL and (D) 0.15 g/mL Ti(OP$_r^i$)$_4$, respectively. (E, F) High-magnification SEM images taken from the samples shown in C and D, respectively. No gold coatings were applied to the samples for all SEM studies. [D. Li and Y. Xia, *Nano Lett.* **3**, 555 (2003).]

4.5. Lithography

Lithography represents another route to the synthesis of nanowires. Various techniques have been explored in the fabrication of nanowires, such as electron beam lithography,[168,169] ion beam lithography, STM lithography, X-ray lithography, proxial-probe lithography and near-field photolithography.[170] Nanowires with diameters less than 10 nm and an aspect ratio of 100 can be readily prepared. Here we just take the fabrication of single crystal silicon nanowires reported by Yin et al.[171] as an example to illustrate the general approach and the products obtained. Figure 4.34 outlines the schematic procedures used for the preparation of

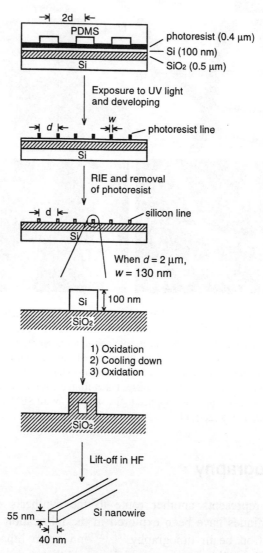

Fig. 4.34. Schematic illustrating procedures used for the preparation of single crystal silicon nanowires. [Y. Yin, B. Gates, and Y. Xia, *Adv. Mater.* **12**, 1426 (2000).]

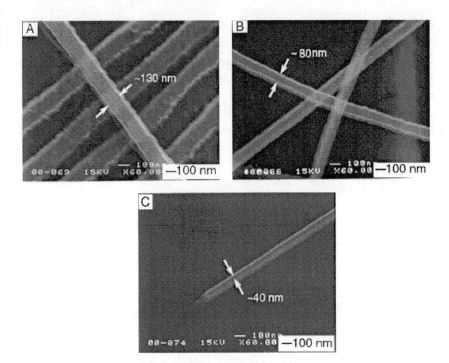

Fig. 4.35. SEM images of silicon nanostructures fabricated using such near-field optical lithography, followed by pattern transfer into silicon with reactive ion etching, oxidation of silicon at 850°C in air for ~1 hr, and finally lift-off in HF solution. [Y. Yin, B. Gates, and Y. Xia, *Adv. Mater.* **12**, 1426 (2000).]

single crystal silicon nanowires.[171] The nanoscale features were defined in a thin film of photoresist by exposing it to a UV light source through a phase shift mask made of a transparent elastomer, such as poly(dimethysiloxane) (PDMS). The light passing through this phase mask was modulated in the near-field such that an array of nulls in the intensity were formed at the edges of the relief structures patterned on the PDMS mask. Therefore, nanoscale features were generated in a thin film of photoresist and the patterns were transferred into the underlying substrate using a reactive ion etching or wet etching process. Silicon nanostructures were separated from underlying substrate by slight over-etching. Figure 4.35 shows SEM images of silicon nanostructures fabricated using such near-field optical lithography, followed by pattern transfer into silicon with reactive ion etching, oxidation of silicon at 850°C in air for ~1 hr, and finally lift-off in HF solution.[171]

4.6. Summary

This chapter summarizes the fundamentals and general approaches for the preparation of one-dimensional nanostructures. For a given fundamental concept, many different approaches can be taken and indeed have been developed. However, not all the synthesis methods have been included in this chapter. The coverage is limited so that the important fundamentals and concepts of various commonly used synthesis techniques are included.

References

1. Y. Xia, P. Yang, Y. Sun, Y. Wu, B. Mayers, B. Gates, Y. Yin, F. Kim, and H. Yan, *Adv. Mater.* **15**, 353 (2003).
2. P. Hartman and W.G. Perdok, *Acta Cryst.* **8**, 49 (1955).
3. A.W. Vere, *Crystal Growth: Principles and Progress*, Plenum, New York, 1987.
4. W. Burton, N. Cabrera, and F.C. Frank, *Phil. Trans. Roy. Soc.* **243**, 299 (1951).
5. P. Hartman, *Z. Kristallogr.*, **121**, 78 (1965).
6. P. Hartman, *Crystal Growth: An Introduction*, North Holland, Amsterdam, 1973.
7. C. Herring, *Structure and Properties of Solid Surfaces*, University of Chicago, Chicago, IL, 1952.
8. W.W. Mullins, *Metal Surfaces: Structure Energetics and Kinetics*, The American Society of Metals, Metals Park, OH, 1962.
9. G.W. Sears, *Acta Metal.* **3**, 361 (1955).
10. G.W. Sears, *Acta Metal.* **3**, 367 (1955).
11. E.I. Givargizov, *Highly Anisotropic Crystals*, D. Reidel, Dordrecht, 1986.
12. G. Bogels, H. Meekes, P. Bennema, and D. Bollen, *J. Phys. Chem.* **B103**, 7577 (1999).
13. W. Dittmar and K. Neumann, in *Growth and Perfection of Crystals*, eds., R.H. Doremus, R.W. Roberts, and D. Turnbull, John Wiley, New York, pp. 121, 1958.
14. R.L. Schwoebel and E. J. Shipsey, *J. Appl. Phys.* **37**, 3682 (1966).
15. R.L Schwoebel, *J. Appl. Phys.* **40**, 614 (1969).
16. Z.Y. Zhang and M.G. Lagally, *Science* **276**, 377 (1997).
17. Z.W. Pan, Z.R. Dai, and Z.L. Wang, *Science* **291**, 1947 (2001).
18. Z.L. Wang, *Adv. Mater.* **15**, 432 (2003).
19. M. Volmer and I. Estermann, *Z. Physik* **7**, 13 (1921).
20. X.Y. Kong and Z.L. Wang, *Nano Lett.* **3**, 1625 (2003).
21. Y. Liu, C. Zheng, W. Wang, C. Yin, and G. Wang, *Adv. Mater.* **13**, 1883 (2001).
22. Y. Yin, G. Zhang, and Y. Xia, *Adv. Func. Mater.* **12**, 293 (2002).
23. X. Jiang, T. Herricks, and Y. Xia, *Nano Lett.* **2**, 1333 (2002).
24. Y. Zhang, N. Wang, S. Gao, R. He, S. Miao, J. Liu, J. Zhu, and X. Zhang, *Chem. Mater.* **14**, 3564 (2002).
25. E.G. Wolfe and T.D. Coskren, *J. Am. Ceram. Soc.* **48**, 279 (1965).
26. S. Hayashi and H. Saito, *J. Cryst. Growth* **24/25**, 345 (1974).
27. W. Shi, H. Peng, Y. Zheng, N. Wang, N. Shang, Z. Pan, C. Lee, and S. Lee, *Adv. Mater.* **12**, 1343 (2000).
28. P. Yang and C.M. Lieber, *Science* **273**, 1836 (1996).

29. B. Gates, Y. Yin, and Y. Xia, *J. Am. Chem. Soc.* **122**, 12582 (2000).
30. B. Wunderlich and H.-C. Shu, *J. Cryst. Growth* **48**, 227 (1980).
31. B. Mayers, B. Gates, Y. Yin, and Y. Xia, *Adv. Mater.* **13**, 1380 (2001).
32. A.A. Kudryavtsev, *The Chemistry and Technology of Selenium and Tellurium*, Collet's, London, 1974.
33. B. Gates, Y. Yin, and Y. Xia, *J. Am. Chem. Soc.* **122**, 582 (1999).
34. Y. Li, Y. Ding, and Z. Wang, *Adv. Mater.* **11**, 847 (1999).
35. W. Wang, C. Xu, G. Wang, Y. Liu, and C. Zheng, *Adv. Mater.* **14**, 837 (2002).
36. Y. Sun, B. Gates, B. Mayers, and Y. Xia, *Nano Lett.* **2**, 165 (2002).
37. K. Govender, D.S. Boyle, P. O'Brien, D. Brinks, D. West, and D. Coleman, *Adv. Mater.* **14**, 1221 (2002).
38. J.J. Urban, W.S. Yun, Q. Gu, and H. Park, *J. Am. Chem. Soc.* **124**, 1186 (2002).
39. J.J. Urban, J.E. Spanier, L. Ouyang, W.S. Yun, and H. Park, *Adv. Mater.* **15**, 423 (2003).
40. H.W. Liao, Y.F. Wang, X.M. Liu, Y.D. Li, and Y.T. Qian, *Chem. Mater.* **12**, 2819 (2000).
41. Q. Chen, W. Zhou, G. Du, and L.-M. Peng, *Adv. Mater.* **14**, 1208 (2002).
42. R.S. Wagner and W.C. Ellis, *Appl. Phys. Lett.* **4**, 89 (1964).
43. R.S. Wagner, W.C. Ellis, K.A. Jackson, and S.M. Arnold, *J. Appl. Phys.* **35**, 2993 (1964).
44. R.S. Wagner, in *Whisker Technology*, ed. by A.P. Levitt, Wiley, New York, 47, 1970.
45. R.S. Wagner and W.C. Ellis, *Trans. Metal. Soc. AIME* **233**, 1053 (1965).
46. G.A. Bootsma and H.J. Gassen, *J. Cryst. Growth* **10**, 223 (1971).
47. C.M. Lieber, *Solid State Commun.* **107**, 106 (1998).
48. J. Hu, T.W. Odom, and C.M. Lieber, *Acc. Chem. Res.* **32**, 435 (1999).
49. A.M. Morales and C.M. Lieber, *Science* **279**, 208 (1998).
50. D.P. Yu, Z.G. Bai, Y. Ding, Q.L. Hang, H.Z. Zhang, J.J. Wang, Y.H. Zou, W. Qian, G.C. Xoing, H.T. Zhou, and S.Q. Feng, *Appl. Phys. Lett.* **72**, 3458 (1998).
51. D.P. Yu, C.S. Lee, I. Bello, X.S. Sun, Y. Tang, G.W. Zhou, Z.G. Bai, and S.Q. Feng, *Solid State Commun.* **105**, 403 (1998).
52. X. Duan and C.M. Lieber, *Adv. Mater.* **12**, 298 (2000).
53. E.I. Givargizov, *J. Vac. Sci. Technol.* **B11**, 449 (1993).
54. Y. Wu and P. Yang, *Chem. Mater.* **12**, 605 (2000).
55. M.S. Gudiksen, J. Wang, and C.M. Lieber, *J. Phys. Chem.* **B105**, 4062 (2001).
56. M.S. Gudiksen and C.M. Lieber, *J. Am. Chem. Soc.* **122**, 8801 (2000).
57. T. Dietz, M. Duncan, M. Liverman, and R.E. Smalley, *J. Chem. Phys.* **73**, 4816 (1980).
58. H.N.V. Temperley, *Proc. Cambridge Phil. Soc.* **48**, 683 (1952).
59. K.A. Jackson, *Growth and Perfection of Crystals*, John Wiley and Sons, New York, 1958.
60. Y. Wang, G. Meng, L. Zhang, C. Liang, and J. Zhang, *Chem. Mater.* **14**, 1773 (2002).
61. Y.Q. Chen, K. Zhang, B. Miao, B. Wang, and J.G. Hou, *Chem. Phys. Lett.* **358**, 396 (2002).
62. K.-W. Chang and J.-J. Wu, *J. Phys. Chem.* **B106**, 7796 (2002).
63. D. Zhang, D.N. McIlroy, Y. Geng, and M.G. Norton, *J. Mater. Sci. Lett.* **18**, 349 (1999).
64. I.-C. Leu, Y.-M. Lu, and M.-H. Hon, *Mater. Chem. Phys.* **56**, 256 (1998).
65. M.H. Huang, Y. Wu, H. Feick, N. Tran, E. Weber, and P. Yang, *Adv. Mater.* **13**, 113 (2001).
66. D.R. Askkeland, *The Science and Engineering of Materials*, PWS, Boston, MA, 1989.

67. Y.C. Choi, W.S. Kim, Y.S. Park, S.M. Lee, D.J. Bae, Y.H. Lee, G.-S. Park, W.B. Choi, N.S. Lee, and J.M. Kim, *Adv. Mater.* **12**, 746 (2000).
68. Z.G. Bai, D.P. Yu, H.Z. Zhang, Y. Ding, Y.P. Wang, X.Z. Gai, Q.L. Hang, G.C. Xoing, and S.Q. Feng, *Chem. Phys. Lett.* **303**, 311 (1999).
69. D.P. Yu, Q.L. Hang, Y. Ding, H.Z. Zhang, Z.G. Bai, J.J. Wang, Y.H. Zou, W. Qian, G.C. Xoing, and S.Q. Feng, *Appl. Phys. Lett.* **73**, 3076 (1998).
70. C.C. Chen and C.C. Yeh, *Adv. Mater.* **12**, 738 (2000).
71. X.F. Duan and C.M. Lieber, *J. Am. Chem. Soc.* **122**, 188 (2000).
72. X. Chen, J. Li, Y. Cao, Y. Lan, H. Li, M. He, C. Wang, Z. Zhang, and Z. Qiao, *Adv. Mater.* **12**, 1432 (2000).
73. T.J. Trentler, K.M. Hickman, S.C. Goel, A.M. Viano, P.C. Gobbons, and W.E. Buhro, *Science* **270**, 1791 (1995).
74. W.E. Buhro, *Polyhedron* **13**, 1131 (1994).
75. H. Yu and W.E. Buhro, *Adv. Mater.* **15**, 416 (2003).
76. M.J. Ludowise, *J. Appl. Phys.* **58**, R31 (1985).
77. J.D. Holmes, K.P. Johnston, C. Doty, and B.A. Korgel, *Science* **287**, 1471 (2000).
78. J. Franks, *Acta Metal.* **6**, 103 (1958).
79. R.M. Fisher, L.S. Darken, and K.G. Carroll, *Acta Metal.* **2**, 368 (1954).
80. J.D. Eshelby, *Phys. Rev.* **91**, 775 (1953).
81. S.E. Koonce and S.M. Arnold, *J. Appl. Phys.* **24**, 365 (1953).
82. R.C. Furneaux, W.R. Rigby, and A.P. Davidson, *Nature* **337**, 147 (1989).
83. R.L. Fleisher, P.B. Price, and R.M. Walker, *Nuclear Tracks in Solids*, University of California Press, Berkeley, CA, 1975.
84. R.J. Tonucci, B.L. Justus, A.J. Campillo, and C.E. Ford, *Science* **258**, 783 (1992).
85. G.E. Possin, *Rev. Sci. Instrum.* **41**, 772 (1970).
86. C. Wu and T. Bein, *Science* **264**, 1757 (1994).
87. S. Fan, M.G. Chapline, N.R. Franklin, T.W. Tombler, A.M. Cassell, and H. Dai, *Science* **283**, 512 (1999).
88. P. Enzel, J.J. Zoller, and T. Bein, *Chem. Commun.* 633 (1992).
89. C. Guerret-Piecourt, Y. Le Bouar, A. Loiseau, and H. Pascard, *Nature* **372**, 761 (1994).
90. P.M. Ajayan, O. Stephan, P. Redlich, and C. Colliex, *Nature* **375**, 564 (1995).
91. A. Despic and V.P. Parkhuitik, *Modern Aspects of Electrochemistry*, Vol. 20 Plenum, New York, 1989.
92. D. AlMawiawi, N. Coombs, and M. Moskovits, *J. Appl. Phys.* **70**, 4421 (1991).
93. C.A. Foss, M.J. Tierney, and C.R. Martin, *J. Phys. Chem.* **96**, 9001 (1992).
94. J.B. Mohler and H.J. Sedusky, *Electroplating for the Metallurgist, Engineer and Chemist*, Chemical Publishing Co. Inc., New York, 1951.
95. A.J. Bard and L.R. Faulkner, *Electrochemical Methods*, John Wiley & Sons, New York, 1980.
96. J.W. Evans and L.C. De Jonghe, *The Production of Inorganic Materials*, Macmillan, New York, 1991.
97. F.R.N. Nabarro and P.J. Jackson, in *Growth and Perfection of Crystals*, eds., R.H. Doremus, B.W. Roberts, and D. Turnbull, John Wiley, New York, p. 13, 1958.
98. T.M. Whitney, J.S. Jiang, P.C. Searson, and C.L. Chien, *Science* **261**, 1316 (1993).
99. W.D. Williams and N. Giordano, *Rev. Sci. Instrum.* **55**, 410 (1984).
100. B.Z. Tang and H. Xu, *Macromolecules* **32**, 2569 (1999).
101. Y. Zhang, G. Li, Y. Wu, B. Zhang, W. Song, and L. Zhang, *Adv. Mater.* **14**, 1227 (2002).
102. G. Yi and W. Schwarzacher, *Appl. Phys. Lett.* **74**, 1746 (1999).

103. J.D. Klein, R.D. Herrick, II, D. Palmer, M.J. Sailor, C.J. Brumlik, and C.R. Martin, *Chem. Mater.* **5**, 902 (1993).
104. C. Schönenberger, B.M.I. van der Zande, L.G.J. Fokkink, M. Henny, C. Schmid, M. Krüger, A. Bachtold, R. Huber, H. Birk, and U. Staufer, *J. Phys. Chem.* **B101**, 5497 (1997).
105. C.J. Brumlik, V.P. Menon, and C.R. Martin, *J. Mater. Res.* **268**, 1174 (1994).
106. C.J. Brumlik and C.R. Martin, *J. Am. Chem. Soc.* **113**, 3174 (1991).
107. C.J. Miller, C.A. Widrig, D.H. Charych, and M. Majda, *J. Phys. Chem.* **92**, 1928 (1988).
108. C.-G. Wu and T. Bein, *Science* **264**, 1757 (1994).
109. P.M. Ajayan, O. Stephan, and Ph. Redlich, *Nature* **375**, 564 (1995).
110. W. Han, S. Fan, Q. Li, and Y. Hu, *Science* **277**, 1287 (1997).
111. G.O. Mallory and J.B. Hajdu (eds.), *Electroless Plating: Fundamentals and Applications*, American Electroplaters and Surface Finishers Society, Orlando, FL, 1990.
112. C.R. Martin, *Chem. Mater.* **8**, 1739 (1996).
113. C.R. Martin, *Science* **266**, 1961 (1994).
114. C.R. Martin, *Adv. Mater.* **3**, 457 (1991).
115. J.C. Hulteen and C.R. Martin, *J. Mater. Chem.* **7**, 1075 (1997).
116. L. Piraux, S. Dubois, and S. Demoustier-Champagne, *Nucl. Instrum. Methods Phys. Res.* **B131**, 357 (1997).
117. I. Zhitomirsky, *Adv. Colloid Interf. Sci.* **97**, 297 (2002).
118. O.O. Van der Biest and L.J. Vandeperre, *Annu. Rev. Mater. Sci.* **29**, 327 (1999).
119. P. Sarkar and P.S. Nicholson, *J. Am. Ceram. Soc.* **79**, 1987 (1996).
120. J.S. Reed, *Introduction to the Principles of Ceramic Processing*, John Wiley & Sons, New York, 1988.
121. R.J. Hunter, *Zeta Potential in Colloid Science: Principles and Applications*, Academic Press, London, 1981.
122. C.J. Brinker and G.W. Scherer, *Sol-Gel Science: the Physics and Chemistry of Sol-Gel Processing*, Academic Press, San Diego, CA, 1990.
123. A.C. Pierre, *Introduction to Sol-Gel Processing*, Kluwer, Norwell, MA, 1998.
124. J.D. Wright and N.A.J.M. Sommerdijk, *Sol-Gel Materials: Chemistry and Applications*, Gordon and Breach, Amsterdam, 2001.
125. D.H. Everett, *Basic Principles of Colloid Science*, the Royal Society of Chemistry, London, 1988.
126. W.D. Callister, *Materials Science and Engineering: An Introduction*, John Wiley & Sons, New York, 1997.
127. S.J. Limmer, S. Seraji, M.J. Forbess, Y. Wu, T.P. Chou, C. Nguyen, and G.Z. Cao, *Adv. Mater.* **13**, 1269 (2001).
128. S.J. Limmer, S. Seraji, M.J. Forbess, Y. Wu, T.P. Chou, C. Nguyen, and G.Z. Cao, *Adv. Func. Mater.* **12**, 59 (2002).
129. S.J. Limmer and G.Z. Cao, *Adv. Mater.* **15**, 427 (2003).
130. Y.C. Wang, I.C. Leu, and M.N. Hon, *J. Mater. Chem.* **12**, 2439 (2002).
131. Z. Miao, D. Xu, J. Ouyang, G. Guo, X. Zhao, and Y. Tang, *Nano Lett.* **2**, 717 (2002).
132. C. Natarajan and G. Nogami, *J. Electrochem. Soc.* **143**, 1547 (1996).
133. B.B. Lakshmi, P.K. Dorhout, and C.R. Martin, *Chem. Mater.* **9**, 857 (1997).
134. R.L. Penn and J.F. Banfield, *Geochim. Cosmochim. Ac.* **63**, 1549 (1999).
135. B.B. Lakshmi, C.J. Patrissi, and C.R. Martin, *Chem. Mater.* **9**, 2544 (1997).

136. J.S. Reed, *Introduction to Principles of Ceramic Processing*, Wiley, New York, 1988.
137. C.A. Huber, T.E. Huber, M. Sadoqi, J.A. Lubin, S. Manalis, and C.B. Prater, *Science* **263**, 800 (1994).
138. Z. Zhang, D. Gekhtman, M.S. Dresselhaus, and J.Y. Ying, *Chem. Mater.* **11**, 1659 (1999).
139. W. Liang and C.R. Martin, *J. Am. Chem. Soc.* **112**, 9666 (1990).
140. S.M. Marinakos, L.C. Brousseau, III, A. Jones, and D.L. Feldheim, *Chem. Mater.* **10**, 1214 (1998).
141. H.D. Sun, Z.K Tang, J. Chen, and G. Li, *Solid State Commun.* **109**, 365 (1999).
142. Z. Cai, J. Lei, W. Liang, V. Menon, and C.R. Martin, *Chem. Mater.* **3**, 960 (1991).
143. Y.-J. Han, J.M. Kim, and G.D. Stucky, *Chem. Mater.* **12**, 2068 (2000).
144. L. Chen, P.J. Klar, W. Heimbrodt, F. Brieler, and M. Fröba, *Appl. Phys. Lett.* **76**, 3531 (2000).
145. K. Matsui, T. Kyotani, and A. Tomita, *Adv. Mater.* **14**, 1216 (2002).
146. R. Leon, D. Margolese, G. Stucky, and P.M. Petroff, *Phys. Rev. B* **52**, R2285 (1995).
147. K.-B. Lee, S.-M. Lee, and J. Cheon, *Adv. Mater.* **13**, 517 (2001).
148. T. Wen, J. Zhang, T.P. Chou, and G.Z. Cao, submitted to *Adv. Mater.*
149. C.-G. Wu and T. Bein, *Science* **264**, 1757 (1994).
150. B. Gates, Y. Wu, Y. Yin, P. Yang, and Y. Xia, *J. Am. Chem. Soc.* **123**, 11500 (2001).
151. Y. Xia, Lecture note of SPIE short course 496, July 7, 2002.
152. H. Dai, E.W. Wong, Y.Z. Lu, S. Fan, and C.M. Lieber, *Nature* **375**, 769 (1995).
153. E.W. Wong, B.W. Maynor, L.D. Burns, and C.M. Lieber, *Chem. Mater.* **8**, 2041 (1996).
154. W. Han, S. Fan, Q. Li, B. Gu, X. Zhang, and D. Yu, *Appl. Phys. Lett.* **71**, 2271 (1997).
155. A. Huczko, *Appl. Phys.* **A70**, 365 (2000).
156. Y. Li, G.S. Cheng, and L.D. Zhang, *J. Mater. Res.* **15**, 2305 (2000).
157. C.M. Zelenski and P.K. Dorhout, *J. Am. Chem. Soc.* **120**, 734 (1998).
158. E. Braun, Y. Eichen, U. Sivan, and G. Ben-Yoseph, *Nature* **391**, 775 (1998).
159. J. Zhan, X. Yang, D. Wang, S. Li, Y. Xie, Y. Xia, and Y. Qian, *Adv. Mater.* **12**, 1348 (2000).
160. A. Frenot and I.S. Chronakis, *Current Opin. Colloid Interf. Sci.* **8**, 64 (2003).
161. D.H. Reneker and I. Chun, *Nanotechnology* **7**, 216 (1996).
162. H. Fong, W. Liu, C.S. Wang, and R.A. Vaia, *Polymer* **43**, 775 (2002).
163. J.A. Mathews, G.E. Wnek, D.G. Simpson, and G.L. Bowlin, *Biomacromolecules* **3**, 232 (2002).
164. G. Larsen, R. Velarde-Ortiz, K. Minchow, A. Barrero, and I.G. Loscertales, *J. Am. Chem. Soc.* **125**, 1154 (2003).
165. H. Dai, J. Gong, H. Kim, and D. Lee, *Nanotechnology* **13**, 674 (2002).
166. D. Li and Y. Xia, *Nano Lett.* **3**, 555 (2003).
167. D. Li and Y. Xia, in *Nanomaterials and Their Optical Applications*, SPIE Proceedings 5224 (2003).
168. K. Kurihara, K. Iwadate, H. Namatsu, M. Nagase, and K. Murase, *J. Vac. Sci. Technol.* **B13**, 2170 (1995).
169. H.I. Liu, D.K. Biegelsen, F.A. Ponce, N.M. Johnson, and R.F. Pease, *Appl. Phys. Lett.* **64**, 1383 (1994).
170. Y. Xia, J.A. Rogers, K.E. Paul, and G.M. Whitesides, *Chem. Rev.* **99**, 1823 (1999).
171. Y. Yin, B. Gates, and Y. Xia, *Adv. Mater.* **12**, 1426 (2000).

Chapter 5

Two-Dimensional Nanostructures: Thin Films

5.1. Introduction

Deposition of thin films has been a subject of intensive study for almost a century, and many methods have been developed and improved. Many such techniques have been developed and widely used in industries, which in turn provides a great driving force for further development and improvement of the deposition techniques. There are many excellent textbooks and monographs available.[1-3] In this chapter, we will briefly introduce the fundamentals and summarize typical experimental approaches of various well-established techniques of film deposition. Film growth methods can be generally divided into two groups: vapor-phase deposition and liquid-based growth. The former includes, for example, evaporation, molecular beam epitaxy (MBE), sputtering, chemical vapor deposition (CVD), and atomic layer deposition (ALD). Examples of the latter are electrochemical deposition, chemical solution deposition (CSD), Langmuir–Blodgett films and self-assembled monolayers (SAMs).

The film deposition involves predominantly heterogeneous processes including heterogeneous chemical reactions, evaporation, adsorption and desorption on growth surfaces, heterogeneous nucleation and surface growth. In addition, most film deposition and characterization processes are conducted under a vacuum. Therefore, in this chapter, before discussing the details of various methods for thin film deposition and growth, a brief discussion will be devoted to the fundamentals of heterogeneous nucleation followed by a general introduction to vacuum science and technology. Other aspects of heterogeneous processes and relevant vacuum issues will be incorporated into the various deposition methods where the subject is relevant.

5.2. Fundamentals of Film Growth

Growth of thin films, as all phase transformation, involves the processes of nucleation and growth on the substrate or growth surfaces. The nucleation process plays a very important role in determining the crystallinity and microstructure of the resultant films. For the deposition of thin films with thickness in the nanometer region, the initial nucleation process is even more important. Nucleation in film formation is a heterogeneous nucleation, and its energy barrier and critical nucleus size have been discussed briefly in Chapter 3. However, the discussion was limited to the simplest situation. The size and the shape of the initial nuclei are assumed to be solely dependent on the change of volume of Gibbs free energy, due to supersaturation, and the combined effect of surface and interface energies governed by Young's equation. No other interactions between the film or nuclei and the substrate were taken into consideration. In practice, the interaction between film and substrate plays a very important role in determining the initial nucleation and the film growth. Many experimental observations revealed that there are three basic nucleation modes:

(1) Island or Volmer–Weber growth,
(2) Layer or Frank–van der Merwe growth, and
(3) Island-layer or Stranski–Krastonov growth.

Figure 5.1 illustrates these three basic modes of initial nucleation in the film growth. Island growth occurs when the growth species are more strongly bonded to each other than to the substrate. Many systems of metals on insulator substrates, alkali halides, graphite and mica substrates display this type of nucleation during the initial film deposition. Subsequent growth results in the islands to coalesce to form a continuous film. The layer growth is the opposite of the island growth, where growth species are equally bound more strongly to the substrate than to each other. First complete monolayer is formed, before the deposition of second layer occurs. The most important examples of layer growth mode are the epitaxial growth of single crystal films. The island-layer growth is an

Two-Dimensional Nanostructures: Thin Films

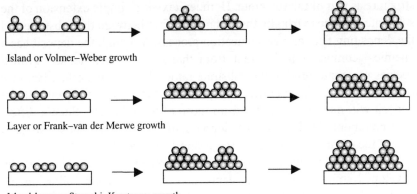

Fig. 5.1. Schematic illustrating three basic modes of initial nucleation in the film growth. Island growth occurs when the growth species are more strongly bonded to each other than to the substrate.

intermediate combination of layer growth and island growth. Such a growth mode typically involves the stress, which is developed during the formation of the nuclei or films.

In Chapter 3, we have arrived at the critical nucleus size, r*, and the corresponding energy barrier, ΔG^*, given by Eqs. (3.47) and (3.49) and shown below:

$$r^* = \left(\frac{2\pi \gamma_{vf}}{\Delta G_v}\right)\left(\frac{\sin^2 \theta \cdot \cos \theta + 2\cos \theta - 2}{2 - 3\cos \theta + \cos^3 \theta}\right) \quad (5.1)$$

$$\Delta G^* = \frac{16\pi \gamma_{vf}}{3(\Delta G_v)^2}\left(\frac{2 - 3\cos \theta + \cos^3 \theta}{4}\right) \quad (5.2)$$

For island growth, the contact angle must be larger than zero, i.e. $\theta > 0$. According to Young's equation, then we have

$$\gamma_{sv} < \gamma_{fs} + \gamma_{vf} \quad (5.3)$$

If the deposit does not wet the substrate at all or $\theta = 180°$, the nucleation is a homogeneous nucleation. For layer growth, the deposit wets the substrate completely and the contact angle equals zero; the corresponding Young's equation becomes:

$$\gamma_{sv} = \gamma_{fs} + \gamma_{vf} \quad (5.4)$$

The most important layer growth is the deposition of single crystal films through either homoepitaxy, in which the depositing film has the same crystal structure and chemical composition as that of the substrate, or heteroepitaxy, in which the depositing film has a close matching crystal structure as that of the substrate. Homoepitaxy is a simple extension of the substrate, and thus virtually there is no interface between the substrate and the depositing film and no nucleation process. Although the deposit has a chemical composition different from that of the substrate, the growth species prefers to bind to the substrate rather than to each other. Because of the difference in chemical composition, the lattice constants of the deposit will most likely differ from those of the substrate. Such a difference commonly leads to the development of stress in the deposit; stress is one of the common reasons for the island-layer growth.

Island-layer growth is a little more complicated and involves *in situ* developed stress. Initially the deposition would proceed following the mode of layer growth. When the deposit is elastically strained due to, for example, lattice mismatch between the deposit and the substrate, strain energy would be developed. As each layer of deposit is added, more stress is developed and so is the strain energy. Such strain energy is proportional to the volume of the deposit, assuming there is no plastic relaxation. Therefore, the change of volume of Gibbs free energy should include the strain energy and Eq. (5.2) is modified accordingly:

$$\Delta G^* = \left(\frac{16\pi\gamma_{vf}}{3(\Delta G_v + \omega)^2}\right)\left(\frac{2 - 3\cos\theta + \cos^3\theta}{4}\right) \quad (5.5)$$

where ω is the strain energy per unit volume generated by the stress in the deposit. Because the sign of ΔG_v is negative, and the sign of ω is positive, the overall energy barrier to nucleation increases. When the stress exceeds a critical point and cannot be released, the strain energy per unit area of deposit is large with respect to γ_{vf}, permitting nuclei to form above the initial layered deposit. In this case, the surface energy of the substrate exceeds the combination of both surface energy of the deposit and the interfacial energy between the substrate and the deposit:

$$\gamma_{sv} > \gamma_{fs} + \gamma_{vf} \quad (5.6)$$

If should be noted that there are other situations when the overall volume of Gibbs free energy may change. For example, initial deposition or nucleation on substrates with cleavage steps and screw dislocations would result in a stress release and, thus, an increased change of the overall Gibbs free energy. As a result, the energy barrier for the initial nucleation is reduced and the critical size of nuclei becomes small. Substrate charge and impurities would affect the ΔG^* through the change of surface, electrostatic and chemical energies in a similar manner.

It should be noted that the aforementioned nucleation models and mechanisms are applicable to the formation of single crystal, polycrystalline and amorphous deposit, and of inorganic, organic and hybrid deposit. Whether the deposit is single crystalline, polycrystalline or amorphous, depends on the growth conditions and the substrate. Deposition temperature and the impinging rate of growth species are the two most important factors and are briefly summarized below:

(1) Growth of single crystal films is most difficult and requires: (i) a single crystal substrate with a close lattice match, (ii) a clean substrate surface so as to avoid possible secondary nucleation, (iii) a high growth temperature so as to ensure sufficient mobility of the growth species and (iv) low impinging rate of growth species so as to ensure sufficient time for surface diffusion and incorporation of growth species into the crystal structure and for structural relaxation before the arrival of next growth species.
(2) Deposition of amorphous films typically occurs (i) when a low growth temperature is applied, there is insufficient surface mobility of growth species and/or (ii) when the influx of growth species onto the growth surface is very high, growth species does not have enough time to find the growth sites with the lowest energy.
(3) The conditions for the growth of polycrystalline crystalline films fall between the conditions of single crystal growth and amorphous film deposition. In general, the deposition temperature is moderate ensuring a reasonable surface mobility of growth species and the impinging flux of growth species is moderately high.

Figure 5.2, as an example, shows the growth conditions for the single crystalline, polycrystalline and amorphous films of silicon by chemical vapor deposition.[4] The above discussion is applicable to single element films; however, growth process is complex in the presence of impurities and additives and in the case of multiple component material systems.

Epitaxy is a very special process, and refers to the formation or growth of single crystal on top of a single crystal substrate or seed. Epitaxial growth can be further divided into homoepitaxy and heteroepitaxy. Homoepitaxy is to grow film on the substrate, in which both are the same material. Homoepitaxial growth is typically used to grow better quality film or introduce dopants into the grown film. Heteroepitaxy refers to the case that films and substrates are different materials. One obvious difference between homoepitaxial films and heteroepitaxial films is the lattice match between films and substrates. There is no lattice mismatch between films and substrates by homoepitaxial growth. On the contrary, there will be a lattice mismatch between films and substrates in heteroepitaxial growth. The lattice mismatch is also called misfit, given by:

$$f = \frac{a_s - a_f}{a_f} \tag{5.7}$$

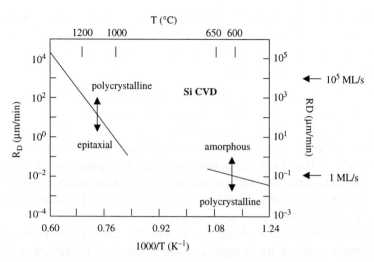

Fig. 5.2. The growth conditions for the single crystalline, polycrystalline and amorphous films of silicon by chemical vapor deposition. [J. Bloem, *Proc. Seventh Conf. CVD*, eds. T.O. Sedgwick and H. Lydtin (ECS PV 79-3), p. 41, 1979]

where a_s is the unstrained lattice constant of the substrate and a_f is the unstrained lattice constant of the film. If $f > 0$, the film is strained in tension, whereas if $f < 0$, the film is trained in compression. Strain energy, E_s, develops in strained films:

$$E_s = 2\mu_f \left(\frac{1+\nu}{1-\nu}\right)\varepsilon^2 \, h \, A \qquad (5.8)$$

where μ_f is the shear modulus of the film, ν, the Poisson's ratio ($<1/2$ for most materials), ε, the plane or lateral strain, h, the thickness, and A, the surface area. It is noted that the strain energy increases with the thickness. The strain energy can be either accommodated by straining both film and substrate when the mismatch is relatively small, or relaxed by formation of dislocations when the mismatch is large. Figure 5.3 schematically illustrates the lattice matched homoepitaxial film and substrate, strained and relaxed heteroepitaxial structures. Both homoepitaxial and heteroepitaxial growth of films has been a well-established technique and found wide applications, particularly in electronic industry.

5.3. Vacuum Science

Most film deposition and processing are carried out in a vacuum. In addition, almost all the characterization of films is performed under a vacuum. Although there is very rich literature on vacuum, it seems that a brief

Two-Dimensional Nanostructures: Thin Films

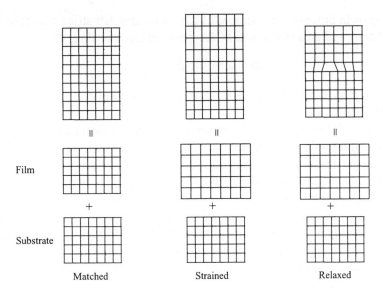

Fig. 5.3. Schematic illustrating the lattice matched homoepitaxial film and substrate, strained and relaxed heteroepitaxial structures.

discussion on relevant subjects is necessary. Specifically, we will introduce some most commonly encountered concepts in thin film deposition and characterization, such as mean free path and flow regimes and their pressure and temperature dependence. Readers who want to learn more fundamentals and technique details of vacuum are recommended to Refs. 5–7

In a gas phase, gas molecules are constantly in motion and colliding among themselves as well as with the container walls. Pressure of a gas is the result of momentum transfer from the gas molecules to the walls, and is the most widely quoted system variable in vacuum technology. The mean distance traveled by molecules between successive collisions is called the mean free path and is an important property of the gas that depends on the pressure, given by:

$$\lambda_{mfp} = \frac{5 \times 10^{-3}}{P} \quad (5.9)$$

where λ_{mfp} is the mean free path in centimeter and P is the pressure in torr. When the pressure is below 10^{-3} torr, the gas molecules in typical film deposition and characterization systems virtually collide only with the walls of the vacuum chamber, i.e. there is no collision among gas molecules.

The gas impingement flux in the film deposition is a measure of the frequency with which gas molecules impinge on or collide with a surface, and is the most important parameter. It is because for film deposition, only molecules impinging onto the growth surface will be able to contribute to the growth process. The number of gas molecules that strike a surface per unit time and area is defined as the gas impingement flux, Φ:

$$\Phi = 3.513 \times 10^{22} \frac{P}{(MT)^{\frac{1}{2}}} \tag{5.10}$$

where P is the pressure in torr, M is the molecular weight and T is temperature.

Figure 5.4 summarizes the molecular density, incident rate, mean free path and monolayer formation time as a function of pressure.[5] As will be discussed further in the following sections, one will see that of the film deposition processes, evaporation requires a vacuum between the high and ultrahigh regimes, whereas sputtering and low pressure chemical vapor deposition are accomplished at the border between the medium and high vacuum ranges. Of the analytical instruments, electron microscopes operate in high vacuum, and surface analytical equipment have the most stringent cleanliness requirements and are operative only under ultrahigh vacuum conditions.

It should be noted that gas flow is different from the restless motion and collision of gas molecules. Gas flow is defined as a net directed movement of gas in a system and occurs when there is a pressure drop. Depending on the geometry of the system involved as well as the pressure, temperature and type of gas in question, gas flow can be divided into three regimes: molecular flow, intermediate flow and viscous flow. Free molecular flow occurs at low gas densities or high vacuum, when the mean free path between intermolecular collisions is larger than the dimensions of the

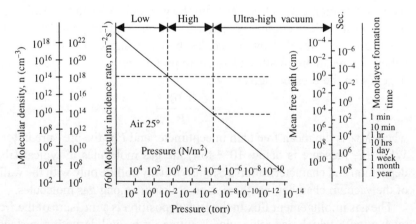

Fig. 5.4. Summary of molecular density, incident rate, mean free path, and monolayer formation time as a function of pressure. [A. Roth, *Vacuum Technology*, North-Holland, Amsterdam, 1976.]

system and the molecules collide with the walls of the system only. At high pressure, intermolecular collisions become predominant since the mean free path is reduced and the gas flow is referred to as in the viscous flow regime. Between free molecular flow and viscous flow, there is a transition regime: intermediate flow. The above gas flow can be defined by the magnitude of the Knudsen number, K_n, given by:

$$K_n = \frac{D}{\lambda_{mfp}} \tag{5.11}$$

where D is the characteristic dimension of the system, e.g. the diameter of a pipe, and λ_{mfp} is the gas mean free path. Figure 5.5 shows the gas flow regimes in a tube as functions of system dimensions and pressure, whereas, the range of Knudsen numbers corresponding to gas flow regimes are summarized in Table 5.1.

Viscous flow is a bit complex and can be further divided into laminar flow, turbulent flow and transition flow. At a low gas flow velocity, the flow is laminar where layered, parallel flow lines may be visualized, no perpendicular velocity is present, and mixing inside the gas is by diffusion only. In this flow, the velocity is zero at the gas-wall interface and gradually increases as moving away from the interface, reaching a maximum at the center when flowing inside a pipe. Flow behavior can be defined by the so-called Reynolds number, Re, which is given below of gas flow inside a pipe:

$$\mathrm{Re} = D \cdot \frac{v \cdot \rho}{\eta} \tag{5.12}$$

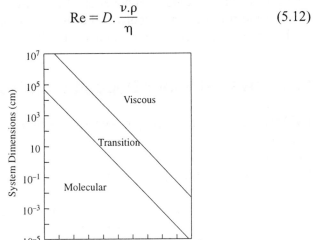

Fig. 5.5. Gas flow regimes in a tube as functions of system dimensions and pressure and the range of Knudsen numbers corresponding to gas flow regimes are summarized in Table 5.1.

Table 5.1. Summary of gas flow regions.

Gas flow regimes	Knudsen number	D·P
Molecular flow	$K_n < 1$	$D \cdot P < 5 \times 10^{-3}$ cm.torr
Intermediate flow	$1 < K_n < 110$	$5 \times 10^{-3} < D.P. < 5 \times 10^{-1}$ cm.torr
Viscous flow	$K_n > 110$	$D \cdot P > 5 \times 10^{-1}$ cm.torr

D is the characteristic dimension of the system and P is the pressure.

where D is the diameter of the pipe, v, the velocity, ρ, the density, and η, the viscosity of the gas. Laminar flow corresponds to a small $Re < 2100$. At a high gas velocity, the flow is turbulent, where the gas is constantly under intermixing, where $Re > 4000$. At $2100 < Re < 4000$, a transition from laminar to turbulent flow occurs and is referred to as transition flow. There is always a laminar flow near to the solid surface in both turbulent and transition flows, since the friction viscous forces a deceleration of the gas at the surface.

Diffusion is one of the mass transfer mechanisms in gases, which also occurs in liquids and solids. Diffusion is the motion of atoms or molecules from regions of higher to lower concentration, thus increasing the entropy of the system. Another mechanism is convection, a bulk gas flow process. Convection arises from the response to gravitational, centrifugal, electric and magnetic forces. Convection can play an important role in high-pressure film deposition. For example, a hotter and less dense gas above a hot substrate would rise, whereas a cooler and denser gas would replace the gap. Such a situation is often encountered in cold wall CVD reactors.

5.4. Physical Vapor Deposition (PVD)

PVD is a process of transferring growth species from a source or target and deposit them on a substrate to form a film. The process proceeds atomistically and mostly involves no chemical reactions. Various methods have been developed for the removal of growth species from the source or target. The thickness of the deposits can vary from angstroms to millimeters. In general, those methods can be divided into two groups: evaporation and sputtering. In evaporation, the growth species are removed from the source by thermal means. In sputtering, atoms or molecules are dislodged from solid target through impact of gaseous ions (plasma). Each group can be further divided into a number of methods, depending on specific techniques applied to activate the source or target atoms or molecules and the deposition conditions applied.

5.4.1. Evaporation

Evaporation is arguably the simplest deposition method, and has been proven particularly useful for the deposition of elemental films. Although formation of thin films by evaporation was known about 150 years ago,[8] it acquired a wide range of applications over 50 years when the industrial scale vacuum techniques were developed.[9] Many excellent books and review articles have been published on evaporated films.[10] A typical evaporation system is schematically shown in Fig. 5.6. The system consists of an evaporation source that vaporizes the desired material and a substrate is located at an appropriate distance facing the evaporation source. Both the source and the substrate are located in a vacuum chamber. The substrate can be heated or electrically biased or rotated during deposition. The desired vapor pressure of source material can be generated by simply heating the source to elevated temperatures, and the concentration of the growth species in the gas phase can be easily controlled by varying the source temperature and the flux of the carrier gas. The equilibrium vapor pressure of an element can be estimated as:

$$\ln P_e = -\frac{\Delta H_e}{R_g T} + C \qquad (5.13)$$

where ΔH_e is the molar heat of evaporation, R_g, gas constant, T, temperature, and C, a constant. However, evaporation of compounds is more

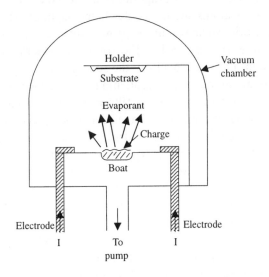

Fig. 5.6. A typical evaporation system consisting of an evaporation source to vaporize the desired material and a substrate located at an appropriate distance facing the evaporation source. Both the source and the substrate are located in a vacuum chamber.

complicated, since compounds may undergo chemical reactions, such as pyrolysis, decomposition and dissociation, and the resulting vapor composition often differs from the source composition during evaporation at elevated temperatures.

The rate of evaporation is dependent on the material in question:

$$\Phi_e = \alpha_e N_A (P_e - P_h)/(2\pi m R_g T)^{1/2} \tag{5.14}$$

where Φ_e is the evaporation rate, α_e, the coefficient of evaporation varying between 0 and 1, N_A, Avogadro's constant, P_e, the vapor pressure, P_h, the hydrostatic pressure acting on the source, m, the molar weight, R_g, the gas constant and T, the temperature. When a mixture of elements or compounds is used as a source for the growth of a complex film, the chemical composition of the vapor phase is most likely to be different from that in the source. Adjusting the composition or molar ratio of the constituents in the source may help. However, the composition of the source would change as the evaporation proceeds, since one element may evaporate much faster than another resulting in the depletion of the first element. As a result, the composition in the vapor phase will change. For a multicomponent system, the chemical composition of evaporated film is likely to be different from that of the source and varies with thickness. Therefore it is in general difficult to deposit complex films using evaporation method.

Deposition of thin films by evaporation is carried out in a low pressure ($10^{-3} \sim 10^{-10}$ torr); atoms and molecules in the vapor phase do not collide with each other prior to arrival at the growth surface, since the mean free path is very large as compared to the source-to-substrate distance. The transport of atoms or molecules from the source to the growth substrate is straightforward along the line of sight, and therefore the conformal coverage is relatively poor and a uniform film over a large area is difficult to obtain. Some special arrangements have been developed to overcome such a shortfall; these include (i) using multiple sources instead of single point source, (ii) rotating the substrates, (iii) loading both source and substrate on the surface of a sphere, and (iv) combination of all the above.

In addition to evaporation of source by resistance heat, other techniques have been developed and have attracted increasing attention and gained more popularity. For example, laser beams have been used to evaporate the material. Absorption characteristics of the material to be evaporated determine the laser wavelength to be used. In order to obtain the high power density required in many cases, pulsed laser beams are generally employed. Such a deposition process is often referred to as laser ablation. Laser ablation has proven to be an effective technique for the deposition of complex films including complex metal oxides such as high T_c superconductor films. One of the great advantages that laser ablation offers is the control of the vapor composition. In principle, the composition of the vapor phase can be controlled as that in the source. The disadvantages of laser ablation include the complex system design, not always possible to find desired laser wavelength for evaporation, and the low energy conversion efficiency. Electron beam evaporation is another technique, but it is limited to the case that the source is electrically conductive. The advantages of electron beam evaporation include a wide range of controlled evaporation rate due to a high power density and low contamination. Arc evaporation is another method commonly used for evaporation of conductive source.

5.4.2. Molecular beam epitaxy (MBE)

MBE can be considered as a special case of evaporation for single crystal film growth, with highly controlled evaporation of a variety of sources in ultrahigh-vacuum of typically $\sim 10^{-10}$ torr.[11-13] Besides the ultrahigh vacuum system, MBE mostly consists of realtime structural and chemical characterization capability, including reflection high energy electron diffraction (RHEED), X-ray photoelectric spectroscopy (XPS), Auger electron spectroscopy (AES). Other analytic instruments may also been attached to the deposition chamber or to a separate analytic chamber, from which the grown films can be transferred to and from the growth chamber without exposing to the ambient. Both ultrahigh vacuum and various structural and chemical characterization facilities are responsible for the fact that the typical MBE reactor can be easily over $1M.

In MBE, the evaporated atoms or molecules from one or more sources do not interact with each other in the vapor phase under such a low pressure. Although some gaseous sources are used in MBE, most molecular beams are generated by heating solid materials placed in source cells, which are referred to as effusion cells or Knudsen cells. A number of effusion cells are radiatically aligned with the substrates as shown in Fig. 5.7. The source materials are most commonly raised to the desired temperatures by resistive heating. The mean free path of atoms or molecules (~ 100 m) far exceeds the distance between the source and the substrate (typically ~ 30 cm) inside the deposition chamber. The atoms or molecules striking on the single crystal substrate results in the formation of the desired epitaxial film. The extremely clean environment, the slow growth rate, and independent control of the evaporation of individual sources enable the precise fabrication of nanostructures and nanomaterials at a

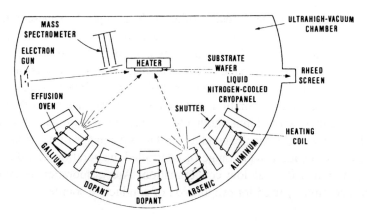

Fig. 5.7. Schematic showing a number of effusion cells radiatically aligned with the substrates.

single atomic layer. Ultrahigh vacuum environment ensures absence of impurity or contamination, and thus a highly pure film can be readily obtained. Individually controlled evaporation of sources permits the precise control of chemical composition of the deposit at any given time. The slow growth rate ensures sufficient surface diffusion and relaxation so that the formation of any crystal defects is kept minimal. The main attributes of MBE include:

(1) A low growth temperature (e.g. 550°C for GaAs) that limits diffusion and maintains hyperabrupt interfaces, which are very important in fabricating two-dimensional nanostructures or multilayer structures such as quantum wells.
(2) A slow growth rate that ensures a well controlled two-dimensional growth at a typical growth rate of 1 μm/h. A very smooth surface and interface is achievable through controlling the growth at the monoatomic layer level.
(3) A simple growth mechanism compared to other film growth techniques ensures better understanding due to the ability of individually controlled evaporation of sources.
(4) A variety of *in situ* analysis capabilities provide invaluable information for the understanding and refinement of the process.

5.4.3. Sputtering

Sputtering is to use energetic ions to knock atoms or molecules out from a target that acts as one electrode and subsequently deposit them on a substrate

Two-Dimensional Nanostructures: Thin Films

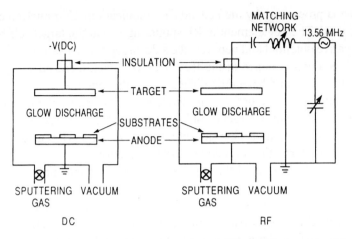

Fig. 5.8. Schematic showing the principles of dc and RF sputtering systems. [M. Ohring, *The Materials Science of Thin Films*, Academic Press, San Diego, CA, 1992.]

acting as another electrode. Although various sputtering techniques have been developed, the fundamentals of the sputtering process are more or less the same. Figure 5.8 schematically illustrates the principles of dc and RF sputtering systems.[1] Let us take the dc discharge as an example to illustrate the process. Target and substrate serve as electrodes and face each other in a typical sputtering chamber. An inert gas, typically argon with a pressure usually ranging from a few to 100 mtorr, is introduced into the system as the medium to initiate and maintain a discharge. When an electric field of several kilovolts per centimeter is introduced or a dc voltage is applied to the electrodes, a glow discharge is initiated and maintained between the electrodes. Free electrons will be accelerated by the electric field and gain sufficient energy to ionize argon atoms. The gas density or pressure must not be too low, or else the electrons will simply strike the anode without having gas phase collision with argon atoms. However, if the gas density or pressure is too high, the electrons will not have gained sufficient energy when they strike gas atoms to cause ionization. Resulting positive ions, Ar^+, in the discharge strike the cathode (the source target) resulting in the ejection of neutral target atoms through momentum transfer. These atoms pass through the discharge and deposit on the opposite electrode (the substrate with growing film). In addition to the growth species, i.e. neutral atoms, other negatively charged species under the electric field will also bombard and interact with the surface of the substrate or grown film.

For the deposition of insulating films, an alternate electric field is applied to generate plasma between two electrodes. Typical RF frequencies employed range from 5 to 30 MHz. However, 13.56 MHz has been reserved for plasma processing by the Federal Communications Commission and is widely used. The key element in RF sputtering is that the target self-biases to a negative potential and behaves like a dc target. Such a self-negative target bias is a consequence of the fact that electrons are considerably more mobile than ions and have little difficulty in following the periodic change in the electric field. To prevent simultaneous sputtering on the grown film or substrate, the sputter target must be an insulator and be capacitively coupled to the RF generator. This capacitor will have a low RF impedance and will allow the formation of a dc bias on the electrodes.

It should also be noted that the types of plasmas encountered in thin film processing techniques and systems are typically formed by partially ionizing a gas at a pressure well below atmospheric. For the most part, these plasmas are very weakly ionized, with an ionization fraction of 10^{-5} to 10^{-1}. Although the above discussion is focused on the deposition of films by sputtering, plasma or glow discharges are widely used in other film processes, such as plasma etching.[14] Other examples include plasma enhanced chemical vapor deposition (PECVD), ion plating and reactive ion etching (RIE). The plasma based film processes differ from other film deposition techniques such as evaporation, since the plasma processes is not thermal and not describable by equilibrium thermodynamics.

Sputtering a mixture of elements or compounds will not result in a change of composition in the target and thus the composition of the vapor phase will be the same as that of the target and remain the same during the deposition. Many modifications have been made to enhance or improve the deposition process and resulted in the establishment of hybrid and modified PVD processes. For example, magnetic field has been introduced into sputtering processes to increase the residence time of growth species in the vapor phase; such sputtering is referred to as magnetron sputtering. Reactive gases have also been introduced into the deposition chamber to form compound films, which are known as reactive sputtering.

5.4.4. *Comparison of evaporation and sputtering*

Some major differences between evaporation and sputtering are briefly summarized below:

(1) The deposition pressure differs noticeably. Evaporation uses low pressures typically ranging from 10^{-3} to 10^{-10} torr, whereas sputtering requires a relatively high pressure typically of ~100 torr. Atoms or molecules in evaporation chamber do not collide with each other,

whereas the atoms and molecules in sputtering do collide with each other prior to arrival at the growth surface.
(2) The evaporation is a process describable by thermodynamical equilibrium, whereas sputtering is not.
(3) The growth surface is not activated in evaporation, whereas the growth surface in sputtering is constantly under electron bombardment and thus is highly energetic.
(4) The evaporated films consist of large grains, whereas the sputtered films consist of smaller grains with better adhesion to the substrates.
(5) Fractionation of multi-component systems is a serious challenge in evaporation, whereas the composition of the target and the film can be the same.

5.5. Chemical Vapor Deposition (CVD)

CVD is the process of chemically reacting a volatile compound of a material to be deposited, with other gases, to produce a nonvolatile solid that deposits atomistically on a suitably placed substrate.[1] CVD process has been very extensively studied and very well documented,[15–17] largely due to the close association with solid-state microelectronics.

5.5.1. Typical chemical reactions

Because of the versatile nature of CVD, the chemistry is very rich, and various types of chemical reactions are involved. Gas phase (homogeneous) reactions and surface (heterogeneous) reactions are intricately mixed. Gas phase reactions become progressively important with increasing temperature and partial pressure of the reactants. An extremely high concentration of reactants will make gas phase reactions predominant, leading to homogeneous nucleation. For deposition of good quality films, homogeneous nucleation should be avoided. The wide variety of chemical reactions can be grouped into: pyrolysis, reduction, oxidation, compound formation, disproportionation and reversible transfer, depending on the precursors used and the deposition conditions applied. Examples of the above chemical reactions are given below:

(A) Pyrolysis or thermal decomposition
$$SiH_4(g) \rightarrow Si(s) + 2H_2(g) \text{ at } 650°C \quad (5.15)$$
$$Ni(CO)_4(g) \rightarrow Ni(s) + 4CO(g) \text{ at } 180°C \quad (5.16)$$

(B) Reduction

$$SiCl_4 (g) + 2H_2(g) \rightarrow Si(s) + 4HCl(g) \text{ at } 1200°C \quad (5.17)$$
$$WF_6(g) + 3H2(g) \rightarrow W(s) + 6HF(g) \text{ at } 300°C \quad (5.18)$$

(C) Oxidation

$$SiH_4(g) + O_2(g) \rightarrow SiO_2(s) + 2H_2(g) \text{ at } 450°C \quad (5.19)$$
$$4PH_3(g) + 5O_2(g) \rightarrow 2P_2O_5(s) + 6H_2(g) \text{ at } 450°C \quad (5.20)$$

(D) Compound formation

$$SiCl_4(g) + CH_4(g) \rightarrow SiC(s) + 4HCl(g) \text{ at } 1400°C \quad (5.21)$$
$$TiCl_4(g) + CH_4(g) \rightarrow TiC(s) + 4HCl(g) \text{ at } 1000°C \quad (5.22)$$

(E) Disproportionation

$$2 \; GeI_2(g) \rightarrow Ge(s) + GeI_4(g) \text{ at } 300°C \quad (5.23)$$

(F) Reversible transfer

$$As_4(g) + As_2(g) + 6GaCl(g) + 3H_2(g)$$
$$\rightarrow 6GaAs(s) + 6HCl(g) \text{ at } 750°C \quad (5.24)$$

The versatile chemical nature of CVD process is further demonstrated by the fact that for deposition of a given film, many different reactants or precursors can be used and different chemical reactions may apply. For example, silica film is attainable through any of the following chemical reactions using various reactants[18–21]:

$$SiH_4(g) + O_2(g) \rightarrow SiO_2(s) + 2H_2(g) \quad (5.25)$$
$$SiH_4(g) + 2N_2O(g) \rightarrow SiO_2(s) + 2H_2(g) + 2N_2(g) \quad (5.26)$$
$$SiH_2Cl_2(g) + 2N_2O(g) \rightarrow SiO_2(s) + 2HCl(g) + 2N_2(g) \quad (5.27)$$
$$Si_2Cl_6(g) + 2N_2O(g) \rightarrow SiO_2(s) + 3Cl_2(g) + 2N_2(g) \quad (5.28)$$
$$Si(OC_2H_5)_4(g) \rightarrow SiO_2(s) + 4C_2H_4(g) + 2H_2O(g) \quad (5.29)$$

From the same precursors and reactants, different films can be deposited when the ratio of reactants and the deposition conditions are varied. For example, both silica and silicon nitride films can be deposited from a mixture of Si_2Cl_6 and N_2O and Fig. 5.9 shows the deposition rates of silica and silicon nitride as functions of the ratio of reactants and deposition conditions.[20]

5.5.2. Reaction kinetics

Although CVD is a nonequilibrium process controlled by chemical kinetics and transport phenomena, equilibrium analysis is still useful in understanding the CVD process. The chemical reaction and phase equilibrium determine the feasibility of a particular process and the final state attainable. In

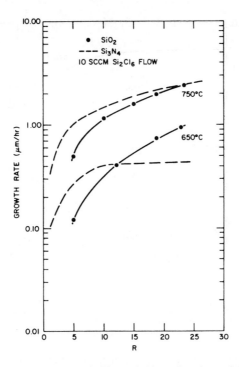

Fig. 5.9. Deposition rates of silica and silicon nitride as functions of the ratio of reactants and deposition conditions. [R.C. Taylor and B.A. Scott, *J. Electrochem. Soc.* **136**, 2382 (1989).]

a given system, multistep complex reactions are often involved. The fundamental reaction pathways and kinetics have been investigated for only a few well-characterized industrially important systems. We will take the reduction of chlorosilane by hydrogen as an example to illustrate the complexity of the reaction pathways and kinetics involved in such a seemingly simple system and deposition process. In this Si–Cl–H system, there exist at least eight gaseous species: $SiCl_4$, $SiCl_3H$, $SiCl_2H_2$, $SiClH_3$, SiH_4, $SiCl_2$, HCl and H_2. These eight gaseous species are in equilibrium under the deposition conditions governed by six equations of chemical equilibrium. Using the available thermodynamic data, composition of gas phase as a function of reactor temperature for a molar ratio of Cl/H = 0.01 and a total pressure of 1 atm, was calculated and presented in Fig. 5.10.[22]

5.5.3. Transport phenomena

Transport phenomena play a critical role in CVD by governing access of film precursors to the substrate and by influencing the degree of desirable

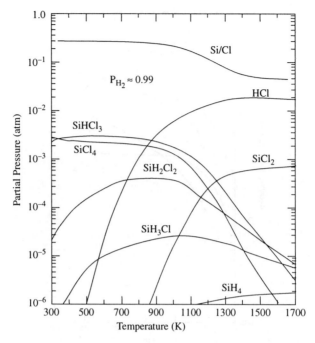

Fig. 5.10. Composition of gas phase as a function of reactor temperature for a molar ratio of Cl/H = 0.01 and a total pressure of 1 atm, calculated using the available thermodynamic data. [E. Sirtl, L.P. Hunt, and D.H. Sawyer, *J. Electrochem. Soc.* **121**, 919 (1974).]

and unwanted gas phase reactions taking place before deposition. The complex reactor geometries and large thermal gradient characteristics of CVD chambers lead to a wide variety of flow structures that affect film thickness, compositional uniformity and impurity levels.[15]

For CVD reactors operating at a low pressure, where the mean free path of gas molecules is 10 times larger than the characteristic length of the reactor, there is no collision between gas molecules and thus the transport of gas is in the free molecular flow regime. For most CVD systems, the characteristic pressure is 0.01 atm and above, and the mean free paths are far larger than the characteristic system dimension. In addition, the gas velocities are low in most CVD reactors, typically of tens of cm/sec, the Reynolds number is typically less than 100 and the flows are laminar. As a result, a stagnant boundary layer of thickness, δ, adjacent to the growth surface is developed during the deposition. In this boundary layer, the composition of growth species decreases from the bulk concentration, P_i, to the surface concentration above the growing film, P_{io}, and the growth species diffuses through the boundary layer prior to depositing onto the growth surface as discussed in Chapter 3 and also illustrated in Fig. 3.6.

Two-Dimensional Nanostructures: Thin Films

When the perfect gas laws are applied since the gas composition in the typical CVD systems is reasonably dilute, the diffusion flux of gas or growth species through the boundary layer is given by:

$$J_i = \frac{D(P_i - P_{io})}{\delta} RT \quad (5.30)$$

Where D is the diffusivity and is dependent on pressure and temperature:

$$D = D_o \left(\frac{P_o}{P}\right)\left(\frac{T}{T_o}\right)^n \quad (5.31)$$

where n is experimentally found to be approximately 1.8. The quantity D_o is the value of D measured at standard temperature T_o (273 K) and pressure P_o (1 atm), and depends on the gas combination in question. Figure 5.11 shows the deposition rate of silicon from four different precursor gases as a function of temperature.[23] This figure also shows that the deposition of silicon films becomes diffusion controlled at high substrate temperatures, whereas surface reaction is a limiting process at relatively low substrate temperatures.

When growth rate is high and the pressure in the reactor chamber is high, diffusion of growth species through the boundary layer can become a rate-limiting process. As Eq. (5.31) indicated, the gas diffusivity varies inversely with pressure, and thus the diffusion flux of gas through the

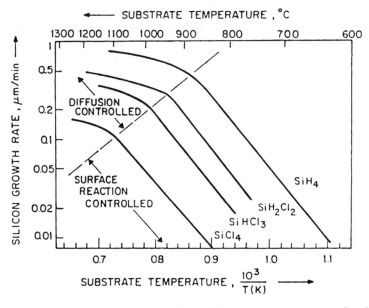

Fig. 5.11. Deposition rate of silicon from four different precursor gases as a function of temperature. [W. Kern, in *Microelectronic Materials and Processes*, ed. R.A. Levy, Kluwer, Boston, MA, p. 203, 1989.]

boundary layer can be enhanced simply by reducing the pressure in the reactor. For deposition of large area films, depletion of growth species or reactants above the growth surface can result in non-uniform deposition of films. To overcome such non-uniformity in deposited films, various reactor designs have been developed to improve the gas-mass transport through the boundary layer. Examples include using low pressure and new designs of reactor chambers and substrate susceptors.

5.5.4. CVD methods

A variety of CVD methods and CVD reactors have been developed, depending on the types of precursors used, the deposition conditions applied and the forms of energy introduced to the system to activate the chemical reactions desired for the deposition of solid films on substrates. For example, when metalorganic compounds are used as precursors, the process is generally referred to as MOCVD (metalorganic CVD), and when plasma is used to promote chemical reactions, this is a plasma enhanced CVD or PECVD. There are many other modified CVD methods, such as LPCVD (low pressure CVD), laser enhanced or assisted CVD, and aerosol-assisted CVD or AACVD.

The CVD reactors are generally divided into hot-wall and cold-wall CVD. Figure 5.12 depicts a few common setups of CVD reactors. Hot-wall

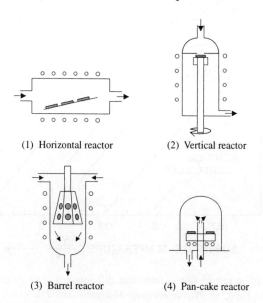

(1) Horizontal reactor (2) Vertical reactor

(3) Barrel reactor (4) Pan-cake reactor

Fig. 5.12. A few common setups of CVD reactors.

CVD reactors are usually tubular in form, and heating is accomplished by surrounding the reactor with resistance elements.[24] In typical cold-wall CVD reactors, substrates are directly heated inductively by graphite susceptors, while chamber walls are air or water-cooled.[25] LPCVD differs from conventional CVD in the low gas pressure of ~0.5 to 1 torr typically used; low pressure is to enhance the mass flux of gaseous reactants and products through the boundary layer between the laminar gas stream and substrates. In PECVD processing, plasma is sustained within chambers where simultaneous CVD reactions occur. Typically, the plasma are excited either by an RF field with frequencies ranging from 100 kHz to 40 MHz at gas pressures between 50 mtorr and 5 torr, or by microwave with a frequency of commonly 2.45 GHz. Often microwave energy is coupled to the natural resonant frequency of the plasma electrons in the presence of a static magnetic field, and such plasma is referred to as electron cyclotron resonance (ECR) plasma.[26] The introduction of plasma results in much enhanced deposition rates, thus permits the growth of films at relatively low substrate temperatures. Figure 5.13 compares the growth rate of polycrystalline silicon films deposited with and without plasma enhancement.[27] MOCVD, also known as organometallic vapor phase epitaxy (OMVPE) differs from other CVD processes by the chemical nature of the precursor gases; metalorganic compounds are employed.[28,29] Laser has also been

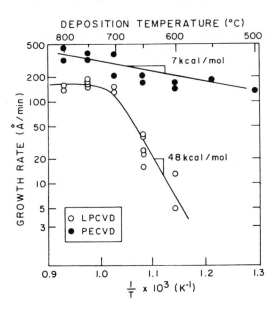

Fig. 5.13. Growth rate of polycrystalline silicon films deposited with and without plasma enhancement. [J.J. Hajjar, R. Reif, and D. Adler, *J. Electron. Mater.* **15**, 279 (1986).]

employed to enhance or assist the chemical reactions or deposition, and two mechanisms are involved: pyrolytic and photolytic processes.[30,31] In the pyrolytic process, the laser heats the substrate to decompose gases above it and enhances rates of chemical reactions, whereas in the photolytic process, laser photons are used to directly dissociate the precursor molecules in the gas phase. Aerosol assisted CVD is developed for the systems where no gaseous precursors are available and the vapor pressures of liquid and solid precursors are too low.[32-34] In this process, liquid precursors are mistified to form liquid droplets that are dispersed in a carrier gas and delivered to the deposition chamber. Inside the deposition chamber, precursor droplets decompose, react and grow films on substrate.

In addition to the growth of thin films on a planar substrate, CVD methods have been modified and developed to deposit solid phase from gaseous precursors on highly porous substrates or inside porous media. Two most noticeable deposition methods are known as electrochemical vapor deposition (EVD) and chemical vapor infiltration (CVI). EVD has been explored for making gas-tight dense solid electrolyte films on porous substrates,[35,36] and the most studied system has been the yttria-stabilized zirconia films on porous alumina substrates for solid oxide fuel cell applications and dense membranes.[35-38] In the EVD process for growing solid oxide electrolyte films, a porous substrate separates metal precursor(s) and oxygen source. Typically chlorides are used as metal precursors, whereas water vapor, oxygen, or air or a mixture of them is used as the source of oxygen. Initially, the two reactants inter-diffuse in the substrate pores and react with each other only when they concur to deposit the corresponding solid oxides. When the deposition conditions are appropriately controlled, the solid deposition can be located at the entrance of pores on the side facing metal precursors, and plug the pores. The location of the solid deposit is mainly dependent on the diffusion rate of the reactants inside the pores as well as the concentrations of the reactants inside the deposition chamber. Under typical deposition conditions, reactant molecules diffusion inside pores is in the Knudsen diffusion region, in which the diffusion rate is inversely proportional to the square root of the molecular weight. Oxygen precursors diffuse much faster than metal precursors, and consequently the deposit occurs normally near the entrance of pores facing the metal precursor chamber. If the deposit solid is an insulator, the deposition by CVD process stops when pores are plugged by the deposit, since no further direct reaction between the two reactants occurs. However, for solid electrolytes, particularly ionic-electronic mixed conductors, the deposition would proceed further by the means of EVD, and the film may grow on the surface exposed to the metal precursor vapor.

In this process, the oxygen or water is reduced at the oxygen/film interface, and the oxygen ions transfer in the film, as the oxygen vacancies diffuse in the opposite direction, and react with the metal precursors at the film/metal precursor interface to continuously form metal oxide.

CVI involves the deposition of solid products onto a porous medium, and the primary focus of CVI is on the filling of voids in porous graphite and fibrous mats to make carbon–carbon composites.[39,40] Various CVI techniques have been developed for infiltrating porous substrates with the main goals to shorten the deposition time and to achieve homogeneous deposition:

(a) Isothermal and isobaric infiltration,
(b) Thermal gradient infiltration,[39]
(c) Pressure gradient infiltration,[39]
(d) Forced flow infiltration,[41]
(e) Pulsed infiltration,[42]
(f) Plasma enhanced infiltration.[41]

Various hydrocarbons have been used as precursors for CVI and typical deposition temperatures range from 850 to 1100°C and the deposition time ranges from 10 to 70 hr, and is rather long as compared to other vapor deposition methods. The long deposition time is due to the relatively low chemical reactivity and gas diffusion into porous media. Furthermore, the gas diffusion will progressively get smaller as more solid is deposit inside the porous substrates. To enhance the gas diffusion, various techniques have been introduced and include forced flow, thermal and pressure gradient. Plasma has been used to enhance the reactivity; however, preferential deposition near surfaces resulted in inhomogeneous filling. The complete filling is difficult and takes very long time, since the gas diffusion becomes very slow in small pores.

5.5.5. Diamond films by CVD

Diamond is a thermodynamically metastable phase at room temperature,[43] so synthetic diamonds are made at high temperatures under high pressures with the aid of transition metal catalysts such as Ni, Fe and Co.[44,45] The growth of diamond films under low pressure (equal to or less than 1 atm) and low temperatures (~800°C) is not a thermodynamically equilibrium process, and differs from other CVD processes. The formation of diamond from gas phase at low pressure was initially reported in late 1960s.[46,47] The typical CVD process of diamond films is illustrated schematically in

Fig. 5.14[48] and can be described as follows. A gaseous mixture of hydrocarbon (typically methane) and hydrogen is fed into an activation zone of the deposition chamber, where activation energy is introduced to the mixture and causes the dissociation of both hydrocarbon and hydrogen molecules to form hydrocarbon free radicals and atomic hydrogen. Many different activation schemes have been found effective in depositing diamond films and include hot-filament, RF and microwave plasma and flames. Upon arrival on the growth surface, a generic set of surface reactions would occur[48]:

$$C_DH + H\cdot \rightarrow C_D\cdot + H_2 \quad (5.32)$$
$$C_D\cdot + \cdot CH3 \rightarrow C_D\text{--}CH_3 \quad (5.33)$$
$$C_D\cdot + C_xH_y \rightarrow C_D\text{--}C_xH_y \quad (5.34)$$

Reaction (5.32) is to activate a surface site by removal of a surface hydrogen atom linked to carbon atom on the diamond surface. An activated surface site readily combines with either a hydrocarbon radical (reaction 5.33) or an

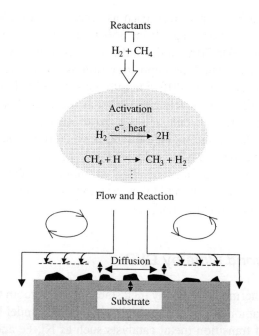

Fig. 5.14. Schematic showing the principal elements in the complex diamond CVD process: flow of reactants into the reactor, activation of the reactants by the thermal and plasma processes, reaction and transport of the species to the growing surface, and surface chemical processes depositing diamond and other forms of carbon. [J.E. Butler and D.G. Goodwin, in *Properties, Growth and Applications of Diamond*, eds. M.H. Nazare and A.J. Neves, INSPEC, London, p. 262, 2001.]

unsaturated hydrocarbon molecule (e.g. C_2H_2, reaction 5.34). A high concentration of atomic hydrogen has proven a key factor in the successful growth of diamond films, and atomic hydrogen is believed to constantly remove the graphite deposits on the diamond growth surface, so as to ensure continued deposition of diamond.[47] Oxygen species have also proven to be important in the deposition of diamond films by atmospheric combustion flames using oxygen and acetylene.[49,50] Other hydrocarbon fuels including ethylene, propylene and methyl acetylene can all be used as precursors for the growth of diamond films.[51-54]

5.6. Atomic Layer Deposition (ALD)

Atomic layer deposition (ALD) is a unique thin film growth method and differs significantly from other thin film deposition methods. The most distinctive feature of ALD has a self-limiting growth nature, each time only one atomic or molecular layer can grow. Therefore, ALD offers the best possibility of controlling the film thickness and surface smoothness in truly nanometer or sub-nanometer range. Excellent reviews on ALD have been published by Ritala and Leskelä.[55,56] In the literature, ALD is also called atomic layer epitaxy (ALE), atomic layer growth (ALG), atomic layer CVD (ALCVD), and molecular layer epitaxy (MLE). In comparison with other thin film deposition techniques, ALD is a relatively new method and was first employed to grow ZnS film.[57] More publications appeared in open literature in early 1980s.[58-60] ALD can be considered as a special modification of the chemical vapor deposition, or a combination of vapor-phase self-assembly and surface reaction. In a typical ALD process, the surface is first activated by chemical reaction. When precursor molecules are introduced into the deposition chamber, they react with the active surface species and form chemical bonds with the substrate. Since the precursor molecules do not react with each other, no more than one molecular layer could be deposited at this stage. Next, the monolayer of precursor molecules that chemically bonded to the substrate is activated again through surface reaction. Either the same or different precursor molecules are subsequently introduced to the deposition chamber and react with the activated monolayer previously deposited. As the steps repeat, more molecular or atomic layers are deposited one layer at a time.

Figure 5.15 schematically illustrates the process of titania film growth by ALD. The substrate is hydroxylated first, prior to the introduction of precursor, titanium tetrachloride. Titanium tetrachloride will

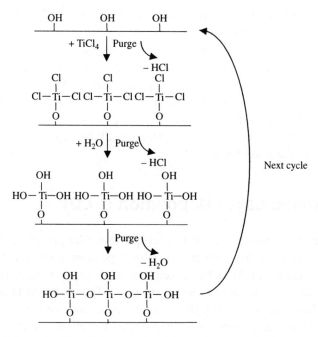

Fig. 5.15. Schematic illustrating the principal reactions and processing steps for the formation of titania film by ALD.

react with the surface hydroxyl groups through a surface condensation reaction:

$$TiCl_4 + HOMe \rightarrow Cl_3Ti-O-Me + HCl \quad (5.35)$$

where Me represents metal or metal oxide substrates. The reaction will stop when all the surface hydroxyl groups reacted with titanium tetrachloride. Then the gaseous by-product, HCl, and excess precursor molecules are purged, and water vapor is subsequently introduced to the system. Titanium trichloride chemically bonded onto the substrate surface undergo hydrolysis reaction:

$$Cl_3Ti-O-Me + H_2O \rightarrow (HO)_3Ti-O-Me + HCl \quad (5.36)$$

Neighboring hydrolyzed Ti precursors subsequently condensate to form Ti–O–Ti linkage:

$$(HO)_3Ti-O-Me + (HO)_3Ti-O-Me$$
$$\rightarrow Me-O-Ti(OH)_2-O-Ti(HO)_2-O-Me + H_2O \quad (5.37)$$

The by-product HCl and excess H_2O will be removed from the reaction chamber. One layer of TiO_2 has been grown by the completion of one cycle of chemical reactions. The surface hydroxyl groups are ready to react with titanium precursor molecules again in the next cycle. By repeating the above steps, second and many more TiO_2 layers can be deposited in a very precisely controlled way.

The growth of ZnS film is another often used classical example for the illustration of the principles of ALD process. $ZnCl_2$ and H_2S are used as precursors. First, $ZnCl_2$ is chemisorbed on the substrate, and then H_2S is introduced to react with $ZnCl_2$ to deposit a monolayer of ZnS on the substrate and HCl is released as a by-product. A wide spectrum of precursor materials and chemical reactions has been studied for the deposition of thin films by ALD. Thin films of various materials including various oxides, nitrides, florides, elements, II–VI, II–VI and III–V compounds, in epitaxial, polycrystalline or amorphous form deposited by ALD are summarized in Table 5.2.[55,56]

The choice of proper precursors is the key issue in a successful design of an ALD process. Table 5.3 summarizes the requirements for ALD precursors.[55,56] A variety of precursors have been used in ALD. For example, elemental zinc and sulfur were used in the first ALD experiments for the growth of ZnS.[57] Metal chlorides were studied soon after the first demonstrations of ALD.[61] Metalloragnic compounds including both organometallic compounds and metal alkoxides are widely used. For non-metals, the simple hydrides have mostly been used: H_2O, H_2O_2, H_2S, H_2Se, H_2Te, NH_3, N_2H_4, PH_3, AsH_3, SbH_3 and HF.

In comparison to other vapor phase deposition methods, ALD offer advantages particularly in the following aspects: (i) precise control of film

Table 5.2. Thin film materials deposited by ALD.[55,56]

II–VI compounds	ZnS, ZnSe, ZnTe, $ZnS_{1-x}Se_x$, CaS, SrS, BaS, $SrS_{1-x}Se_xCdS$, CdTe, MnTe, HgTe, $Hg_{1-x}Cd_xTe$, $Cd_{1-x}Mn_xTe$
II–VI based phosphors	ZnS:M (M = Mn, Tb, Tm), CaS:M (M = Eu, Ce, Tb, Pb), SrS:M (M = Ce, Tb, Pb, Mn, Cu)
III–V compounds	GaAs, AlAs, AlP, InP, GaP, InAs, $Al_xGa_{1-x}As$, $Ga_xIn_{1-x}As$, $Ga_xIn_{1-x}P$
Nitrides	AlN, GaN, InN, SiN_x, TiN, TaN, Ta_3N_5, NbN, MoN, W_2N, Ti–Si–N
Oxides	Al_2O_3, TiO_2, ZrO_2, HfO_2, Ta_2O_5, Nb_2O_5, Y_2O_3, MgO, CeO_2, SiO_2, La_2O_3, $SrTiO_3$, $BaTiO_3$, $Bi_xTi_yO_z$, In_2O_3, In_2O_3:Sn, In_2O_3:F, In_2O_3:Zr, SnO_2, SnO_2:Sb, ZnO, ZnO:Al, Ga_2O_3, NiO, CoO_x, $YBa_2Cu_3O_{7-x}$, $LaCoO_3$, $LaNiO_3$
Fluorides	CaF_2, SrF_2, ZnF_2
Elements	Si, Ge, Cu, Mo, Ta, W
Others	La_2S_3, PbS, In_2S_3, $CuGaS_2$, SiC

Table 5.3. Requirements for ALD precursors.[55]

Requirement	Comments
Volatility	For efficient transportation, a rough limit of 0.1 torr at the applicable maximum source temperature
	Preferably liquids or gases
No self-decomposition	Would destroy the self-limiting film growth mechanism
Aggressive and complete reactions	Ensure fast completion of the surface reactions and thereby short cycle times
	Lead to high film purity
	No problems of gas phase reactions
No etching of the film or substrate material	No competing reaction pathways
	Would prevent the film growth
No dissolution to the film	Would destroy the self-limiting film growth mechanism
Un-reactive byproduct	To avoid corrosion
	Byproduct re-adsorption may decrease the growth rate
Sufficient purity	To meet the requirements specific to each process
Inexpensive	
Easy to synthesize & handle	
Nontoxic and environmentally friendly	

thickness and (ii) conformal coverage. Precise control of film thickness is due to the nature of self-limiting process, and the thickness of a film can be set digitally by counting the number of reaction cycles. Conformal coverage is due to the fact that the film deposition is immune to variations caused by non-uniform distribution of vapor or temperature in the reaction zone. Figure 5.16 shows the X-ray diffraction spectra and the cross-sectional SEM image of 160 nm Ta(Al)N(C) film on patterned silicon wafer.[62] The film is polycrystalline and shows perfect conformality. The deposition temperature was 350°C and precursors used were $TaCl_5$, trimethylaluminum (TMA) and NH_3. However, it should be noted that excellent conformal coverage can only be achieved when the precursor doses and pulse time are sufficient for reaching the saturated state at each step at all surfaces and no extensive precursor decomposition takes place. ALD has demonstrated its capability of depositing multilayer structures or nanolaminates and Fig. 5.17, as an example, shows such a schematic representation of nanolaminates prepared onto glass substrates by ALD.[63]

ALD is an established technique for the production of large area electroluminescent displays,[64] and is a likely future method for the production

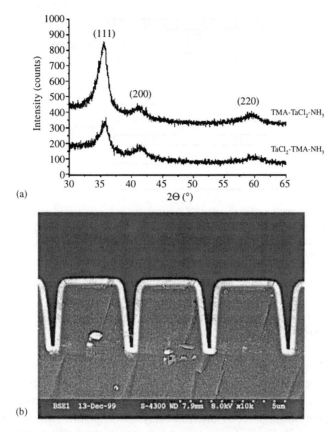

Fig. 5.16. (a) X-ray diffraction spectra and (b) cross-sectional SEM image of 160 nm Ta(Al)N(C) film on patterned silicon wafer. [P. Allén, M. Juppo, M. Ritala, T. Sajavaara, J. Keinonen, and M. Leskelä, *J. Electrochem. Soc.* **148**, G566 (2001).]

of very thin films needed in microelectronics.[65] However, many other potential applications of ALD are discouraged by its low deposition rate, typically <0.2 nm (less than half a monolayer) per cycle. For silica deposition, completing a cycle of reactions typically requires more than 1 min.[66,67] Some recent efforts have been directed towards the development of rapid ALD deposition method. For example, highly conformal layers of amorphous silicon dioxide and aluminum oxide nanolaminates were deposited at rates of 12 nm or <32 monolayers per cycle, and the method has been referred to as "alternating layer deposition".[68] The exact mechanism for such a multilayer deposition in each cycle is unknown, but obviously different from the self-limiting growth discussed above. The precursor employed in this experiment, tris(tert-butoxy)silanol, can react with each other, and thus the growth is not self-limiting.

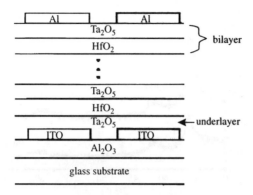

Fig. 5.17. Schematic representation of the nanolaminates prepared onto $5 \times 5\,\text{cm}^2$ glass subtrates by ALD. The Al_2O_3 layer serves as an ion barrier against sodium diffusion from soda lime glass substrate. [K. Kukli, J. Ihanus, M. Ritala, and M. Leskelä, *Appl. Phys. Lett.* **68**, 3737 (1996).]

5.7. Superlattices

Superlattices in this chapter are specifically referred to as thin film structures composed of periodically alternating single crystal film layers; however, it should be noted that the term superlattice was originally used to describe homogeneous ordered alloys. Composite film supperlattices are capable of displaying a broad spectrum of conventional properties as well as a number of interesting quantum effects. When both layers are relatively thick, properties of bulk materials are observed due to the frequently synergistic extensions of the laws of property mixtures that are operative. However, when the layers are very thin, quantum effects emerge, since the wavefunctions of charge carriers in adjacent thin layers penetrate the barriers and couple with one another. Such structures are mostly fabricated by MBE; however, CVD methods are also capable of making superlattices. ALD is another unique technique in the fabrication of superlattice structures. Organic superlattices can be fabricated using LB technique or by self-assembly, which are to be discussed in the following sections. Some of semiconductor superlattice systems are listed in Table 5.4.[69] Semiconductor superlattices can be categorized into compositional superlattices and modulation doping, i.e. selective periodic doping, superlattice. The fabrication of semiconductor superlattices is basically the controlled synthesis of band gap structures, which is also known as band gap engineering.[70-72] Esaki and Tsu were the pioneers in the

Table 5.4. Examples of superlattice systems.[69]

Film materials	Lattice mismatch	Deposition methods
$GaAs-As_xGa_{1-x}As$	0.16% for x = 1	MBE, MOCVD
$In_{1-x}Ga_xAs-GaSb_{1-y}As_y$	0.61%	MBE
GaSb–AlSb	0.66%	MBE
$InP-Ga_xIn_{1-x}As_yP_{1-y}$		MBE
$InP-In_{1-x}Ga_xAs$	0%, x = 0.47	MBE, MOCVD, LPE
$GaP-GaP_{1-x}As_x$	1.86%	MOCVD
$GaAs-GaAs_{1-x}P_x$	1.79%, x = 0.5	MOCVD, CVD
Ge–GaAs	0.08%	MBE
$Si-Si_{1-x}Ge_x$	0.92%, x = 0.22	MBE, CVD
CdTe–HgTe	0.74%	MBE
MnSe–ZnSe	4.7%	MBE
$PbTe-Pb_{1-x}Sn_xTe$	0.44%, x = 0.2	CVD

synthesis of semiconductor thin film superlattices in 1970.[73] Figure 5.18 shows the TEM images of $InGaO_3(ZnO)_5$ superlattice structure.[74]

So far all the methods discussed are vapor phase deposition methods. Thin films can also be made through wet chemical processes. There are many methods developed and examples include electrochemical deposition, sol-gel processing and self-assembly. In comparison with vacuum deposition methods, solution based film deposition methods offer a wide range of advantages including the mild processing conditions so that they are applicable and widely used for the fabrication of thin films of temperature sensitive materials. Mild processing conditions also lead to stress-free films.

5.8. Self-Assembly

Self-assembly is a generic term used to describe a process that ordered arrangement of molecules and small components such as small particles occurred spontaneously under the influence of certain forces such as chemical reactions, electrostatic attraction and capillary forces. In this section, we will focus our discussion on the formation of monolayer or multiple layers of molecules through self-assembly. In general, chemical bonds are formed between the assembled molecules and the substrate surface, as well as between molecules in the adjacent layers. Therefore, the major driving force here is the reduction of overall chemical potential. Further discussion on self-assembly of nanoparticles and nanowires will be presented in Chapter 7. A variety of interactions or forces have been

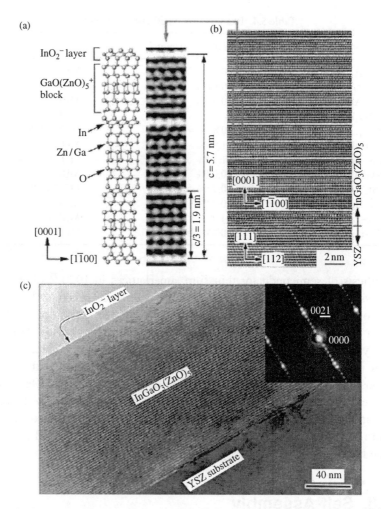

Fig. 5.18. Structure of $InGaO_3(ZnO)_5$. (a) Schematic of the crystal structure. A HRTEM lattice image is shown for comparison. The InO_2^- layer (In_3+ ion locates at an octahedral site coordinated by oxygens) and the $GaO+(ZnO)_5$ block ($Ga3+$ and $Zn2+$ ions share trigonal-bipyramidal and tetrahedral sites) are alternately stacked along the 0001 direction at a period of 1.9 nm (d0003). (b and c) Cross-sectional HRTEM images of a $InGaO_3(ZnO)_5$ thin film grown on YSZ(111) by reactive solid-phase epitaxy. Periodic stacking of the InO_2- layer and the $GaO+(ZnO)_5$ block is clearly visible, which is also confirmed in the electron diffraction image [(C), inset]. Single-crystalline film is formed over the entire observation area. The topmost layer of the film is the InO_2- layer. [K. Nomura, H. Ohta, K. Ueda, T. Kamiya, M. Hirano, and H. Hosono, *Science* **300**, 1269 (2003).]

explored as driving forces for the self-assembly of nanometer subjects as the fundamental building blocks.

Self-assembled monolayers are molecular assemblies that are formed spontaneously by the immersion of an appropriate substrate into a solution of an active surfactant in an organic solvent.[75,76] A typical self-assembling surfactant molecule can be divided into three parts as sketched in Fig. 5.19. The first part is the head group that provides the most exothermic process, i.e. chemisorption on the substrate surface. The very strong molecular-substrate interactions result in an apparent pinning of the head group to a specific site on the surface through a chemical bond, such as covalent Si–O and S–Au bonds, and ionic $-CO_2^-Ag^+$ bond. The second part is the alkyl chain, and the exothermic energies associated with its interchain van der Waals interactions are an order of magnitude smaller than the chemisorption of head groups on substrates. The third molecular part is the terminal functionality; these surface functional groups in SA monolayers are thermally disordered at room temperature. The most important process in self-assembly is the chemisorption, and the associated energy is at the order of tens of kcal/mol (e.g. ~40–45 kcal/mol for thiolate on gold[77,78]). As a result of the exothermic head group-substrate interactions, molecules try to occupy every available binding site on the surface and adsorbed molecules may diffuse along the surface. In general, SA monolayers are considered ordered and closely packed molecular assemblies that have a two-dimensional crystalline-like structure, though there exist a lot of defects.

The driving force for the self-assembly includes: electrostatic force, hydrophobicity and hydrophilicity, capillary force and chemisorption. In the following discussion, we will focus on the formation of SA monolayers

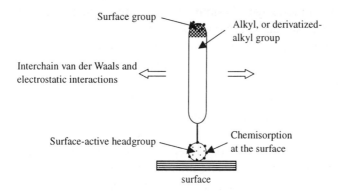

Fig. 5.19. A typical self-assembling surfactant molecule consisting of three parts: surface group, alkyl or derivatized alkyl group, and surface-active headgroup.

that chemisorb on the substrates. There are several types of self-assembly methods for the organic monolayers and these include (i) organosilicon on hydroxylated surfaces, such as SiO_2 on Si, Al_2O_3 on Al, glass, etc,[79-81] (ii) alkanethiols on gold, silver and copper,[82-85] (iii) dialkyl sulfides on gold,[86] (iv) dialkyl disulfides on gold,[87] (v) alcohols and amines on platinum,[86] and (vi) carboxylic acids on aluminum oxide and silver.[88,89] Another way to group the self-assembly methods could be based on the types of chemical bonds formed between the head groups and substrates. There are (i) covalent Si–O bond between organosilicon on hydroxylated substrates that include metals and oxides, (ii) polar covalent S–Me bond between alkanethiols, sulfides and noble metals such as gold, silver, platinum and copper, and (iii) ionic bond between carboxylic acids, amines, alcohols on metal or ionic compound substrates.

One of the important applications of self-assembly that has been extensively studied is the introduction of various desired functionalities and surface chemistry to the inorganic materials. In the synthesis and fabrication of nanomaterials and nanostructures, particularly the core-shell structures, self-assembled organic monolayers are widely used to link different materials together.

5.8.1. Monolayers of organosilicon or alkylsilane derivatives

Typical formulas of alkylsilanes are $RSiX_3$, R_2SiX_2 or R_3SiX, where X is chloride or alkoxy group and R is a carbon chain that can bear different functionalities, such as amine or pyridyl. The chemistry of organosilicon derivatives has been discussed in a great detail by Plueddemann.[90] The formation of monolayers is simply by reacting alkylsilane derivatives with hydroxylated surfaces such as SiO_2, TiO_2.

In a typical procedure, a hydroxylated surface is introduced into a solution (e.g. $\sim 5 \times 10^{-3}$ M) of alkyltrichlorosilane in an organic solvent (e.g. a mixture of 80/20 Isopar-G/CCl_4) for a few minutes (e.g. 2–3 min.). A longer immersion time is required for surfactants with long alkyl chains. A reduction in surfactant concentration in solution takes longer time to form a complete monolayer as illustrated in Fig. 5.20, which presents the results of stearic acid ($C_{17}H_{35}COOH$) monolayers on glass slides.[91] The ability to form a complete monolayer is obviously dependent on the substrate, or the interactions between the monolayer molecules and the substrate surface. After immersion, the substrate is rinsed with methanol, DI water and then dried. Organic solvent is in general required for the

Two-Dimensional Nanostructures: Thin Films 209

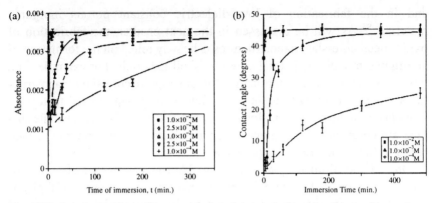

Fig. 5.20. A reduction in surfactant concentration in solution takes longer time to form a complete monolayer, as illustrated by the results of stearic acid ($C_{17}H_{35}COOH$) monolyares on glass slides. [S.H. Chen and C.F. Frank, *Langmuir* 5, 978 (1989).]

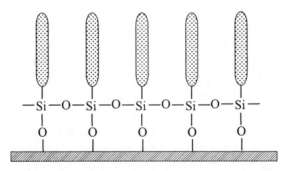

Fig. 5.21. Alkylsilanes with more than one chloride or alkoxy groups, surface polymerization capable of forming silicon–oxygen–silicon bonds between adjacent molecules as commonly invoked deliberately by the addition of moisture.

self-assembly for the alkylsilane derivatives, since silane groups undergo hydrolysis and condensation reaction when in contact with water, resulting in aggregation. In general, monolayers of alkylsilanes may be inherently more disordered than those of alkanethiols, where molecules have more freedom to establish a long-range order. For alkylsilanes with more than one chloride or alkoxy groups, surface polymerization is commonly invoked deliberately by the addition of moisture, so as to form silicon–oxygen–silicon bonds between adjacent molecules as sketched in Fig. 5.21.

Monolayers of organosilicon were studied for applications in enzyme immobilization as early as in the late 1960s,[92] surface silanization for the preparation of hydrophobic surfaces for LB films.[93] This has also been studied for the preparation of inorganic aerogels under ambient pressure[94] and in the fabrication of low dielectric constant porous inorganic materials.[95] As will be discussed in the next chapter, the fabrication of oxide-metal core-shell nanostructures is heavily relied on the formation of an organic monolayer linking core and shell materials. For example, in a typical approach to the fabrication of silica-gold core-shell nanostructures, organosilicon with amine as a functional group is used to form a monolayer on the surface of silica nanoparticles by self-assembly. The surface amine groups then attract gold nanoclusters in the solution, which result in the formation of a gold shell.

One of the ultimate goals of using SA films is the construction of multilayer films that contain functional groups that possess useful physical properties in a layer-by-layer fashion. Examples of those functional groups include electron donor or electron acceptor groups, nonlinear optical chromophores, moieties with unpaired spins. The construction of an SA multilayer requires that the monolayer surface be modified to be a hydroxylated surface, so that another SA monolayer can be formed through surface condensation. Such hydroxylated surfaces can be prepared by a chemical reaction and the conversion of a nonpolar terminal group to a hydroxyl group. Examples include a reduction of a surface ester group, a hydrolysis of a protected surface hydroxy group, and a hydroboration-oxidation of a terminal double bond.[96,97] Oxygen plasma etching followed with immersion in DI-water also effectively makes the surface hydroxylated.[98] A subsequent monolayer is added onto the activated or hydroxylated monolayer through the same self-assembly procedure and multilayers can be built just by repetition of this process. Figure 5.22 shows such a SA multilayer structure. However, it should be noted that in the construction of multilayers, the quality of monolayers formed by self-assembly may rapidly degrade as the thickness of the film increases.[80,99]

5.8.2. Monolayers of alkanethiols and sulfides

Monolayers of alkanethiols on gold surfaces are an extensively studied SA system since 1983.[100] Sulfur compounds can form strong chemical bonds to gold,[101,102] silver,[84] copper,[103] and platinum[86] surfaces. When a fresh, clean, hydrophilic gold substrate is immersed into a dilute solution (e.g. 10^{-3}M) of the organosulfur compound in an organic solvent, a closely packed and oriented monolayer would form. However, immersion times vary from a few minutes to a few hours for alkanethiols, or as long as several days for sulfides and disulfides. For the self-assembly of alkanethiol monolayers, 10^{-3} M is a convenient concentration widely used for most

Two-Dimensional Nanostructures: Thin Films

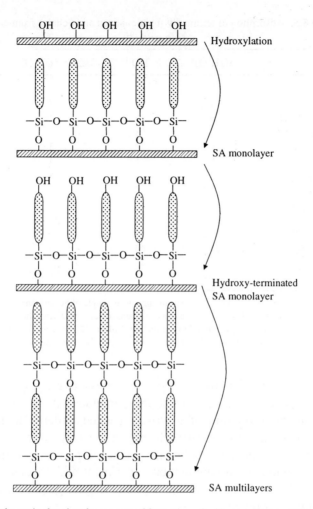

Fig. 5.22. Schematic showing the process of formation of self-assembled multilayer structure.

experimental work, a higher concentration such as 10^{-2} M can be used for simple alkanethiols. Although ethanol has been used in most experiments as the preferred solvent, other solvents may be used. One important consideration in choosing a solvent is the solubility of alkanethiol derivatives. Bain et al.[101] showed that there is no considerable solvent effect on the formation of monolayers of alkanethiols. However, it is recommended to use a solvent that does not show a tendency to incorporate into the two-dimensional system and examples include ethanol, THF, acetonitrile, etc.

Table 5.5 summarizes the effect of various head groups on monolayer formation on a gold surface.[104] The advancing contact angles of water and

Table 5.5. Adsorption of terminally functionalized alkyl chains from ethanol onto gold.[104]

	θ (H$_2$O)[a]	θ (HD)[b]	Thickness (Å) Obsd[c]	Calcd[d]
CH$_3$(CH$_2$)$_{17}$NH$_2$	90	12	6	22–24
CH$_3$(CH$_2$)$_{16}$OH	95	33	9	21–23
CH$_3$(CH2)$_{16}$CO$_2$H	92	38	7	22–24
CH$_3$(CH$_2$)$_{16}$CONH$_2$	74	18	7	22–24
CH$_3$(CH$_2$)$_{16}$CN	69	0	3	22–24
CH$_3$(CH$_2$)$_{21}$Br	84	31	4	28–31
CH$_3$(CH$_2$)$_{14}$CO$_2$Et	82	28	6	h
[CH$_3$(CH$_2$)$_9$CC]$_2$Hg	70	0	4	17–19
[CH$_3$(CH$_2$)$_{15}$]$_3$P[e]	111	44	21	21–23
CH$_3$(CH$_2$)$_{22}$NC	102	28	30	29–33
CH$_3$(CH$_2$)$_{15}$SH[f]	112	47	20	22–24
[CH$_3$(CH$_2$)$_{15}$S]$_2$	110	44	23	22–24
[CH$_3$(CH$_2$)$_{15}$]$_2$S[g]	112	45	20	22–24
CH$_3$(CH$_2$)$_{15}$OCS$_2$Na	108	45	21	24–26

Note: [a]Advancing contact angle of water. [b]Advancing contact angle of hexadecane. [c]Computed from ellipsometric data using $n = 1.45$. [d]Assumed that the chains are closely packed, trans-extended and tilted between 30° and 0° from the normal to the surface. [e]Adsorbed from acetonitrile. [f]Reference 105. [g]Reference 106. [h]An ester group is too large to form a closely packed monolayer.

hexadecane are studied, and the thickness measured using ellipsometer. It is clear that sulfur, phosphorus strongly interact with the gold surface, resulting in the formation of a closely packed, ordered monolayer. It should be noted that the isonitrile forms only poorly packed monolayers as compared with those formed by the thiols and phosphines. In the same study, they conducted a competition experiment and concluded that the thiol group forms the strongest interaction with the gold surface over all the head groups studied.

5.8.3. Monolayers of carboxylic acids, amines and alcohols

Spontaneous adsorption and self-arrangement of long chain alkanoic acids on oxide[88,107] and metal[89] substrates have been another widely studied self-assembly system. The most commonly used head groups include –COOH, –OH and –NH$_2$, which ionize in the solution first and then form ionic bond with substrates. Although the interaction between head groups and substrate plays the most important role in self-assembly and thus determines the quality of the resultant SA monolayers, the alkyl chains also play an important role. In addition to the interchain van der Waals and electrostatic interactions, alkyl chains may provide space for better arrangement of head groups resulting in the formation of closely packed SA monolayers or restrict packing and ordering in the self-assembly, depending on the molecular structures of alkyl chains.[108,110]

SA monolayers have been exploited for applications of surface chemistry modification, introduction of functional groups on the surface, construction of multilayer structures. SA monolayers have also been used to enhance the adhesion at the interfaces.[90] Various functional groups can also be incorporated into or partially substitute alkyl chains in surfactant molecules. SA monolayers have also be used in the synthesis and fabrication of core-shell nanostructures with silane groups linking to oxides and amines linking to metals.[109]

Self-assembly is a wet chemical route to the synthesis of thin films, mostly organic or inorganic–organic hybrid films. This method is often used for the surface modification by formation of a single layer of molecules, which is commonly referred to as self-assembled monolayer (SAM). This method has also been explored to assemble nanostructured materials, such as nanoparticles into an ordered macroscale structures, such as arrays or photonic bandgap crystals. Self-assembly of nanoparticles and nanowires is one of the topics to be discussed in Chapter 7. Arguably, all spontaneous growth processes of formation of materials such as single crystal growth or thin film deposition can be considered as self-assembly process. In those processes, growth species self-assemble at low energy sites. Growth species here for self-assembly are commonly atoms. For more conventional definition of self-assembly, the growth species are commonly molecules. However, nanoparticles or even micron-sized particles are also used as growth species for self-assembly.

5.9. Langmuir–Blodgett Films

Langmuir–Blodgett films (LB films) are monolayers and multilayers of amphiphilic molecules transferred from the liquid–gas interface (commonly water–air interface) onto a solid substrate and the process is generally referred to as Langmuir–Blodgett technique (LB technique).[110] Langmuir carried out the first systematic study on monolayers of amphiphilic molecules at the water–air interface and the first study on a deposition of multilayers of long-chain carboxylic acid onto a solid substrate was carried out.[111]

Before discussing in more detail about the LB films, let us briefly review what is the amphiphile. The amphiphile is a molecule that is insoluble in water, with one end that is hydrophilic, and therefore is preferentially immersed in the water and the other that is hydrophobic and preferentially resides in the air or in the nonpolar solvent. A classical example of an amphiphile is stearic acid, $C_{17}H_{35}CO_2H$. In this molecule, the long hydrocarbon tail, $C_{17}H_{35}$ — is hydrophobic, and the carboxylic acid head group, $-CO_2H$ is hydrophilic. Since the amphiphiles have one end that is hydrophilic and the other that is hydrophobic, they like to locate in interfaces such as between air and water, or between oil and water. This is the reason they are also called surfactants. However, it should be noted that the solubility of an amphiphilic molecule in water depends on the balance between the alkyl chain length and the strength of its hydrophilic head. Certain strength of the hydrophilic head is required to form LB films. If the hydrophilicity is too weak, no LB film can be formed. However, if the strength of the hydrophilic head is too strong, the amphiphilic molecule is too soluble in water to allow the formation of a monolayer. Table 5.6 summarizes the properties of different head groups.[112] The soluble amphiphile molecules may form micelles in water when their concentration exceeds their critical micellar concentration, which will be discussed further in the synthesis of ordered mesoporous materials in the next chapter.

The LB technique is unique, since monolayers can be transferred to many different substrates. Most LB depositions have involved hydrophilic substrates where the monolayers are transferred in the retraction mode.[113] Glass, quartz and other metal substrates with an oxidized surface are used as substrate, but silicon wafer with a surface of silicon dioxide is the most commonly used substrate. Gold is an oxide-free substrate and also commonly used to deposit LB films. However, gold has a high surface energy ($\sim 1000\,mJ/m^2$) and is easily contaminated, which results in an uneven

Table 5.6. The effect of different functional groups on LB film formation of C_{16}-compounds.[112]

Very weak (no film)	Weak (unstable film)	Strong (stable LB film)	Very strong (soluble)
Hydrocarbon	$-CH_2OCH_3$	$-CH_2OH$	$-SO_3^-$
$-CH_2I$	$-C_6H_4OCH_3$	$-COOH$	$-OSO_3^-$
$-CH_2Br$	$-COOCH_3$	$-CN$	$-C_6H_4SO_4^-$
$-CH_2Cl$		$-CONH_2$	$-NR_4^+$
$-NO_2$		$-CH=NOH$	
		$-C_6H_4OH$	
		$-CH_2COCH_3$	
		$-NHCONH_2$	
		$-NHCOCH_3$	

Two-Dimensional Nanostructures: Thin Films   215

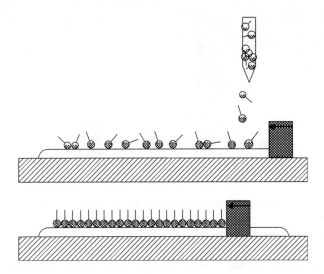

Fig. 5.23. Schematic showing the formation of Langmuir films, which denote the molecular films at the water–air interface, a drop of a dilute solution of an amphiphilic molecule in a volatile solvent, such as $CHCl_3$, is spread on the water–air interface of a trough.

quality of LB films. Cleanliness of the substrate surface is crucial to high quality LB films. In addition, the purity of the organic amphiphiles under study is of great importance, since any contamination in the amphiphile will be incorporated into the monolayer.

Figure 5.23 schematically shows the formation of Langmuir films, which denote the molecular films at the water–air interface, a drop of a dilute solution of an amphiphilic molecule in a volatile solvent, such as $CHCl_3$, is spread on the water–air interface of a trough. As the solvent evaporates, the amphiphilic molecules are dispersed on the interface. The barrier moves and compresses the molecules on the water–air interface; the intermolecular distance decreases and the surface pressure increases. A phase transition may occur, which is assigned to a transition from the "gas" to the "liquid" state. In the liquid state, the monolayer is coherent, except the molecules occupy a larger area than in the condensed phase. When the barrier compresses the film further, a second phase transition can be observed from the "liquid" to the "solid" state. In this condensed phase, the molecules are closely packed and uniformly oriented.

Two methods are commonly used to transfer monolayers from the water–air interface onto a solid substrate. The more conventional method is the vertical deposition as sketched in Fig. 5.24. When a substrate is moved through the monolayer at the water–air interface, the monolayer can be transferred during emersion (retraction or upstroke) or immersion

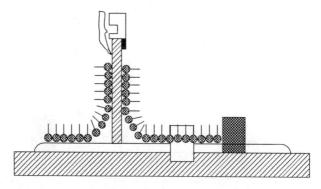

Fig. 5.24. The more conventional vertical deposition method for the formation of LB films on substrates.

(dipping or down stroke). A monolayer usually will be transferred during retraction when the substrate surface is hydrophilic, and the hydrophilic head groups interact with the surface. However, if the substrate surface is hydrophobic, the monolayer will be transferred in the immersion, and the hydrophobic alkyl chains interact with the surface. If the deposition process starts with a hydrophilic substrate, it becomes hydrophobic after the first monolayer transfer, and thus the second monolayer will be transferred in the immersion. Multiple layer films can be synthesized just by repeating the process. Figure 5.25 shows the film thickness proportionally increased with the number of layers.[114]

Another method to build LB multilayer structure is the horizontal lifting, also referred to as Schaefer's method. Schaefer's method is useful for the deposition of very rigid films. In this method as sketched in Fig. 5.26, a compressed monolayer is first formed at the water and air interface, a flat substrate is placed horizontally on the monolayer film. When the substrate is lifted and separated from the water surface, the monolayer is transferred onto the substrate.

Thermal stability and order–disorder transition are two important issues for any practical applications of LB films. Although a lot of research has been done in the past two decades, many issues remain unsolved and our understanding on the structures and stability of LB films is still very limited.

Self-assembly and LB technique offer the possibility of design and the construction of stable organic superlattices. For example, SA can be applied to assemble electron donor and electron acceptor groups, separated by well-defined distances — that can exchange electrons following optical excitation. This may allow the construction, for example, of an

Two-Dimensional Nanostructures: Thin Films

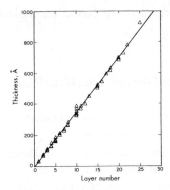

Fig. 5.25. Film thickness proportionally increased with the number of layers. [N. Tillman, A. Ulman, and T.L. Penner, *Langmuir* **5**, 101 (1989).]

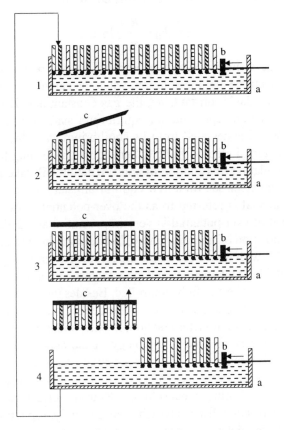

Fig. 5.26. Schaefer's method useful for the deposition of very rigid films, in which, a compressed monolayer is first formed at the water and air interface, a flat substrate is placed horizontally on the monolayer film.

electronic shift register memory based on molecular electron transfer reactions.[115]

5.10. Electrochemical Deposition

Electrochemical deposition or electrodeposition is a very well established thin film growth method. In the previous chapter, we discussed the growth of nanowires using this method, and some fundamentals of the process have been introduced already. In this section, our focus will be on the deposition of films. The key parameters in the electrodeposition of elemental films can be conveniently grouped into thermodynamic and kinetic considerations.

As discussed in the previous chapter, the electrochemical potential of a metal electrode, E, is given by the Nernst equation:

$$E = E_0 + \frac{R_g T}{n_i F} \ln a_i \quad (5.38)$$

where E_0 is the standard electrode potential, or the potential difference between the electrode and the solution, when the activity, a_i, of the ions is unity, F, the Faraday's constant, R_g, the gas constant and T, temperature. The Nernst equation represents an equilibrium state. When the electrochemical potential is deviated from its equilibrium value by, for example, applying an external electric field, either reduction (leading to deposition of solid) or oxidation (dissolution of solid) reaction will take place on the surface or metal electrode till a new equilibrium state is reached. The difference in potential is referred to as the over-potential or over-voltage. A careful control of over-potential is very important to avoid electrolysis of solvent or deposition of impurity phase. In addition, the interactions of the solute ion M^{m+} with the solvent, or with complex-forming ligands should be considered. These interactions and other factors such as the ionic strength of the solution must be carefully controlled. Besides thermodynamics, there are many kinetic factors that influence the deposition of elemental films. The rate of the electron transfer reaction, i.e. the oxidation–reduction kinetics, influences the nature and morphology of the deposit. The nucleation rate of crystals is a function of the over potential,[116] and also influences the nature of the deposit. In the case of a diffusion-limited deposition, the rate of mass transport of solute species to the electrode surface has great effect on the rate of deposition that can be achieved. Electrolyte agitation can lessen the diffusion layer thickness and favor rapid deposition, but maximum stable growth is generally produced in solutions of relatively high solute activity, high diffusion coefficient (low solution viscosity), and low growth velocity.[117] Dissociation kinetics of solvated or complex ions influences the metal ion

Two-Dimensional Nanostructures: Thin Films

activity at the electrode surface and may limit the deposition rate that can be achieved for desired deposition morphology.

Electrodeposition of alloys and compounds is far more complex.[118-120] In alloy and compound electrodeposition, the equilibrium potentials of the alloy or compound components, the activities of the ions in solution, and the stability of the resultant deposit all are important thermodynamic considerations. For a compound M_nN_m, the conditions necessary to obtain the simultaneous deposition of two different kinds of ions at the cathode is:

$$E_M + \eta^m = E_N + \eta^n \quad (5.39)$$

where E_M and E_N are the respective equilibrium potentials of M and N, η^m and η^n are the over potentials required for electrodepositing M and N, respectively. Considering the fact that the activities of the metals M and N in the compound or alloy are determined by their concentrations in the solution and by the thermodynamic stability of the deposit, and often vary during deposition, it is very difficult to control deposit stoichiometry. In addition, control of ionic strength and solute concentration are important for uniform deposition.

For the growth of films by electrodeposition, a few practical concerns deserve a brief discussion here:

(1) Though aqueous solutions are often used, nonaqueous solvent or molten salts are also used. Electrolysis of water is one of the main reasons that nonaqueous solvent or molten salts are used.
(2) The electrical conductivity of the deposit must be high enough to permit the deposition of successive layers. The electrodeposition is therefore applied only for the growth of metal, semiconductors and conductive polymer films.
(3) Deposition can be accomplished at constant current or constant potential, or by other means, such as involving pulsed current or voltage.
(4) Post treatment may be employed to improve the characteristics of the deposits.

5.11. Sol-Gel Films

Sol-gel processing is widely used in the synthesis of inorganic and organic–inorganic hybrid materials and capable of producing nanoparticles, nanorods, thin films and monolith. In the previous chapters, we discussed the fabrication of nanoparticles and nanorods using sol-gel processing. A general introduction to sol-gel processing was presented in Chapter 3. For more detailed information, the readers are recommended to excellent books by Brinker and Scherer,[121] Pierre,[112] and Wright and

Sommerdijk.[123] Sol-gel methods for oxide coatings were reviewed by Francis.[124] Here we will focus our discussion only on the fundamentals and methods of the formation of sol-gel thin films. Prior to sol-gel transition or gelation, sol is a highly diluted suspension of nanoclusters in a solvent, and typically sol-gel films are made by coating sols onto substrates. Although some two-dozen methods are available for applying liquid coatings to substrates, the best choice depends on several factors including solution viscosity, desired coating thickness and coating speed.[125] Most commonly used methods for sol-gel film deposition are spin- and dip-coatings,[126,127] though spray and ultrasonically pulverized spray were also used.[128,129]

In dip-coating, a substrate is immersed in a solution and withdrawn at a constant speed. As the substrate is withdrawn upward, a layer of solution is entrained, and a combination of viscous drag and gravitational forces determines the film thickness, H[130]:

$$H = c_1 \left(\frac{\eta U_0}{\rho g}\right)^{\frac{1}{2}} \qquad (5.40)$$

where η is the viscosity, U_0 the withdrawal speed, ρ the density of the coating sol, and c_1 is a constant. Figure 5.27 illustrates various stages of the dip-coating process.[131] It should be noted that the equation does not account for the evaporation of solvent and continuous condensation between nanoclusters dispersed in the sol as illustrated in Fig. 5.28.[132] However, the relationship between the thickness and the coating variables is the same and supported by

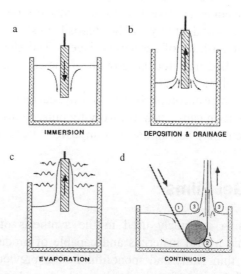

Fig. 5.27. Stages of the dip-coating process: (a–e) batch and (f) continuous. [L.E. Scriven, in *Better Ceramics Through Chemistry III*, eds. C.J. Brinker, D.E. Clark, and D.R. Ulrich, The Materials Research Society, Pittsburgh, PA, p. 717, 1988.]

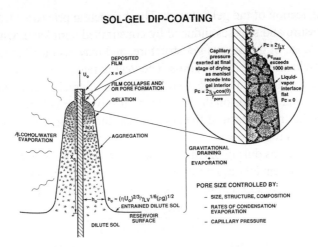

Fig. 5.28. Schematic showing the competing processes of evaporation of solvent and continuous condensation between nanoclusters dispersed in the sol during dip-coating. [C.J. Brinker and A.J. Hurd, *J. Phys. III* (Fr.) **4**, 1231 (1994).]

the experimental results,[133] but the proportionality constant is different. The thickness of a dip-coated film is commonly in the range of 50–500 nm,[134] though a thinner film of ~8 nm per coating was also reported.[135]

Spin-coating is used routinely in microelectronics to deposit photoresists and specialty polymers and has been well studied.[136,137] A typical spin coating consists of four stages: delivery of solution or sol onto the substrate center, spin-up, spin-off and evaporation (overlaps with all stages). After delivering the liquid to the substrate, centrifugal forces drive the liquid across the substrate (spin-up). The excess liquid leaves the substrate during spin off. When flow in the thin coating is no longer possible, evaporation takes over to further reduce the film thickness. A uniform film can be obtained when the viscosity of the liquid is not dependent on shear rate (i.e. Newtonian) and the evaporation rate is independent of position. The thickness of a spin-coated film, H, is given by[138]

$$H = \left(1 - \frac{\rho_A^o}{\rho_A}\right)\left(\frac{3\eta e}{2\rho_A^o \omega^2}\right) \qquad (5.41)$$

where ρ_A is the mass of volatile solvent per unit volume, ρ_A^o its initial volume, ω the angular velocity, η the liquid viscosity, and e the evaporation rate, which is related to the mass transfer coefficient. It is clear from the equation that the film thickness can be controlled by adjusting the solution properties and the deposition conditions.

In the process of creating a sol-gel coating, the removal of solvent or drying of the coating proceeds simultaneously with continuous condensation and solidification of the gel network. The competing processes lead to capillary pressure and stresses induced by constrained shrinkage, which result in the collapse of the porous gel structure, and may also lead to the formation of cracks in the resultant films. The drying rate plays a very important role in the development of stress and formation of cracks particularly in the later stages and depends on the rate at which solvent or volatile components diffuse to the free surface of the coating and the rate at which the vapor is transported away in the gas.

Stress develops during drying of a solidified coating due to constrained shrinkage. Solvent loss after solidification is a common source of stress in solvent-cast polymer coatings and Croll defines such a stress as[139,140]:

$$\sigma = \frac{E(\sigma)\,(\phi_s - \phi_r)}{(1-\nu)3(1-\phi_r)} \quad (5.35)$$

where $E(\sigma)$ is a nonlinear elastic modulus and ν the Poisson's ratio of the coating, ϕ_s and ϕ_r are the volume fractions of solvent at solidification and residual after drying, respectively. The relationship shows that solvent content at solidification should be minimized to lower the stress in the coating. In the formation of sol-gel coating, it is very important to limit the condensation reaction rate during the removal of solvent upon drying, so that the volume fraction of solvent at solidification is kept small. To relieve stresses, the material can relax internally by molecular motion or it can deform. Internal relaxation slows as the material approaches an elastic solid and deformation is restricted by adherence to the substrate. Since the stress-free state shrinks during solidification and adherence to the substrate confines shrinkage in the coating to the thickness direction, in-plane tensile stresses result. Cracking is another form of stress relief. For sol-gel coatings, the formation of cracks limits the coating thickness commonly less than 1 micron. A critical coating thickness, T_c, has been defined.[141,142]

$$T_c = \frac{EG_c}{A\sigma^2} \quad (5.36)$$

where E is the Young's modulus of the film, A is a dimenionless proportionality constant, and G_c the energy required to form two new crack surfaces. The concept of critical thickness is supported by experimental reports. For example, a critical thickness of 600 nm was reported in Ceria sol-gel films, and cracks formed above this thickness.[143]

It should also been noted that sol-gel coatings are commonly porous and amorphous. For many applications, subsequent heat treatment is required to achieve full densification and convert amorphous to crystalline. Mismatch of thermal expansion coefficients of sol-gel coatings and substrates is another important source of stress, and a residual stress in sol-gel coatings can be as high as 350 MPa.[144]

Organic–inorganic hybrids are a new type of materials, which are not present in nature, and sol-gel is the obliged route to synthesize them.[122,145] The organic and inorganic components can interpenetrate each other on a nanometer scale. Depending on the interaction between organic and inorganic components, hybrids are divided into two classes: (i) hybrids consisting of organic molecules, oligomers or low molecular weight polymers embedded in an inorganic matrix to which they are held by weak hydrogen bond or van der Waals force and (ii) in those, the organic and inorganic components are bonded to each other by strong covalent or partially covalent chemical bonds. The organic component can significantly modify the mechanical properties of the inorganic component.[146] The porosity can also be controlled as well as the hydrophilic and hydrophobic balance.[147] Hybrids with new optical[148,149] or electrical[150] properties can be tailored. Some hybrids can display new electrochemical reactions as well as special chemical or biochemical reactivity.[151,152]

Porosity is another important property of sol-gel film. Although for many applications, heat-treatment at elevated temperatures is employed to remove the porosity, the inherited porosity enables sol-gel film for many applications such as matrix of catalyst, host of sensing organic or biocomponents, electrode in solar cells. Porosity itself also renders other unique physical properties such as low dielectric constant, low thermal conductivity, etc. Organic molecules such as surfactants and diblock polymers have been used to form templates in the synthesis of ordered mesoporous materials, which will be another subject of discussion in the next chapter.

There are many other chemical solution deposition (CSD) methods. Fundamentals discussed above are generally applicable to other CSD methods. For example, the competing processes during drying, the development of stresses and the formation of cracks are similar to that in sol-gel films.

5.12. Summary

A variety of methods for thin film deposition has been summarized and briefly discussed. Although all the methods can make films with thickness less than 100 nm, they do offer varied degree of control of thickness and surface smoothness. Both MBE and ALD offer the most precise control of deposition at the single atomic level, and the best quality of the grown film. However, they suffer from the complicated deposition instrumentation and slow growth rate. Self-assembly is another method offering a single atomic level control; however, it is in general limited to the fabrication of organic or inorganic–organic hybrid thin films.

References

1. M. Ohring, *The Materials Science of Thin Films*, Academic Press, San Diego, CA, 1992.
2. J.L. Vossen and W. Kern (eds.), *Thin Film Processes II*, Academic Press, San Diego, CA, 1991.
3. H.S. Nalwa (ed.), *Handbook of Thin Film Materials, Vol. 1: Deposition and Processing of Thin Films*, Academic Press, San Diego, CA, 2002.
4. J. Bloem, *Proc. Seventh Conference on CVD*, eds. T.O. Sedgwick and H. Lydtin, (ECS, PV 79-3), p. 41, 1979.
5. A. Roth, *Vacuum Technology*, North-Holland, Amsterdam, 1976.
6. S. Dushman, *Scientific Foundations of Vacuum Techniques*, Wiley, NY, 1962.
7. R. Glang, in *Handbook of Thin Film Technology*, eds. L.I. Maissel and R. Glang, McGraw-Hill, NY, 1970.
8. M. Faraday, *Phil. Trans.* **147**, 145 (1857).
9. L. Holland, *Vacuum Deposition of Thin Films*, Chapman and Hall, London, 1957.
10. C.V. Deshpandey and R.F. Bunshah, in *Thin Film Processes II*, eds. J.L. Vossen and W. Kern, Academic Press, San Diego, CA, 1991.
11. M.A. Herman and H. Sitter, *Molecular Beam Epitaxy — Fundamentals and Current Status*, Springer-Verlag, Berlin, 1989.
12. E. Kasper and J.C. Bean (eds.), *Silicon-Molecular Beam Epitaxy I and II*, CRC Press, Boca Raton, FL, 1988.
13. E.H.C. Parker (ed.), *The Technology and Physics of Molecular Beam Epitaxy*, Plenum Press, NY, 1985.
14. J.A. Mucha and D.W. Hess, in *Introduction to Microlithography*, eds. L.F. Thompson, C.G. Willson, and M.J. Bowden, The American Chemical Society, Washington, DC, p. 215, 1983.
15. K.F. Jensen and W. Kern, in *Thin Film Processes II*, eds. J.L. Vossen and W. Kern, Academic Press, San Diego, CA, 1991.
16. K.L. Choy, *Prog. Mater. Sci.* **48**, 57 (2003).
17. P. Ser, P. Kalck, and R. Feurer, *Chem. Rev.* **102**, 3085 (2002).
18. N. Goldsmith and W. Kern, *RCA Rev.* **28**, 153 (1967).
19. R. Rosler, *Solid State Technol.* **20**, 63 (1977).
20. R.C. Taylor and B.A. Scott, *J. Electrochem. Soc.* **136**, 2382 (1989).
21. E.L. Jordon, *J. Electrochem. Soc.* **108**, 478 (1961).
22. E. Sirtl, L.P. Hunt, and D.H. Sawyer, *J. Electrochem. Soc.* **121**, 919 (1974).
23. W. Kern, in *Microelectronic Materials and Processes*, ed. R.A. Levy, Kluwer, Boston, MA, 1989.
24. A.C. Adams, in *VLSI Technology*, 2nd edition, S.M. Sze, McGraw-Hill, NY, 1988.
25. S.M. Sze, *Semiconductor Devices: Physics and Technology*, John Wiley and Sons, NY, 1985.
26. S. Matuso, *Handbook of Thin Film Deposition Processes and Techniques*, Noyes, Park Ridge, NJ, 1982.
27. J.J. Hajjar, R. Reif, and D. Adler, *J. Electronic Mater.* **15**, 279 (1986).
28. R.D. Dupuis, *Science* **226**, 623 (1984).
29. G.B. Stringfellow, *Organo Vapor-Phase Epitaxy: Theory and Practice*, Academic Press, NY, 1989.
30. R.M. Osgood and H.H. Gilgen, *Ann. Rev. Mater. Sci.* **15**, 549 (1985).

31. R.L. Abber, in *Handbook of Thin-Film Deposition Processes and Techniques*, ed. K.K. Schuegraf, Noyes, Park Ridge, NJ, 1988.
32. L.D. McMillan, C.A. de Araujo, J.D. Cuchlaro, M.C. Scott, and J.F. Scott, *Integ. Ferroelec.* **2**, 351 (1992).
33. C.F. Xia, T.L. Ward, and P. Atanasova, *J. Mater. Res.* **13**, 173 (1998).
34. P.C. Van Buskirk, J.F. Roeder, and S. Bilodeau, *Integ. Ferroelec.* **10**, 9 (1995).
35. A.O. Isenberg, in *Electrode Materials and Processes for Energy Conversion and Storage*, eds. J.D.E. McIntyre, S. Srinivasan, and F.G. Will, *Electrochem. Soc. Proc.* **77-86**, 572 (1977).
36. M.F. Carolan and J.M. Michaels, *Solid State Ionics* **25**, 207 (1987).
37. Y.S. Lin, L.G.J. de Haart, K.J. de Vries, and A.J. Burggraaf, *J. Electrochem. Soc.* **137**, 3960 (1990).
38. G.Z. Cao, H.W. Brinkman, J. Meijerink, K.J. de Vries, and A.J. Burggraaf, *J. Am. Ceram. Soc.* **76**, 2201 (1993).
39. W.V. Kotlensky, *Chem. Phys. Carbon* **9**, 173 (1973).
40. P. Delhaes, in *Proc. Fourteenth Conf. Chemical Vapor Deposition, Electrochem. Soc. Proc.* **97-25**, 486 (1997).
41. S. Vaidyaraman, W.J. Lackey, G.B. Freeman, P.K. Agrawal, and M.D. Langman, *J. Mater. Res.* **10**, 1469 (1995).
42. P. Dupel, X. Bourrat, and R. Pailler, *Carbon* **33**, 1193 (1995).
43. R. Berman, in *Physical Properties of Diamond*, ed. R. Berman, Clarendon Press, Oxford, p. 371, 1965.
44. H.P. Bovenkerk, F.P. Bundy, H.T. Hall, H.M. Strong, and R.H. Wentorf, *Nature* **184**, 1094 (1959).
45. J. Wilks and E. Wilks, *Properties and Applications of Diamonds*, Butterworth-Heinemann, Oxford, 1991.
46. B.V. Derjaguin and D.V. Fedoseev, *Sci. Am.* **233**, 102 (1975).
47. J.C. Angus, H.A. Will, and W.S. Stanko, *J. Appl. Phys.* **39**, 2915 (1968).
48. J.E. Butler and D.G. Goodwin, in *Properties, Growth and Applications of Diamond*, eds. M.H. Nazare and A.J. Neves, INSPEC, London, p. 262, 2001.
49. L.M. Hanssen, W.A. Carrington, J.E. Butler, and K.A. Snail, *Mater. Lett.* **7**, 289 (1988).
50. D.E. Rosner, *Ann. Rev. Mater. Sci.* **2**, 573 (1972).
51. J.J. Schermer, F.K. de Theije, and W.A.L.M. Elst, *J. Cryst. Growth* **243**, 302 (2002).
52. S.J. Harris, H.S. Shin, and D.G. Goodwin, *Appl. Phys. Lett.* **66**, 891 (1995).
53. K.L. Yarina, D.S. Dandy, E. Jensen, and J.E. Butler, *Diamond Relat. Mater.* **7**, 1491 (1998).
54. D.M. Gruen, *Ann. Rev. Mater. Sci.* **29**, 211 (1999).
55. M. Ritala and M. Leskelä, in *Handbook of Thin Film Materials, Vol. 1: Deposition and Processing of Thin Films*, ed. H.S. Nalwa, Academic Press, San Diego, CA, pp. 103, 2002.
56. M. Ritala and M. Leskelä, *Nanotechnology* **10**, 19 (1999).
57. T. Suntola and J. Antson, US Patent No. 4,058,430, 1977.
58. M. Ahonen and M. Pessa, *Thin Solid Films* **65**, 301 (1980).
59. M. Pessa, R. Mäkelä, and T. Suntola, *Appl. Phys. Lett.* **38**, 131 (1981).
60. T. Suntola and J. Hyvärinen, *Ann. Rev. Mater. Sci.* **15**, 177 (1985).
61. T. Suntola, J. Antson, A. Pakkala, and S. Lindfors, *SID 80 Dig.* **11**, 108 (1980).
62. P. Allén, M. Juppo, M. Ritala, T. Sajavaara, J. Keinonen, and M. Leskelä, *J. Electrochem. Soc.* **148**, G566 (2001).

63. K. Kukli, J. Ihanus, M. Ritala, and M. Leskelä, *Appl. Phys. Lett.* **68**, 3737 (1996).
64. T. Suntola and M. Simpson (eds.), *Atomic Layer Epitaxy*, Blackie, London, 1990.
65. A.I. Kingon, J.P. Maria, and S.K. Streiffer, *Nature* **406**, 1032 (2000).
66. J.D. Ferguson, A.W. Weimer, and S.M. George, *Appl. Surf. Sci.* **162–163**, 280 (2000).
67. S. Morishita, W. Gasser, K. Usami, and M. Matsumura, *J. Non-Cryst. Solids* **187**, 66 (1995).
68. D. Hausmann, J. Becker, S. Wang, and R.G. Gordon, *Science* **298**, 402 (2002).
69. L. Esaki, in *Symp. Recent Topics in Semiconductor Physics*, eds. H. Kamimura and Y. Toyozawa, World Scientific, Singapore, 1982.
70. F. Capasso, *Science* **235**, 172 (1987).
71. F. Capasso and S. Datta, *Phys. Today* **43**(2), 74 (1990).
72. H. Ichinose, Y. Ishida, and H. Sakaki, *JOEL News* **26E**(1), 8 (1988).
73. L. Esaki and R. Tsu, *IBM J. Res. Dev.* **14**, 61 (1970).
74. K. Nomura, H. Ohta, K. Ueda, T. Kamiya, M. Hirano, and H. Hosono, *Science* **300**, 1269 (2003).
75. W.C. Bigelow, D.L. Pickett, and W.A. Zisman, *J. Colloid Interface Sci.* **1**, 513 (1946).
76. W.A. Zisman, *Adv. Chem. Ser.* **43**, 1 (1964).
77. L.H. Dubois, B.R. Zegarski, and R.G. Nuzzo, *Proc. Natl. Acad. Sci.* **84**, 4739 (1987).
78. L.H. Dubois, B.R. Zegarski, and R.G. Nuzzo, *J. Am. Chem. Soc.* **112**, 570 (1990).
79. R. Maoz and J. Sagiv, *Langmuir* **3**, 1045 (1987).
80. L. Netzer and J. Sagiv, *J. Am. Chem. Soc.* **105**, 674 (1983).
81. N. Tillman, A. Ulman, and T.L. Penner, *Langmuir* **5**, 101 (1989).
82. I. Rubinstein, S. Teinberg, Y. Tor, A. Shanzer, and J. Sagiv, *Nature* **332**, 426 (1988).
83. G.M. Whitesides and P.E. Laibinis, *Langmuir* **6**, 87 (1990).
84. A. Ulman, *J. Mater. Ed.* **11**, 205 (1989).
85. L.C.F. Blackman, M.J.S. Dewar, and H. Hampson, *J. Appl. Chem.* **7**, 160 (1957).
86. E.B. Troughton, C.D. Bain, G.M. Whitesides, R.G. Nuzzo, D.L. Allara, and M.D. Porter, *Langmuir* **4**, 365 (1988).
87. R.G. Nuzzo, F.A. Fusco, and D.L. Allara, *J. Am. Chem. Soc.* **109**, 2358 (1987).
88. H. Ogawa, T. Chihera, and K. Taya, *J. Am. Chem. Soc.* **107**, 1365 (1985).
89. N.E. Schlotter, M.D. Porter, T.B. Bright, and D.L. Allara, *Chem. Phys. Lett.* **132**, 93 (1986).
90. E.P. Plueddemann, *Silane Coupling Agents*, Plenum Press, NY, 1982.
91. S.H. Chen and C.F. Frank, *Langmuir* **5**, 978 (1989).
92. H.H. Weetall, *Science* **160**, 615 (1969).
93. R.R. Highfield, R.K. Thomas, P.G. Cummins, D.P. Gregory, J. Mingis, J.B. Hayter, and O. Schärpf, *Thin Solid Films* **99**, 165 (1983).
94. S.S. Prakash, C.J. Brinker, A.J. Hurd, and S.M. Rao, *Nature* **374**, 439 (1995).
95. S. Seraji, Y. Wu, M.J. Forbess, S.J. Limmer, T.P. Chou, and G.Z. Cao, *Adv. Mater.* **12**, 1695 (2000).
96. R. Maoz and J, Sagiv, *Langmuir* **3**, 1045 (1987).
97. N. Tillman, A. Ulman, and T.L. Penner, *Langmuir* **5**, 101 (1989).
98. T.P. Chou and G.Z. Cao, *J. Sol-Gel Sci. Technol.* **27**, 31–41 (2003).
99. L. Netzer, R Iscovoci, and J. Sagiv, *Thin Solid Films* **99**, 235 (1983).
100. R.G. Nuzzo and D.L. Allara, *J. Am. Chem. Soc.* **105**, 4481 (1983).
101. C.D. Bain, E.B. Troughton, Y.T. Tao, J. Evall, G.M. Whitesides, and R.G. Nuzzo, *J. Am. Chem. Soc.* **111**, 321 (1989).
102. M.D. Porter, T.B. Bright, D.L. Allara, and C.E.D. Chidsey, *J. Am. Chem. Soc.* **109**, 3559 (1987).

103. K.R. Stewart, G.M. Whitesides, H.P. Godfried, and I.F. Silvera, *Surf. Sci.* **57**, 1381 (1986).
104. C.D. Bain, J. Evall, and G.M. Whitesides, *J. Am. Chem. Soc.* **111**, 7155 (1989).
105. N. Tillman, A. Ulman and T.L. Penner, *Langmuir* **5**, 101 (1989).
106. L. Netzer, R. Iscovici, and J. Sagiv, *Thin Solid Films* **100**, 67 (1983).
107. D.L. Allara and R.G. Nuzzo, *Langmuir* **1**, 45 (1985).
108. D.Y. Huang and Y.T. Tao, *Bull. Inst. Chem. Acad. Sinica* **33**, 73 (1986).
109. L.M. Liz-Marzán, M. Giersig, and P. Mulvaney, *J. Chem. Soc. Chem. Commun.* **731** (1996).
110. A. Ulman, *An Introduction to Ultrathin Organic Films: From Langmuir–Blodgett to Self-Assembly*, Academic Press, San Diego, CA, 1991.
111. G.L. Gaines, *Insoluble Monolayers Liquid–Gas Interfaces*, Interscience, NY, 1966.
112. N.K. Adam, *The Physics and Chemistry of Surfaces*, 3rd edition, Oxford University Press, London, 1941.
113. I.R. Peterson, G. Veale, and C.M. Montgomery, *J. Colloid Interface Sci.* **109**, 527 (1986).
114. N. Tillman, A. Ulman, and T.L. Penner, *Langmuir* **5**, 101 (1989).
115. J.J. Hopfield, J.N. Onuchic, and D.N. Beratan, *Science* **241**, 817 (1988).
116. E.B. Budevski, in *Comprehensive Treatise of Electrochemistry*, Vol. 7, eds. B.E. Conway, J. O'M. Bockris, and E. Yeagers, S.U.M. Khan and R.E. White, Plenum, NY, p. 399, 1983.
117. D. Elwell, *J. Cryst. Growth* **52**, 741 (1981).
118. G.F. Fulop and R.M. Taylor, *Ann. Rev. Mater. Sci.* **15**, 197 (1985).
119. K.M. Gorbunova and Y.M. Polukarov, in *Advances in Electrochemistry and Electrochemical Engineering*, Vol. 5, eds. C.W. Tobias and P. Delahay, Wiley, NY, p. 249, 1967.
120. A.R. Despic, in *Comprehensive Treatise of Electrochemistry*, Vol. 7, eds. B.E. Conway, J. O'M. Bockris, and E. Yeagers, S.U.M. Khan and R.E. White, Plenum, NY, p. 451, 1983.
121. C.J. Brinker and G.W. Scherer, *Sol-Gel Science: The Physics and Chemistry of Sol-Gel Processing*, Academic Press, San Diego, CA, 1990.
122. A.C. Pierre, *Introduction to Sol-Gel Processing*, Kluwer, Norwell, MA, 1998.
123. J.D. Wright and N.A.J.M. Sommerdijk, *Sol-Gel Materials*, Gordon and Breach Science Publishers, Amsterdam, 2001.
124. L.F. Francis, *Mater. Manufac. Proc.* **12**, 963 (1997).
125. E.D. Cohen, in *Modern Coating and Drying Technology*, eds. E.D. Cohen and E.B. Gutoff, VCH, NY, p.1, 1992.
126. C.J. Brinker, A.J. Hurd, and K.J. Ward, in *Ultrastructure Processing of Advanced Ceramics*, eds. L.L. Hench and D.R. Ulrich, John Wiley & Sons, NY, p. 223 (1988).
127. C.J. Brinker, A.J. Hurd, P.R. Schunk, G.C. Frye, and C.S. Ashley, *J. Non-Cryst. Solids* **147–148**, 424 (1992).
128. P. Hinz and H. Dislicj, *J. Non-Cryst. Solids* **82**, 411 (1986).
129. P. Marage, M. Langlet, and J.C. Joubert, *Thin Solid Films* **238**, 218 (1994).
130. R.P. Spiers, C.V. Subbaraman, and W.L. Wilkinson, *Chem. Eng. Sci.* **29**, 389 (1974).
131. L.E. Scriven, in *Better Ceramics Through Chemistry III*, eds. C.J. Brinker, D.E. Clark, and D.R. Ulrich, The Materials Research Society, Pittsburgh, PA, p. 717, 1988.
132. C.J. Brinker and A.J. Hurd, *J. Phys. III* (Fr.) **4**, 1231 (1994).
133. M. Guglielmi, P. Colombo, F. Peron, and L.M. Degliespoti, *J. Mater. Sci.* **27**, 5052 (1992).

134. S. Sakka, K. Kamiya, K. Makita, and Y. Yamamoto, *J. Non-Cryst. Solids* **63**, 223 (1984).
135. J.G. Cheng, X.J. Meng, J. Tang, S.L. Guo, J.H. Chu, M. Wang, H. Wang, and Z. Wang, *J. Am. Ceram. Soc.* **83**, 2616 (2000).
136. A.G. Emslie, F.T. Bonner, and G. Peck, *J. Appl. Phys.* **29**, 858 (1958).
137. D. Meyerhofer, *J. Appl. Phys.* **49**, 3993 (1978).
138. D.E. Bomside, C.W. Macosko, and L.E. Scriven, *J. Imag. Technol.* **13**, 122 (1987).
139. S.G. Croll, *J. Coatings Technol.* **51**, 64 (1979).
140. S.G. Croll, *J. Appl. Polymer Sci.* **23**, 847 (1979).
141. F.F. Lange, in *Chemical Processing of Advanced Materials*, eds. L.L. Hench and J.K. West, John Wiley & Sons, New York, p. 611, 1992.
142. M.S. Hu, M.D. Thouless, and A.G. Evans, *Acta Metallurgica* **36**, 1301 (1988).
143. A. Atkinson and R.M. Guppy, *J. Mater. Sci.* **26**, 3869 (1991).
144. T.J. Garino and M. Harrington, *Mater. Res. Soc. Symp. Proc.* **243**, 341 (1992).
145. C. Sanchez and F. Ribot, *New J. Chem.* **18**, 1007 (1994).
146. A. Morikawa, Y. Iyoku, M. Kakimoto, and Y. Imai, *J. Mater. Chem.* **2**, 679 (1992).
147. K. Izumi, H. Tanaka, M. Murakami, T. Degushi, A. Morita, N. Toghe, and T. Minami, *J. Non-Cryst. Solids* **121**, 344 (1990).
148. D. Avnir, D. Levy, and R. Reisfeld, *J. Phys. Chem.* **88**, 5956 (1984).
149. B. Dunn and J.I. Zink, *J. Mater. Chem.* **1**, 903 (1991).
150. S.J. Kramer, M.W. Colby, J.D. Mackenzie, B.R. Mattes, and R.B. Kaner, in *Chemical Processing of Advanced Materials*, eds. L.L. Hench and J.K. West, Wiley, New York, p. 737, 1992.
151. L.M. Ellerby, C.R. Nishida, F. Nishida, S.A. Yamanaka, B. Dunn, J.S. Valentine, and J.I. Zink, *Science* **255**, 1113 (1992).
152. P. Audebert, C. Demaille, and C. Sanchez, *Chem. Mater.* **5**, 911 (1993).

Chapter 6

Special Nanomaterials

6.1. Introduction

In the previous chapters, we have introduced the fundamentals and general methods for the synthesis and fabrication of various nanostructures and nanomaterials including nanoparticles, nanowires and thin films. However, there are a number of important nanomaterials not included in these discussions, since their syntheses are unique and difficult to group into previous chapters. Examples of such nanomaterials are carbon fullerenes and nanotubes, ordered mesoporous materials, organic-inorganic hybrids, intercalation compounds, and oxide-metal core-shell structures. In addition, bulk materials with nanosized building blocks, such as nanograined ceramics and nanocomposites, have not been discussed so far. In this chapter, we will discuss the synthesis of these special nanomaterials. Most of these nanomaterials are unique, do not exist in nature, and are truly "man-made" relatively recently, therefore, a brief introduction to the materials, such as their peculiar structures and properties, has also been included in the discussion. All the discussion has been kept very general, but detailed references are given so that the readers can easily find literature to gain more insight to those subjects when needed.

6.2. Carbon Fullerenes and Nanotubes

Carbon is a unique material, and can be a good metallic conductor in the form of graphite, a wide band gap semiconductor in the form of diamond, or a polymer when reacted with hydrogen. Carbon provides examples of materials showing the entire range of intrinsic nanometer scaled structures from fullerenes, which are zero-dimensional nanoparticles, to carbon nanotubes, one-dimensional nanowires to graphite, a two-dimensional layered anisotropic material, to fullerene solids, a three-dimensional bulk materials with the fullerene molecules as the fundamental building block of the crystalline phase. In this section, we will briefly discuss the synthesis and some properties of fullerenes, fullerene crystals and carbon nanotubes. For more general research information about carbon science or detailed information on carbon fullerenes and carbon nanotubes, the readers are referred to excellent review articles and books, such as that by Dresselhaus[1,2] and references therein.

6.2.1. Carbon fullerenes

Carbon fullerene commonly refers to a molecule with 60 carbon atoms, C_{60}, and with an icosahedral symmetry,[3] but also includes larger molecular weight fullerenes $C_n(n > 60)$. Examples of larger molecular weight fullerenes are C_{70}, C_{76}, C_{78}, C_{80}, and higher mass fullerenes, which possess different geometric structure,[4–6] e.g. C_{70} has a rugby ball-shaped symmetry. Figure 6.1 shows the structure and geometry of some fullerene molecules.[7] The name of fullerene was given to this family of carbon molecules because of the resemblance of these molecules to the geodesic dome designed and built by R. Buckminster Fuller,[8] whereas the name of buckminster fullerene or buckyball was specifically given to the C_{60} molecules, which are the most widely studied in the fullerene family and deserve a little more discussion on its structure and properties.

The 60 carbon atoms in C_{60} are located at the vertices of a regular truncated icosahedron and every carbon site on C_{60} is equivalent to every other site. The average nearest neighbor C–C distant in C_{60} (1.44 Å)[9] is almost identical to that in graphite (1.42 Å). Each carbon atom in C_{60} is trigonally bonded to other carbon atoms, the same as that in graphite, and most of the faces on the regular truncated icosahedron are hexagons. There are 20 hexagonal faces and 12 additional pentagonal faces in each C_{60} molecule, which has a molecule diameter of 7.10 Å.[3,10] Although each carbon atom in C_{60} is equivalent to every other carbon atom, the three bonds emanating

Special Nanomaterials

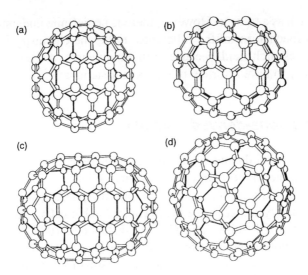

Fig. 6.1. (a) The icosahedral C_{60} molecule. (b) The C_{70} molecule as a rugby ball-shaped molecule. (c) The C_{80} molecule as an extended rugby ball-shaped molecule. (d) The C_{80} molecule as an icosahedron. [M.S. Dresselhaus and G. Gresselhaus, *Ann. Rev. Mater. Sci.* **25**, 487 (1995).]

from each atom are not equivalent. Each carbon atom has four valence electrons for the formation of three chemical bonds, and there will be two single bonds and one double bond. The hexagonal faces consist of alternating single and double bonds, whereas the pentagonal faces are defined by single bonds. In addition, the length of single bonds is 1.46 Å, longer than the average bond length, 1.44 Å, while the double bonds are shorter, 1.40 Å.[11,12] The structures of other fullerene molecules can be considered as a modification of C_{60} by varying the number of hexagonal faces as far as the Euler's theorem is not violated, which states that a closed surface consisting of hexagons and pentagons has exactly 12 pentagons and an arbitrary number of hexagons.[13] For example, C_{70} structure can be envisioned by adding a belt of five hexagons around the equatorial plane of the C_{60} molecule normal to one of the five-fold axis.

Fullerenes are usually synthesized by using an arc discharge between graphite electrodes in approximately 200 torr of He gas, first demonstrated in 1990 by Krätschmer and coworkers.[14] The heat generated at the contact point between the electrodes evaporates carbon to form soot and fullerenes, which condense on the water-cooled walls of the reactor. This discharge produces a carbon soot that can contain up to ~15% fullerenes: C_{60} (~13%) and C_{70} (~2%). The fullerenes are next separated from the soot, according to their mass, by use of liquid chromatography and using a solvent such as a toluene. However, there is no definite understanding of the growth mechanism of the fullerenes. Fullerene chemistry has been a very active research field, because of the uniqueness of the C_{60} molecule and its ability to have a variety of chemical reactions.[15,16]

6.2.2. Fullerene-derived crystals

In a solid state, fullerene molecules crystallize into crystal structures through weak intermolecular forces and each fullerene molecule serves as a fundamental building block of the crystalline phase. For example, the C_{60} molecules crystallize into a face-centered cubic (FCC) structure with lattice constant of 14.17 Å and a C_{60}–C_{60} distance of 10.02 Å.[17] The molecules are almost freely rotating with three degree of rotation at room temperature, as shown by nuclear magnetic resonance methods. The crystalline forms of fullerenes are often called fullerites.[18] Single crystals can be grown either from solution using solvents such as CS_2 and toluene or by vacuum sublimation, though the sublimation yields better crystals and is generally the method of choice.[19]

6.2.3. Carbon nanotubes

There are excellent reviews and books on the synthesis and the physical properties of carbon nanotubes,[20–23] and therefore, in this section, only a brief summary of the fundamentals and general approaches for the synthesis of carbon nanotubes is presented. There are single-wall carbon nanotube or SWCNT, and multi-wall carbon nanotube or MWCNT. The fundamental carbon nanotube is a single-wall structure and can be understood by referring to Fig. 6.2.[1] In this figure we see that points O and A are crystallographically equivalent on the graphene sheet, where X-axis is placed parallel to one side of the honeycomb lattice. The points O and A can be connected by a vector $C_h = na_1 + ma_2$, where a_1 and a_2 are unit vectors for the honeycomb lattice of the graphene sheet. Next we can draw normals to C_h at points O and A to obtain lines OB and AB'. If we now superimpose OB onto AB', we obtain a cylinder of carbon atoms that constitutes a carbon nanotube when properly capped at both ends with half of a fullerene. Such a single-wall carbon nanotube is uniquely determined by the integers (n,m). However, from an experimental standpoint, it is more convenient to denote each carbon nanotube by its diameter $d_t = C_h/\pi$ and the chiral angle θ. Depending on the chiral angle, a single-wall carbon

Special Nanomaterials

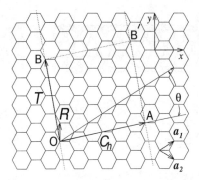

Fig. 6.2. The chiral vector OA or $C_h = na_1 + ma_2$ is defined on the honeycomb lattice of carbon atoms by unit vectors a1 and a2 and the chiral angle θ with respect to the zigzag axis. Along the zigzag axis, $\theta = 0°$. Also shown is the lattice vector OB = T of the one-dimensional nanotube unit cell. The rotation angle ψ and the translation τ (not shown) constitute the basic symmetry operation $R = (\psi|\tau)$ for the carbon nanotube. The diagram is constructed for $(n,m) = (4,2)$. The area defined by the rectangle (OAB'B) is the area of the one-dimensional unit cell of the nanotube. [M.S. Dresselhaus, *Ann. Rev. Mater. Sci.* **27**, 1 (1997).]

nanotube can have three basic geometries—armchair with $\theta = 30°$, zigzag with $\theta = 0°$, and chiral with $0 < \theta < 30°$, as shown in Fig. 6.3.[24]

Multi-wall carbon nanotube consists of several nested coaxial single-wall tubules. The arrangement of the carbon atoms in the hexagonal network of the multi-wall carbon nanotube is often helicoidal, resulting in the formation of chiral tubes.[31] However, there appears to be no particular ordering between individual cylindrical planes forming the multi-wall carbon nanotube such as can be found in graphite where the planes are stacked relative to each other in an ABAB configuration. In other words, a given multi-wall carbon nanotube will typically be composed of a mixture of cylindrical tubes having different helicity or no hecility, thereby resembling turbostratic graphite. Typical dimensions of multi-wall carbon nanotube are outer diameter: 2–20 nm, inner diameter: 1–3 nm, and length: 1–100 μm. The intertubular distance is 0.340 nm, which is slightly larger than the interplanar distance in graphite.

Carbon nanotubes can be prepared by arc evaporation,[25] laser ablation,[26] pyrolysis,[27] PECVD,[28] and electrochemical methods.[29,30] Carbon nanotubes were first synthesized by Iijima in 1991 in the carbon cathode by arc discharge.[31] However, the experimental discovery of single-wall carbon nanotubes came in 1993,[32,33] whereas the discovery in 1996 of a much more efficient synthesis route, involving laser vaporization of graphite to prepare arrays of ordered single-wall nanotubes,[34] offered major new opportunities for quantitative experimental studies of carbon nanotubes.

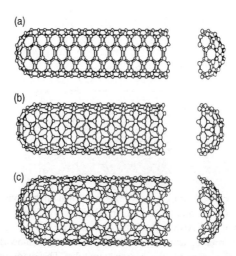

Fig. 6.3. Schematic models for single-wall carbon nanotubes with the nanotube axis normal to (a) the $\theta = 30°$ direction (an armchair (n,n) nanotube); (b) the $\theta = 0°$ direction (a zigzag $(n,0)$ nanotube); and (c) a general direction OB (see Fig. 6.2) with $0° < \theta < 30°$ [a chiral (n,m) nanotube]. The actual nanotubes shown in the figure correspond to (n,m) values of (a) (5,5), (b) (9,0), and (c) (10,5). [M.S. Dresselhaus, G. Dresselhaus, and R. Saito, *Carbon* **33**, 883 (1995).]

The formation of the carbon nanotubes in most cases requires an "open end" where the carbon atoms arriving from the gas phase could coherently land and incorporate into the structure. Growth of nested multi-wall nanotubes can be stabilized by the strained "lip–lip" bonding between the coaxial edges, highly fluctuating and, therefore, accessible for new atoms. In general the open end can be maintained either by a high electric field, by the entropy opposing the orderly cap termination, or by the presence of a metal cluster.

The presence of an electric field in the arc-discharge is believed to promote the growth of carbon nanotubes.[35,37] Nanotubes form only where the current flows, on the larger negative electrodes. Typical rate of the cathode deposition is about a millimeter per minute, with the current and voltage in the range of 100 A and 20 V respectively, which maintains a high temperature of 2000–3000°C. For example, Ebbesen and Ajayan[36] used carbon arc evaporation to produce carbon nanotubes in high yields. In their experiment, an arc plasma is generated between the two carbon electrodes in an inert atmosphere, e.g. He, by applying a DC current density of ~ 150 A/cm^2 with a voltage of ~ 20 V. The extremely high growth temperatures required in the arc discharge experiments can cause the grown carbon nanotubes to sinter, and the sintering of carbon nanotubes is believed to be the predominant source of defects.[37]

An addition of a small amount of transition metal powder such as Co, Ni or Fe, favors a growth of single-wall nanotubes.[32,33] Thess et al.[34] grew uniform diameter (10,10) nanotubes with high yield, by condensation of laser-vaporized carbon catalyst mixture at a lower temperature of ~1200°C. It is believed that the alloy cluster anneals all unfavorable structures into hexagons, which in turn welcome the newcomers and promote the continuous growth of straight nanotubes. Figure 6.4 illustrates the energetics of growth, the relative binding energies in nanotubes, graphite and the feedstock components.[18]

Growth of aligned carbon nanotubes was first demonstrated by CVD directly on Fe nanoparticles embedded in mesoporous silica.[38] The diameter, growth rate and density of vertically aligned carbon nanotubes are found to be dependent on the size of the catalyst.[39] Plasma induced well-aligned carbon nanotubes can be grown on contoured surfaces and with a growth direction always perpendicular to the local substrate surface as shown in Fig. 6.5.[40] The alignment is primarily induced by the electrical self-bias field imposed on the substrate surface from the plasma environment. It was found that switching the plasma off effectively turns the alignment mechanism off, leading to a smooth transition between the plasma-grown straight nanotubes and the thermally grown "curly" nanotubes as shown in Fig. 6.6.[40] DC-bias has found to enhance the nucleation and growth of aligned carbon nanotubes.[41]

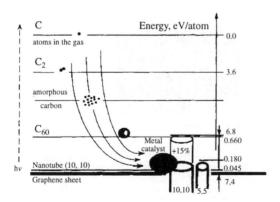

Fig. 6.4. "Food chain" illustrates how the metal (Ni/Co) cluster is able to eat essentially any carbon material it encounters and feed the digested carbon bits to the growing end of the nanotube. The vertical axis shows the cohesive energy per atom for the different forms of carbon consumed in nanotube growth. The energy cost for curving the graphene sheet into the cylinder of the (10,10) tube is only 0.045 eV, or 0.08 eV nm^2/d^2 for a tube of any diameter d. The elastic stretching of a tube by 15% adds approximately 0.66 eV per atom above the graphene. [R.E. Smalley and B.I. Yakobson, Solid State Commun. 107, 597 (1998).]

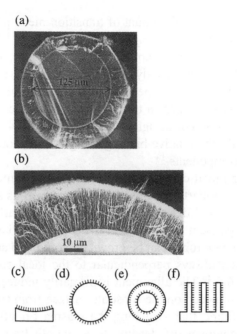

Fig. 6.5. (a) An SEM micrograph showing the radially grown nanotubes on the surface of a 125 μm-diameter optical fiber. (b) A close-up micrograph showing the conformally perpendicular nature of the nanotube grown on the fiber. (c)–(f) are examples of nonplanar, complex surfaces where nanotubes can be conformally grown perpendicular to the local surface. [C. Bower, W. Zhu, S. Jin, and O. Zhou, *Appl. Phys. Lett.* **77**, 830 (2000).]

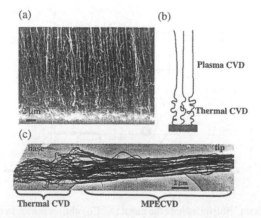

Fig. 6.6. (a) An SEM micrograph and (b) a schematic showing the straight/curled nanotube structure produced by an alternating plasma and thermal process (a 2 min plasma process followed by a 70 min thermal process), indicating both the field induced alignment effect and the base growth mechanism. (c) TEM micrograph showing a bundle of nanotubes with the upper portion straight and the lower portion curled. [C. Bower, W. Zhu, S. Jin, and O. Zhou, *Appl. Phys. Lett.* **77**, 830 (2000).]

It should be noted that the catalyst growth mechanism of carbon nanotubes is similar to that of VLS growth of nanowires or nanorods discussed in Chapter 4. Baker and Harris proposed this model for the catalytic carbon filament growth.[42] Atomic carbon dissolves into the metal droplets, then diffuses to and deposits at the growth surface, resulting in the growth of carbon nanotubes. The catalyst growth offers an additional advantage; it is relatively easy to prepare patterned carbon nanotube films by standard lithographic techniques[43,44] and to grow aligned carbon nanotubes with or without the substrate.[45,46] Methods of CVD growth of carbon nanotubes, assisted by the transition metal catalysts, are also considered as the method for the mass production.[47] CVD methods also allow the growth of carbon nanotubes at much lower growth temperatures such as 700 or 800°C.[48,49] The as-grown nanotubes generally show poor crystallinity, but can be much improved by a heat treatment at 2500–3000°C in argon.[50]

For the catalytic growth, two models have been proposed to explain the experimental observations: the base growth and tip growth, which were originally developed for the catalytic growth of carbon filaments.[51] Both models are used to explain the growth of carbon nanotubes. In the case of PECVD and pyrolysis growth, the catalytic particles are usually found at the tip and explained by the tip growth model.[52–55] The base model has been used to explain the vertically aligned carbon nanotube growth by thermal CVD using iron as catalyst.[56–59] However, experiments showed that the vertical growth of aligned carbon nanotubes does not necessarily follow the base-growth model.[60] The growth of aligned carbon nanotubes is possible through both tip-growth and base-growth models, depending on the catalyst and substrate used in the deposition method. Furthermore, the diffusion of precursor molecules to the catalyst at the bottom of the growing nanotubes would be difficult, particularly considering the high density and large length (up to 100 μm) of the grown carbon nanotubes. However, no research has been done to address this issue yet.

Another proposed mechanism for the carbon nanotube growth assumes that the nanotubes are always capped.[61] The growth is nucleated at active sites of a vapor-grown carbon fiber and the growth involves C_2 dimer absorption near a pentagon at the cap of the nanotube. Subsequent restructuring would result in the formation of an additional carbon hexagon, which is added into the nanotube and leads to the growth of the tube.

In almost all the synthesis methods, carbon nanotubes are found along with other carbon materials, such as amorphous carbon and carbon nanoparticles. Purification is generally required and refers to the isolation of carbon nanotubes from other entities. Three basic methods have been used for purification: gas phase, liquid phase and intercalation methods.[62] The gas phase purification method removes nanoparticles and amorphous carbon in the presence of nanotubes by an oxidation process.[63,64] The gas phase purification process also tends to burn off many of the nanotubes, particularly the smaller diameter nanotubes. Liquid phase removal of nanoparticles and other unwanted carbons is achieved using a potassium permanganate, $KMnO_4$ treatment.[65] This method retains most of carbon nanotubes, a higher yield than gas phase purification, but with shorter length. Carbon nanoparticles and other carbon species can be intercalated by reacting with $CuCl_2$–KCl, whereas the nanotubes would not intercalate since they have closed cage structures. Subsequent chemical reactions can remove the intercalated species.[66]

Properties of carbon nanotubes have been extensively studied. Langer et al.[67] were the first to study the transport properties of carbon nanotubes, and further measurements were done by many research groups.[68–70] Carbon nanotubes are excellent candidates for stiff and robust structures, since the carbon–carbon bond in graphite is one of the strongest in nature. TEM observation revealed that carbon nanotubes are flexible and do not break upon bending.[71] Thermal conductivity of carbon nanotubes could be extremely high, considering the fact that thermal conductivity of diamond and graphite (in-plane) are extremely high,[72] and thermal conductivity of individual carbon nanotubes was found much higher than that of graphite and bulk nanotubes.[73] Carbon nanotubes have a wide spectrum of potential applications. Examples include use in catalysis,[74] storage of hydrogen and other gases,[75] biological cell electrodes,[76] quantum resistors,[77] nanoscale electronic and mechanical devices,[78] electron field emission tips,[79] scanning probe tip,[80] flow sensors[81] and nanocomposites.[82]

6.3. Micro and Mesoporous Materials

According to the classification made by IUPAC,[83] porous solids can be grouped into three categories, depending on their pore diameter: microporous ($d < 2$ nm), mesoporous (2 nm $< d < 50$ nm), and macroporous ($d > 50$ nm) materials. Almost all of zeolites and their derivatives are microporous, whereas surfactant templated mesoporous materials and most xerogels and aerogels are mesoporous materials. In this section, we will briefly introduce these meso and microporous materials and their respective synthesis techniques. This field has been extensively covered with excellent review articles.[84,85]

6.3.1. Ordered mesoporous structures

Ordered mesoporous materials are made with a combination of using self-assembled surfactants as template and simultaneous sol-gel condensation around template. Mesoporous materials may have many important technological applications as supports, adsorbents, sieves or nanoscale chemical reactors. Such materials have uniformly sized and shaped pores with diameters ranging from 3 nm to several tens nanometers and microns long, and often have a very large pore volume (up to 70%) and very high surface area (>700 m^2/g). Before we discuss the details of the synthesis of ordered mesoporous materials, a brief introduction to surfactants and the formation of micelles is given.

Surfactants are organic molecules, which comprise two parts with different polarity.[86] One part is a hydrocarbon chain (often referred to as polymer tail), which is nonpolar and hence hydrophobic and lipophilic, whereas the other is polar and hydrophilic (often called hydrophilic head). Because of such a molecular structure, surfactants tend to enrich at the surface of a solution or interface between aqueous and hydrocarbon solvents, so that the hydrophilic head can turn towards the aqueous solution, resulting in a reduction of surface or interface energy. Such concentration segregation is spontaneous and thermodynamically favorable. Surfactant molecules can be generally classified into four families, and they are known as anionic, cationic, nonionic and amphoteric surfactants, which are briefly discussed below:

(1) Typical anionic surfactants are sulfonated compound with a general formula R–SO$_3$Na, and sulfated compounds of R–OSO$_3$Na, with R being an alkyl chain consisting of 11 to 21 carbon atoms.
(2) Cationic surfactants commonly comprise of an alkyl hydrophobic tail and a methyl-ammonium ionic compound head, such as cetyl trimethyl ammonium bromide (CTAB), $C_{16}H_{33}N(CH_3)_3Br$ and cetyl trimethyl ammonium chloride (CTAC), $C_{16}H_{33}N(CH_3)_3Cl$.
(3) Nonionic surfactants do not dissociate into ions when dissolved in a solvent as both anionic and cationic surfactant. Their hydrophilic head is a polar group such as ether, R–O–R, alcohol, R–OH, carbonyl, R–CO–R, and amine, R–NH–R.
(4) Amphoteric surfactants have properties similar to either nonionic surfactants or ionic surfactants. Examples are betaines and phospholipids.

When surfactants dissolve into a solvent forming a solution, the surface energy of the solution will decrease rapidly and linearly with an increasing concentration. This decrease is due to the preferential enrichment and the ordered arrangement of surfactant molecules on the solution surface i.e. hydrophilic heads inside the aqueous solution and/or away from non-polar solution or air. However, such a decrease stops when a critical concentration is reached, and the surface energy remains constant with further increase in the surfactant concentration, as shown in Fig. 6.7. This figure also shows that surface energy of a solution changes with the addition of general organic or inorganic solutes. The critical concentration in Fig. 6.7 is termed as the critical micellar concentration, or CMC. Below the CMC, the surface energy decreases due to an increased coverage of surfactant molecules on the surface as the concentration increases. At the CMC, the surface has been fully covered with the surfactant molecules. Above the CMC, further addition of surfactant molecules leads to phase segregation and formation of colloidal aggregates, or micelles.[87] The initial micelles are spherical and individually dispersed in the solution, and would transfer to a cylindrical rod shape with further increased surfactant concentration. Continued increase of surfactant concentration results in an ordered parallel hexagonal packing of cylindrical micelles. At a still higher concentration, lamellar micelles would form. Inverse micelles would form at an even higher concentration. Figure 6.8 are schematics of various micelles formed at various surfactant concentrations above the CMC.

Micelles, particularly hexagonal or cubic packing of cylindrical micelles have been used as templates to synthesize ordered mesoporous materials through sol-gel processing.[88] The formation of this new family of materials was first reported in 1992.[89,90] The first ordered mesoporous

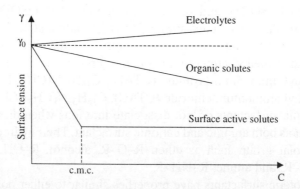

Fig. 6.7. Effect of different solutes on the surface tension of a solution. Surfactants, or surface active molecules will preferably allocate at the surface, resulting in a decrease in surface tension with an increasing concentration till the critical micellar concentration, or CMC is reached. Further increase in surfactant concentration will not reduce the surface tension.

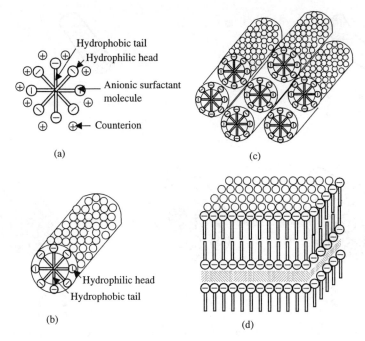

Fig. 6.8. (a) Spherical micelle forms first as the concentration of surfactants is above the CMC. (b) Individual cylindrical micelle forms as the concentration of surfactants increases further. (c) Further increased concentration of surfactants results in the formation of hexagonally packed cylindrical micelles, (d) Lamellar micelles would form when the concentration of surfactants rises even further.

materials synthesized were denoted as MCM-41 and MCM-48. MCM-41 is an aluminosilicate possessing hexagonally arranged one-dimensional pores with diameters ranging from 1.5 to 10 nm, and MCM-48 is an aluminosilicate with a three-dimensional pore system and diameters of order of 3 nm. It should be noted that the inorganic portion of mesoporous materials MCM-41 and MCM-48 are amorphous aluminosilicates.

The process is conceptually straightforward and can be briefly described below. Surfactants with a certain molecule length are dissolved into a polar solvent with a concentration exceeding its CMC, mostly at a concentration, at which hexagonal or cubic packing of cylindrical micelles is formed. At the same time, the precursors for the formation of desired oxide(s) are also dissolved into the same solvent, together with other necessary chemicals such as a catalyst. Inside the solution, several processes proceed simultaneously. Surfactants segregate and form micelles, whereas oxide precursors undergo hydrolysis and condensation around the micelles simultaneously, as schematically shown in Fig. 6.9.

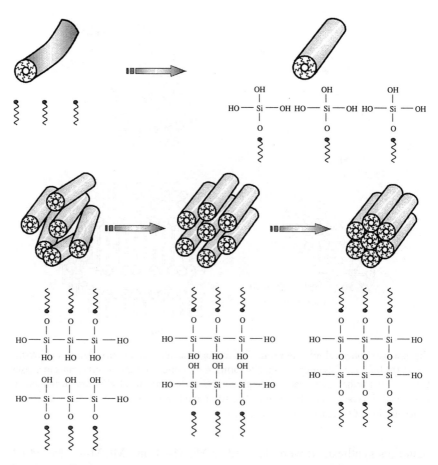

Fig. 6.9. Schematic showing the process of the formation of mesoporous materials. Surfactant molecules form cylindrical micelles with hexagonal packing, while inorganic precursors form a framework around the micelles through hydrolysis and condensation reactions.

Various organic molecules including surfactants and block copolymers have been used to direct the formation of ordered mesoporous materials.[91–94] Various oxides other than silica and aluminosilicates are found to form ordered mesoporous structures.[95–101] A lot of research has been conducted in the synthesis of ordered mesoporous complex metal oxides,[95,102,103] also called mixed metal oxides, which possess many important physical properties conducive to a wide range of applications, particularly as heterogeneous catalysts in modern chemical industry. The greatest challenge in the synthesis of ordered mesoporous complex metal oxides is the same as the formation of nanoparticles and nanowires of complex metal oxides by sol-gel processing, which is to ensure the formation of homogeneous desired stoichiometric composition through heterocondensation. All the general considerations that have been discussed previously are applicable here. However, the situation here is even more complex, since the presence of surfactants in the solution would complicate the reaction kinetics of hydrolysis and condensation reactions. Some surfactants would act as catalysts to promote hydrolysis and condensation reactions. The presence of relatively large surfactant molecules and micelles in the solution would certainly have a steric effect on the diffusion process. Although all these surfactant effects are present in the synthesis of single metal oxide mesoporous materials, a given surfactant may have varied degree effects on different precursors. Therefore, the influences of surfactants on the hydrolysis and condensation reactions in the formation of ordered mesoporous complex metal oxides should be carefully considered. Table 6.1 summarizes some physical properties of mesoporous complex oxides and Fig. 6.10 shows TEM images of various mesoporous materials.[95]

Optically transparent and electronically conductive complex oxide, indium tin oxide (ITO), has also been studied to form mesoporous structure.[104] In the fabrication of mesoporous ITO, a prime impediment is used to control the competing hydrolysis and condensation reactions, which is achieved by employing atrane complexes as precursors to slow the kinetics of hydrolysis. Indium acetate and tin isopropoxide with desired stoichiometric ratios were dissolved in a 10-fold molar excess of triethanolamine under an inert nitrogen atmosphere. Approximately 10 vol% dry formamide was also added to lower the viscosity. After the solution was mixed for 4 hr, CTAB in a 3.5 : 1 molar ratio with respect to the total metal concentration was admixed to the solution, and the pH was adjusted to 8 with 4 M sodium hydroxide. The mixture was held at 80°C for 96 hr prior to filtering off the product. The resultant ITO powder with a In : Sn molar

Table 6.1. Physical properties of mesoporous complex metal oxides.

Oxide	Pore size (nm)	BET surface area (m^2/g)	BET surface area (m^2/cm^3)	Porosity (%)
SiAlO$_{3.5}$	6	310	986	59
Si$_2$AlO$_{5.5}$	10	330	965	55
SiTiO$_4$	5	495	1638	63
Al$_2$TiO$_5$	8	270	1093	59
ZrTiO$_4$	8	130	670	46
ZrW$_2$O$_8$	5	170	1144	51

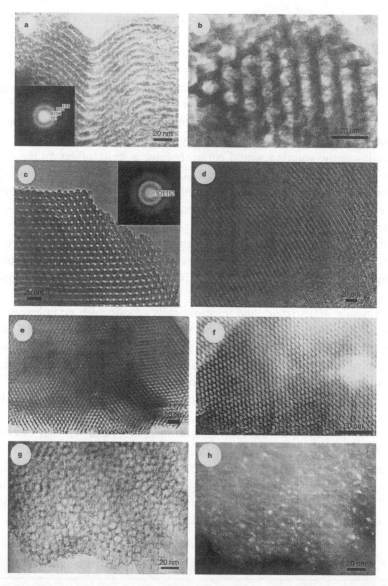

Fig. 6.10. TEM micrographs of two-dimensional hexagonal mesoporous TiO_2 (a, b), ZrO_2 (c, d), Nb_2O_5 (e) and $SiAlO_{3.5}$ (f). (a, d) are recorded along the [110] zone axis and (b, c, e, f) along the [001] zone axis, respectively, of each material. Insets in (a) and (c) are selected-area electron diffraction patterns obtained on the image area. (g) Bright-field TEM image of a thin slice of the mesoporous TiO_2 sample. (h) Dark-field image obtained on the same area of the same TiO_2 sample. The bright spots in the image correspond to TiO_2 nanocrystals. The images were recorded with a 200 kV JEOL-2000 TEM. All samples were calcined at 400°C for 5 hr in air to remove the block copolymer surfactant species. [P.D. Yang, D.Y. Zhao, D.I. Margoless, B.F. Chemelka, and G.D. Stucky, *Nature* **396**, 152 (1998).]

ratio of 1 : 1 has a BET surface area of 273 m²/g and a pore diameter of ~2 nm as determined by nitrogen adsorption isotherms. XRD indicates the formation of crystalline ITO after calcinations at unspecified temperatures, and TEM image shows a worm-hole topography. However, electrical conductivity measurements taken on a water-free pressed pellet showed an average value of $\sigma = 1.2 \times 10^{-3}$ S/cm at room temperature, which is about 3 orders of magnitude lower than that of ITO thin films under the same condition.

In addition to manipulating chemical compositions, crystal and microstructures, physical and chemical properties can also be introduced into order mesoporous materials through various surface modifications, including coating, grafting and self-assembly.[105–111]

Typical mesoporous materials are in the form of powders (or bulk mesoporous materials) and films. Bulk mesoporous materials comprise a collection of macroscopically sized grains (up to several hundreds micrometers). In each grain, there is crystallographically ordered mesoporous structure, however, all grains are randomly packed. This hinders diffusional accessibility to the mesopore structure, and thus limits the applications of ordered mesoporous materials in practice. Several groups have been successful in aligning mesoporous films parallel to a substrate surface over large areas, or within microchannels.[112–117] However, there is limited accessibility to the pores due to parallel alignment to the surface rather than the ideal perpendicular alignment. Efforts have also been made in achieving the alignment of mesoporous silica perpendicular to a surface (i.e. dead end pores), but this was done with a strong magnetic field on a small sample size and has very limited practical possibilities.[118] The synthesis of oriented or hierarchically structured mesoporous materials has also been reported.[119–121]

6.3.2. Random mesoporous structures

Mesoporous structures can be created by a variety of other methods. Examples include leaching a phase separated glass,[122] anodic oxidation of thin metal foils in an acidic electrolyte,[123] radiation-track etching,[124] and sol-gel processing.[125] In this section, the discussion will be focused on sol-gel derived mesoporous materials. Depending on the conditions applied for the removal of solvent during drying, two types of mesoporous materials can be obtained. One is xerogel, when solvents are removed under ambient conditions. Another is aerogel, which refers to mesoporous material with very high porosity and surface area and is generally made with supercritical drying. Both xerogels and aerogels are highly porous with typical average pore size of several nanometers. However, aerogels have higher porosity ranging from 75% to 99%, whereas xerogels typically have a porosity of 50%, but can have less than 1%.

Xerogels. The formation of porous structure by sol-gel processing is conceptually straightforward. During a sol-gel processing, precursor molecules undergo hydrolysis and condensation reactions, leading to the formation of nanoclusters. Aging will allow such nanoclusters to form gel, which consists of three-dimensional and interpenetrated networks of both solvent and solid. When the solvent is subsequently removed during drying, the gel network would partially collapse due to the capillary force, P_c, as given by the Laplace equation[126]:

$$P_c = -\gamma_{LV} \cos\theta \left(\frac{1}{R_1} + \frac{1}{R_2}\right) \qquad (6.1)$$

where γ_{LV} is the surface energy of liquid–vapor interface, θ is the wetting angle of liquid on solid surface, R_1 and R_2 are the principal radii of a curved liquid–vapor surface. For a spherical interface, $R_1 = R_2$. The collapse of the solid gel network driven by the capillary force would result in an appreciable loss of porosity and surface area. However, such a process in general would not result in the formation of dense structure. It is because the collapse of the gel network would promote the surface condensation and result in strengthening the gel network. When the strength of the gel network is sufficiently large to resist the capillary force, the collapse of gel network stops and porosity would be preserved. Similar processes occur in both monolith formation where the sol is allowed to gel through aging, and film formation where solvent evaporates prior to gelation, though kinetics and the strength of the gel network are significantly different. Table 6.2 listed some properties of porous oxide synthesized by sol-gel processing.[127] Typical pore size of sol-gel derived porous materials ranges from subnanometer to several tens nanometers depending on the sol-gel processing conditions and subsequent thermal treatment. For a given system, a higher thermal treatment temperature results in a larger pore size. The initial pore size is largely dependent on the size of nanoclusters formed in the sol and how well is the packing of these nanoclusters. Smallest pores are generally obtained from silica system. When silicon alkoxide precursors are hydrolyzed and condensed with acid as a catalyst, a linear silica chain would form. Such linearly structured silica chain would collapse almost completely upon removal of solvent, leading to the formation of relatively dense material. When the base is used as a catalyst, a highly branched nanocluster structure would form and subsequently lead to the formation of highly porous materials. Organic components are also often incorporated into the gel network to facilitate the pore

Table 6.2. Structural properties of sol-gel derived porous materials.[127]

Materials	Sintering Temp	Sintering time	Pore diameter	Porosity	BET surface area
γ-AlOOH	200	34	2.5	41	315
	300	5	5.6	47	131
γ-Al_2O_3	500	34	3.2	50	240
	550	5	6.1	59	147
	800	34	4.8	50	154
θ-Al_2O_3	900	34	5.4	48	99
α-Al_2O_3	1000	34	78	41	15
TiO_2	300	3	3.8	30	119
	400	3	4.6	30	87
	450	3	3.8	22	80
	600	3	20	21	10
CeO_2	300	3	2	15	41
	400	3	2	5	11
	600	3		1	1
Al_2O_3-CeO_2	450	3	2.4	39	164
	600	3	2.6	46	133
Al_2O_3-TiO_2	450	3	2.5	38–48	220–260
Al_2O_3-ZrO_2	450	5	2.6	43	216
	750	5	2.6	44	179
	1000	5	≥20		

size and porosity. For example, alkyl chains were incorporated into silica network to form relatively dense organic-silica hybrids. Porous structures were obtained when organic components were pyrolyzed. It should be noted that the porous structure formed by sol-gel processing is random and pores are tortuous, though the size distribution of pores is relatively narrow.

Aerogels were first made in the early 1930s[128] and have been studied for various applications since 1960s.[129,130] The chemistry of aerogels and their applications has summarized in an excellent review paper.[131] Typically wet gels are aged for a certain period of time to strengthen the gel network, and then brought to temperature and pressure above the supercritical point of the solvent in an autoclave, under which the solvent is removed from the gel network. Above the supercritical point, the difference between solid and liquid disappears, and thus the capillary force ceases existence. As a result, the highly porous structure of the gel network is preserved. Such prepared aerogels can have porosity as high as 99% and surface area exceeding 1000 m^2/g. Supercritical drying process consists of heating the wet gel in an autoclave to both pressure and temperature above the critical point of the solvent, and then slowly evacuating the liquid phase by reducing the pressure while maintaining the temperature above the critical point. Figure 6.11 shows two common supercritical drying paths with a solvent of CO_2, and Table 6.3 listed some critical point parameters of common solvents.[132]

All materials that can be synthesized as wet gels by sol-gel processing can form aerogels with supercritical drying. In addition to silica aerogels, examples of other commonly studied inorganic aerogels are TiO_2,[133] Al_2O_3,[134] Cr_2O_3[135] and mixed silica–alumina.[136] Solvent exchange has been widely used to reduce the temperature and pressure required to reach a supercritical condition. Highly porous structure can also be obtained using ambient drying. To prevent the collapse of the original porous structure of gel network, there are two approaches. One is to eliminate the

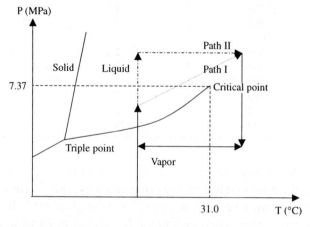

Fig. 6.11. An example of a possible supercritical drying path of CO_2. There are two practical approaches: (I) increase the pressure above the vapor pressure at room temperature and then increase both temperature and pressure simultaneously by heating, and (II) increase the pressure above the supercritical point of the solvent and then heat the sample above supercritical temperature while maintaining the pressure constant.

Table 6.3. Critical point parameters of common solvents.[132]

Solvents	Formula	T_c (°C)	P_c (MPa)
Water	H_2O	374.1	22.04
Carbon dioxide	CO_2	31.0	7.37
Freon 116	$(CF_3)_2$	19.7	2.97
Acetone	$(CH_3)_2O$	235.0	4.66
Nitrous oxide	N_2O	36.4	7.24
Methanol	CH_3OH	239.4	8.09
Ethanol	C_2H_5OH	243.0	6.3

capillary force, which is the fundamental concept of using supercritical drying and is discussed above. Another is to manipulate the inbalance between the huge capillary force and the small mechanical strength of the gel network, so that the gel network is strong enough to resist the capillary force during the removal of solvent. Organic components is incorporated into inorganic gel network to change the surface chemistry of the silica gel network and, thus, to minimize the capillary force and prevent the collapse of gel network. Organic components can be introduced through either copolymerization with organic components introduced in the form of organic precursor,[137,138] or self-assembly with solvent exchange.[139] The incorporation of organic components into silica gel network resulted in the formation of highly porous silica under ambient conditions, with a porosity of 75% or higher and a specific surface area of 1000 m²/g. Organic aerogels can be made by polymerizing organic precursors and subsequent supercritical drying of aged wet gels. The most extensively studied organic aerogels are the resorcinol-formaldehyde (RF) and formaldehyde (MF) aerogels.[140,141] Carbon aerogels are formed by pyrolysis of organic aerogels, typically at temperatures above 500°C. Carbon aerogels retain the high surface area and pore volume of their parent organic aerogels.[142]

6.3.3. Crystalline microporous materials: zeolites

Zeolites are crystalline aluminosilicates and were first discovered in 1756.[143,144] There are 34 naturally occurring zeolites and nearly 100 synthetic type zeolites. A zeolite has a three-dimensional framework structure with uniformly sized pores of molecular dimensions, typically ranging from ~0.3 to 1 nm in diameter, and pore volumes vary from about 0.1 to 0.35 cc/g. Zeolites have a broad diverse spectrum of applications, and examples include catalysts, adsorbents and molecular sieves. Many review articles and books have been published.[145–147] Details of the structures and specific names of various zeolites have been summarized in literature.[148–150] Only a brief description is given below.

Zeolites are tectoaluminosilicates with a formal composition $M_{2/n}O \cdot Al_2O_3 \cdot xSiO_2 \cdot yH_2O$ (n = valence state of the mobile cation, M^+ and $x \geq 2$), in that they are composed of TO_4 tetrahedra (T = tetrahedral atom, i.e. Si, Al), each oxygen atom is shared between adjacent tetrahedral, which leads to the framework ratio of O/T being equal to 2 for all zeolites.[151] A dimensional framework is formed by 4-corner connecting TO_4 tetrahedra. When a zeolite is made of pure silica without any defects, each oxygen atom at the corner is shared by two SiO_4 tetrahedra and the charge is balanced. When silicon is replaced by aluminum, alkali metal ions, such as K^+, Na^+, alkaline earth ions, such as Ba^{2+}, Ca^{2+}, and protons, H^+ are typically introduced to balance the charges. Such a framework formed is relatively open and characterized by the presence of channels and cavities. The size of the pores and the dimensionality of the channel system are determined by arrangement of TO_4 tetrahedra. More specifically, the pore sizes are determined by the size of the rings that are formed by connecting various numbers of TO_4 tetrahedra or T atoms. An 8-ring is designated to a ring comprised of 8 TO_4 tetrahedra and is considered to be a small pore opening (0.41 nm in diameter), a 10-ring medium one (0.55 nm), and a 12-ring large one (0.74 nm), when rings are free of distortion. Depending on the arrangement or the connection of various rings, different structures or pore openings, such as cages, channels, chains and sheets, can be formed. Figure 6.12 shows some of these subunits, in which each cross point is designated to a TO_4 tetrahedron for clarity.[150] In this figure, the designations in terms of the n-rings defining the faces of these subunits are also included. For example, a cancrinite cage subunit is defined by six 4-rings

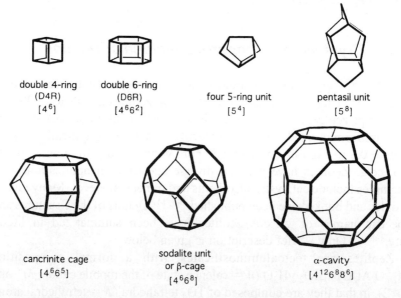

Fig. 6.12. Some subunits and cages that recur in several framework types of zeolites; each cross point is designated to a TO_4 tetrahedron where T is a metal such as silicon or aluminum. [L.B. McCusker and C. Baerlocher, in *Introduction to Zeolite Science and Practice*, 2nd edition, eds., H. van Bekkum, E.M. Flanigen, P.A. Jacobs, and J.C. Jansen, Elsevier, Amsterdam, p. 37, 2001.]

Special Nanomaterials

and five 6-rings, and is thus designated a $[4^6 6^5]$ cage. A nomenclature similar to that used for cages has also been developed to describe channels, chains and sheets. Different frameworks are formed by stacking various subunits and/or with different stacking sequences. There are 133 confirmed zeolite framework types. Figure 6.13 shows two schematics of zeolite frameworks: SOD and LTA framework types.[150]

Zeolites are normally prepared by hydrothermal synthesis techniques.[152,153] A typical synthesis procedure involves the use of water, a silica source, an alumina source, a mineralizing agent and a structure-directing agent. The sources of silica are numerous and include colloidal silica, fumed silica, precipitated silica and silicon alkoxides. Typical alumina sources include sodium aluminate, boehmite, aluminum hydroxide, aluminum nitrate and alumina. The typical mineralizing agent is hydroxyl ion, OH⁻ and fluorine ion, F⁻. The structure-directing agent is a soluble organic species, such as quaternary ammonium ion, which assists in the formation of the silica framework and ultimately resides within the intracrystalline voids. Alkali metal ions can also play a structure-directing role in the crystallization process. Table 6.4 lists the reactants, synthesis temperatures, and the physical and chemical properties of zeolites Na-A and TPA-ZSM-5.[153] Figure 6.14 gives SEM images of a few zeolites.[153]

The syntheses can be sensitive to the reagent type, the order of addition, the degree of mixing, the crystallization temperature and time and the composition. There are numerous complex chemical reactions and organic–inorganic interactions occurring during the synthesis process.

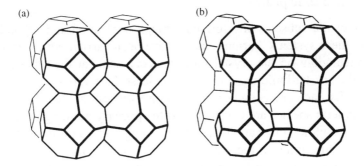

Fig. 6.13. (a) The SOD framework type, having a body centered cubic arrangement of β or sodalite cages (see Fig. 6.12). (b) The LTA framework type with a primitive cubic arrangement of α-cages joined through single 8-rings (producing a sodalite cage in the center). [L.B. McCusker and C. Baerlocher, in *Introduction to Zeolite Science and Practice*, 2nd edition, eds. H. van Bekkum, E.M. Flanigen, P.A. Jacobs, and J.C. Jansen, Elsevier, Amsterdam, p. 37, 2001.]

Table 6.4. Synthesis mixtures (in molar ratio), physical and chemical properties of zeolites Na-A and TPA-ZSM-5.[153]

Materials & properties	Na-A	TPA-SZM-5
SiO_2	1	1
Al_2O_3	0.5	<0.14
Na_2O	1	0.16
TPA_2O	–	0.3
H_2O	17	49
T (°C)	<100	<150
Pore structure	3D, cages linked via windows	2D, intersecting channels
Density (g/cm^3)	1.28	1.77
Pore volume (cm^3/g)	0.37	0.18
Lattice stabilization	Na^+, H_2O	TPA^+
Si/Al ratio	1	0.12
Bronsted activity	Low	High
Affinity	Hydrophilic	Hydrophobic

Depending on the mixture composition, the extent of reactions and the synthesis temperature, at least four types of liquids can be yielded[153]:

(i) Clear solution that consists of molecular, monomeric and ionic species only,

(ii) Sol or colloidal consisting of amorphous clusters with open structure (also called dispersed low density gel),

(iii) Sol or colloidal with dispersed amorphous clusters with dense structure (also referred to as separated high density gel), and

(iv) Sol or colloidal with metastable crystalline solid nanoparticles (also called solid phase).

Zeolites are subsequently formed through nucleation and crystallization from these systems. Various studies have been carried out to establish an understanding on the crystal growth mechanisms at the molecular level and the crystal-building units.[152,153] At least three types of crystal building units have been suggested for the growth of zeolites: (i) tetrahedral monomeric species are considered as the primary building units, (ii) secondary building units are the crystal building units, and (iii) clathrates are the building units in the nucleation and crystallization of zeolites. Two recent synthesis models are outlined below. Figure 6.15 illustrates the mechanism of structure direction and crystal growth in the synthesis of TPA-Si-ZSM-5 proposed by Burkett and Davis.[154] During the synthesis, the inorganic–organic composite clusters are first formed by overlapping the hydrophobic hydration spheres of the inorganic and organic components and subsequent releasing of ordered water to establish favorable

Fig. 6.14. SEM images showing the crystalline nature of zeolites. Single crystals of (a) zeolite A, (b) analcime, and (c) natrolite. (d) A batch of zeolite L and (e) typical needle aggregates of zeolite mordenite. [J.C. Jansen, in *Introduction to Zeolite Science and Practice*, 2nd edition, eds. H. van Bekkum, E.M. Flanigen, P.A. Jacobs, and J.C. Jansen, Elsevier, Amsterdam, p. 175, 2001.]

van der Waals interactions. Such inorganic–organic composite clusters serve as growth species for both initial nucleation and subsequent growth of zeolite crystals. The nucleation occurs through epitaxial aggregation of these composite clusters, whereas the crystal growth proceeds through diffusion of the same species to the growing surface to give a layer-by-layer growth mechanism. Another mechanism, called "nanoslab" hypothesis,[155] builds on the mechanism discussed above. The difference is that the inorganic–organic composite clusters form "nanoslabs" through epitaxial aggregation first. Such formed "nanoslabs" aggregates with other "nanoslabs" to form bigger slabs as shown in Fig. 6.16.[155]

Effects of structure-directing agent. When different organic molecules as structure-directing agents are included in an otherwise identical synthesis mixture, zeolites with completely different crystal structures can

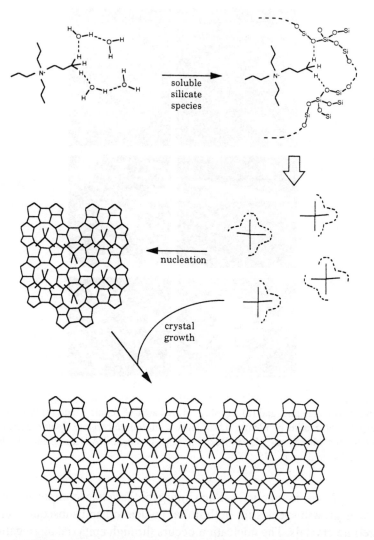

Fig. 6.15. Mechanism of structure-direction crystal growth involving organic–inorganic composites in the synthesis of pure-silica ZSM-5 zeolite using TPA$^+$ as structure-directing agent. [S.L. Burkett and M.E. Davis, *J. Phys. Chem.* **98**, 4647 (1994).]

be formed. For example, when N,N,N-trimethyl 1-adamantammonium hydroxide is used as a structure-directing agent, zeolite SSZ-24 was formed, while ZSM-5 was produced by using tetrapropylammonium hydroxide as structure-directing agent. In addition, the choice of a structure-directing agent can affect the synthesis rate.[156]

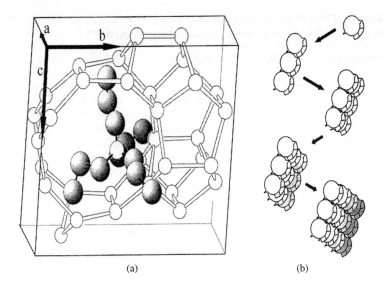

(a)   (b)

Fig. 6.16. The "nanoslab" hypothesis: (a) the precursor unit containing one TPA cation and (b) schematic representation of nanoslab formation by aggregation of precursor units. [C.E.A. Kirschhock, R. Ravishankar, L. Van Looveren, P.A. Jacobs, and J.A. Martens, *J. Phys. Chem.* **B103**, 4972 (1999).]

The geometry of the structure-directing agent has a direct impact on the geometry of the zeolite synthesized. For example, SSZ-26 is a zeolite with intersecting 10- and 12-ring pores,[157] and was synthesized with *a priori* design using a propellane-based structure-directing agent.[158] It has been demonstrated experimentally and through molecular force field calculations that the geometry of the pore sections of SSZ-26 matches very well with that of the organic structure-directing molecules and one structure-directing molecule is present at each channel intersection.[159] ZSM-18 is an aluminosilicate zeolite containing 3-member rings[160] and was synthesized using structure-directing agent what was designed using molecular modeling.[161] An excellent fit exists between the zeolite cage and the organic structure-directing agent.

Effects of heteroatoms. The addition of small quantities of tetrahedral cations, such as Al, Zn, B etc., to the synthesis mixtures has dramatic effects and results in significantly different zeolite structures when using identical structure-directing agents.[162] Table 6.5 compares some systems.[144] For example, when other synthesis parameters are kept the same with tetraethylammonium cation, TEA$^+$, as a structure-directing agent, ZSM-12 is formed when the ratio of SiO_2 to Al_2O_3 is greater than 50. When a small amount of alumina is added, zeolite beta is formed. Further

Table 6.5. Effect of aluminum, boron and zinc on the structure of zeolites or other compounds obtained using organic structural-directing agents.[144]

Organic agent	SiO_2	SiO_2/Al_2O_3 <50	SiO_2/B_2O_3 <30	SiO_2/ZnO <100
$C_8H_{20}N$	ZSM-12	Zeolite Beta	Zeolite Beta	VPI-8
$C_{16}H_{32}N_4$	ZSM-12	Zeolite Beta	Zeolite Beta	VPI-8
$C_{13}H_{24}N$*	ZSM-12	Mordenite	Zeolite Beta	Layered Mater.
$C_{13}H_{24}N$*	SSZ-24	SSZ-25	SSZ-33	–
$C_{13}H_{24}N$*	SSZ-31	Mordenite	SSZ-33	VPI-8
$C_{12}H_{20}N$	SSZ-31	SSZ-37	SSZ-33	–

* with different molecular structures

addition of alumina to reach a ratio of 15 of SiO_2/Al_2O_3, ZSM-20 is then synthesized. The substitution of divalent and trivalent tetrahedral cations for Si^{4+} in the synthesis mixtures results in a negatively charged zeolite framework, which will coordinate more strongly with both the organic structure-directing cations and the inorganic cations, such as alkali metal cations. In addition, the change of both the cation-oxygen bond lengths and the cation-oxygen-cation bond angles would have appreciable influences on the formation of building units.[163]

Effects of alkali metal cations. The presence of alkali metal cations is required for the vast majority of zeolite syntheses at basic conditions.[164] A small concentration of alkali metal cations in aqueous solutions significantly increases the dissolution rate of quartz, up to 15 times as much as the rate in deionized water.[165,166] It is generally accepted that the presence of alkali metal cations can accelerate the rate of nucleation and crystal growth of high-silica zeolites.[164,167] However, it was also found that too much alkali metal cations may result in competition with the organic structure-directing agent for interactions with silica to result in layer-structured products.[168]

Organic–inorganic hybrid zeolites. Recently Yamamoto et al.[169] succeeded in synthesis of organic–inorganic hybrid zeolites that contain an organic framework by partially superseding a lattice oxygen atom by a methylene group. Such hybrid zeolites are significantly different from the zeolites containing pendant organic groups.[170] The use of methylene-bridged organosilane as a silicon source gives zeolite materials containing an organic group as lattice, with several zeolitic phases such as the MFI and the LTA structures. In such hybrid zeolites, some of siloxane bonds, Si–O–Si, are replaced by methylene frameworks, Si–CH$_2$–Si.

6.4. Core-Shell Structures

In Chapter 3, we have discussed the synthesis of heteroepitaxial semiconductor core-shell structure. Although the chemical compositions of the core and shell are different, they possess similar crystal structure and lattice constants. Therefore, the formation of the shell material on the surface of grown nanometer sized particle (the core) is an extension of particle growth with different chemical composition. The core-shell structures to be discussed in this section are significantly different. First, the core and shell often have totally different crystal structures. For example, one can be single crystal and another amorphous. Secondly, the physical properties of core and shell often differ significantly from one another; one may be metallic and another dielectric. Furthermore, the synthesis processes of cores and shells in each core-shell structure are significantly different. Although a variety of core-shell structures can be fabricated by various techniques, such as coating, self-assembly, and vapor phase deposition, the discussion in this section will be focused mainly on the core-shell structures of novel metal-oxide, novel metal-polymer, and oxide-polymer systems mostly by solution methods. Further, a monolayer of molecules assembled on the surface of nanoparticles will not be included in the following discussion. Polymer monolayers are often used to induce the diffusion-controlled growth and stabilize the nanoparticles, which has been already discussed in Chapter 3. Self-assembly of molecular monolayers has been one of the topics discussed in the previous chapter.

6.4.1. Metal-oxide structures

We shall take gold–silica core-shell structure as an example to illustrate the typical experimental approaches.[171,172] Gold surface has very little electrostatic affinity for silica, since gold does not form a passivation oxide layer in solution, and thus no silica layer will form directly on the particle surface. Furthermore, there are usually adsorbed organic monolayers on the surface to stabilize the particles against coagulation. These stabilizers also render the gold surface vitreophobic. A variety of thioalkane and aminoalkane derivatives may be used to stabilize gold nanoparticles.[173] However, for the formation of core-shell structures, the stabilizers are not only needed to stabilize the gold nanoparticle by forming a monolayer on the surface, but also required to interact with silica shell. One approach is to use organic stabilizers with two functionalities at two ends. One would link to gold particle surface and the other to silica shell. The simplest way to link to silica is to use silane coupling agents.[174] (3-aminopropyl)trimethoxysilane (APS) has been the most widely used complexing agent to link gold core with silica shell.

Figure 6.17 sketched the principal procedures of fabricating gold–silica core-shell structures. There are typically three steps. The first step is to form the gold cores with desired particle size and size distribution. The second step is to modify the surface of gold particle from vitreophobic to vitreophilic through introducing an organic monolayer. The third step involves the deposition of oxide shell. In the first developed fabrication process,[171,172] gold colloidal dispersion is first prepared using the sodium citrate reduction method,[175] resulting in the formation of a stable colloidal solution with gold nanoparticles of ~15 nm and 10% dispersity.

In the second step, a freshly prepared aqueous solution of APS (2.5 mL, 1 mM) is added to 500 mL of gold colloidal solution under vigorous stirring for 15 min. A complete coverage of one monolayer of APS is formed on the gold particle surface. During this process, the previously adsorbed, negatively charged citrate groups are displaced by APS molecules, with the silanol groups pointing into solution. The process is driven by the large complexation constant for gold amines. The silane groups in APS molecules in aqueous solution undergo rapid hydrolysis and convert to silanol groups, which may react with one another through condensation reactions to form three-dimensional network. However, the rate of condensation reaction is rather slow at low concentration.[174] It should also be noted that during the self-assembly of APS on the surface of gold particles, the pH needs to be maintained above the isoelectric point of silica, which is 2–3,[176] so that the silanol groups is negatively charged. In addition, the pH is required to ensure the adequate negative surface charge on the gold

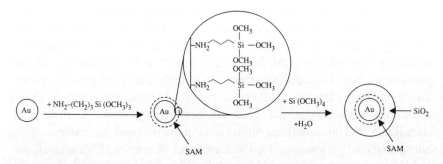

Fig. 6.17. Principal procedures for the formation of gold–silica core-shell structures. (a) Formation of monosized gold particles, (b) modifying the surface of gold nanoparticles by introducing a monolayer of organic molecules through self-assembly, and (c) deposition of silica shell. [L.M. Liz-Marzán, M. Giersig, and P. Mulvaney, *Langmuir* **12**, 4329 (1996).]

nanoparticles, so that the positively charged amino groups are attracted to the gold surface.

In the third step, a silica sol prepared by slowly reducing the pH of a 0.54 wt% sodium silicate solution to 10–11 is added to the gold colloidal solution (with a resulting pH of ~8.5) under vigorous stirring for at least 24 hours. A layer of silica of 2–4 nm thick is formed on the modified surface of the gold nanoparticles. In this step, slow condensation or polymerization reaction is promoted by controlling the pH, so that the formation of a thin, dense and relatively homogeneous silica layer around the gold particle can be produced.[176,177] Further growth of the silica layer was achieved by transferring the core-shell nanostructures to ethanol solution and by controlling the growth condition such that further growth of silica layer would be diffusion predominant, which is often referred to as Stöber method.[178] Figure 6.18 shows TEM images of gold–silica core-shell nanostructure.[172]

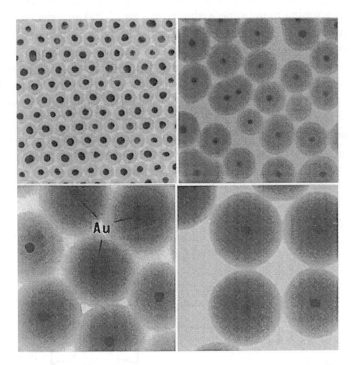

Fig. 6.18. TEM Images of silica-coated gold particles produced during the extensive growth of the silica shell around 15 nm Au particles with TES in 4:1 ethanol/water mixtures. The shell thicknesses are (a, top left) 10 nm, (b, top right) 23 nm, (c, bottom left) 58 nm, and (d, bottom right) 83 nm. [L.M. Liz-Marzán, M. Giersig, and P. Mulvaney, *Langmuir* **12**, 4329 (1996).]

6.4.2. Metal–polymer structures

Emulsion polymerization is one of the widely used strategies for the creation of metal–polymer core-shell structures.[179] For example, silver-polystyrene/methacrylate core-shell structures were prepared by emulsion polymerization of styrene and/or methacrylic acid in oleic acid. In this system, silver particles are coated with a uniform and well-defined layer with thickness ranging 2–10 nm.[180] The thickness of the layer can be readily controlled by changing the concentrations of monomers. Further etching tests in concentrated chloride solutions revealed that the polymer coatings have a strong protection effect.

Another example of the formation of metal–polymer core-shell structures is membrane-based synthesis.[181–183] In this method, the metal particles are first trapped and aligned inside the pores of membranes by vacuum filtration and then the polymerization of conducting polymers inside the pores are followed as schematically illustrated in Fig. 6.19.[183] A porous alumina membrane with a pore size of 200 nm was used to trap gold nanoparticles, $Fe(ClO_4)_3$ was used as polymerization initiator and poured on top of the membrane. Several drops of pyrrole or N-methylpyrrole were placed underneath the membrane. Upon diffusion of the monomer molecules as a vapor inside the pores, it contacted with the initiators and formed polymer. The deposition of polymer was found to preferentially occur on the surface of the gold nanoparticles. The thickness of polymer shell can be controlled by the polymerization time and can be easily varied from 5 nm to 100 nm. However, further polymerization time resulted in the formation of aggregated core-shell structures. Figure 6.20 shows the

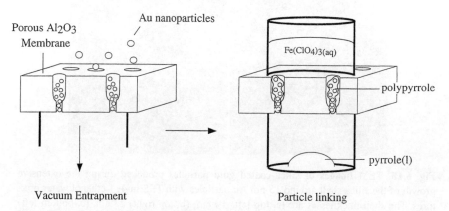

Fig. 6.19. Schematic representing the fabrication procedures of metal-polymer core-shell structures. [S.M. Marinakos, D.A. Shultz, and D.L. Feldheim, *Adv. Mater.* **11**, 34 (1999).]

Special Nanomaterials

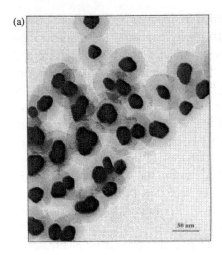
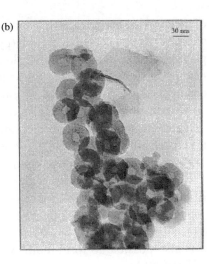

Fig. 6.20. TEM images of the gold-poly(pyrrole) and composite poly(N-methylpyrrole)/poly(pyrrole) core-shell structures: (a) ~30 nm diameter gold particles coated with Ppy and (b) polymer shell after the removal of Au with a mixture of 0.002 M $K_4[Fe(CN)_6]$ and 0.1 M KCN. [S.M. Marinakos, D.A.Shultz, and D.L. Feldheim, *Adv. Mater.* **11**, 34 (1999).]

TEM images of the gold-poly(pyrrole) and composite poly(N-methylpyrrole)/poly(pyrrole) core-shell structures.[183]

6.4.3. Oxide–polymer structures

The synthetic routes to produce polymer-coated oxide particles can be grouped into two main classes: polymerization at the particle surface or adsorption onto the particles.[184,185] Polymerization-based methods include monomer adsorption onto particles followed by subsequent polymerization[181,182,186] and emulsion polymerization.[179,180] In the adsorption and polymerization of monomer approach, the polymerization can be activated by either the addition of an initiator or the oxide itself. For example, the coating of aluminum hydrous oxide modified silica particles with poly(divinylbenzene) (PDVB) layers was prepared by pre-treatment of the silica particles with coupling agents such as 4-vinylpyridine or 1-vinyl-2-pyrrolidone, followed by subsequent admixing of divinylbenzene and a radical polymerization initiator.[187] The similar approach can be used to synthesize polymer layers of poly(vinylbenzene chloride) (PVBC), copolymers of PDVB–PVBC, and double shells of PDVB and PVBC.[188] Polymerization of adsorbed monomers can also be initiated by the surface sites of oxide nanoparticles. For example, poly(pyrrole) coatings on a range of metal oxide particles have been formed in this way.[186,189] α-Fe_2O_3, SiO_2, and CeO_2 were coated with poly(pyrrole) by exposing the oxides to the polymerization medium of pyrrole in an ethanol and water mixture and heating to 100°C.[189] Further, it was found that the thickness of the polymer coatings can be controlled by varying the contact time of the core with the polymerization solution and also depends on the inorganic core composition and the additives in the solution. Figure 6.21 is the TEM images of silica coated with poly(pyrrole).[189] Polymer layers on inorganic nanoparticles can also be obtained through emulsion polymerization.

Self-assembly has been widely studied for the construction of thin films.[190,191] Self-assembled thin polymer layer has been used to stabilize the colloidal particles by direct adsorption of polymers from solution onto their surface,[192] which has been discussed briefly in Chapters 2 and 3. It is also possible to form multilayers of polyelectrolytes by electrostatic self-assembly.

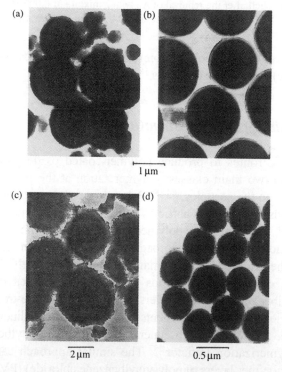

Fig. 6.21. TEM (a,b) CuO, (c) NiO, and (d) SiO2 particles coated with polypyrrole prepared with the same mass of metal oxides (1.0 mg cm^{-3}), pyrrole (0.039 g cm^{-3}), and ethanol (5%), using 0.0016 g cm^{-3} PVA in (b) and no PVA in other cases. [C.L. Huang and E. Metijevic, J. Mater. Res. 10, 1327 (1995).]

6.5. Organic–Inorganic Hybrids

Organic–inorganic hybrids are materials in which organic and inorganic components interpenetrate each other in nanometer scale and both form percolated three-dimensional networks commonly by sol-gel processing. Such organic–inorganic hybrids have also been termed Ormosils (organically modified silicates) or Ormocers (organically modified ceramics) in literature. Hybrids are generally divided into two classes: (i) hybrids that consist of organic molecules, oligomers or low molecular weight polymers embedded in an inorganic matrix to which they are held by weak hydrogen bonds or van der Waals forces, and (ii) hybrids in that the organic and inorganic components are linked to each other through covalent bonds. Class I hybrids can be considered as molecular scale nanocomposites where organic components are physically trapped in an inorganic matrix; whereas class II hybrids can be considered as a huge molecule that links organic and inorganic components through true chemical bonds.

6.5.1. Class I hybrids

There are a few routes developed for the synthesis of class I hybrids, including hydrolysis and condensation of alkoxides inside soluble organic polymers, mixing alkoxides and organic compounds in a common solvent, and impregnating a porous oxide gel with organic compounds. All three techniques have been widely explored for the formation of various organic–inorganic hybrids. For example, hybrids comprising organic dyes embedded in inorganic matrix, such as silica, aluminosilicate and transition metal oxides,[193,194] composed of polymers in inorganic matrix, such as poly(N-vinyl pyrrolidone)-silica[195] and poly (methylmethacrylate)-silica[196] are made by hydrolysis-condensation of alkoxides together with soluble organic polymers. Simultaneous gelation of the organic and inorganic components by mixing alkoxides and organic components in a common solvent is a method to ensure the formation of interpenetrated three-dimensional networks of both organic and inorganic components. However, the challenge is to prevent phase segregation and precipitation of organic components during hydrolysis and condensation processing, some precursor modification is desired.[197] Various silica-based hybrids with organics including polyparaphenylene and polyaniline were synthesized using this approach.[198] Infiltration of organic components into highly porous inorganic gel networks is yet another method to make class I hybrids such as PMMA-silica.[199]

Ordered hybrids can also be made by intercalation of organic compounds in ordered inorganic hosts, which include clay silicates, metal phosphates, layered metal oxides, halides or chalcogenides.[200] For example, alkyl amines can be intercalated in between vanadium oxide layers that was made by hydrolyzing and condensing $VO(OPr^n)_3$ in n-propanol.[201] Intercalating materials will be discussed further later in Sect. 6.6.

6.5.2. Class II hybrids

Class II hybrids comprise organic and inorganic components chemically bonded with each other and truly differ from organic–inorganic nanocomposites. In general, such hybrids are synthesized by hydrolyzing and polymerizing organic and inorganic precursors simultaneously. Inorganic precursors are referred to inorganic salts, such as $SiCl_4$ and $ErCl_3$, organic salts, such as $Cd(acac)_2$, and alkoxides, such as $Al(OR)_3$ and $Ti(OR)_4$ where R is alkyl group. All the coordination groups associated with the metal cations in inorganic precursors are hydrolysable, i.e. readily replaceable by hydroxyl and/or oxo groups during hydrolysis and condensation process. Organic precursors consist of at least one unhydrolyzable coordination group and examples are $Si(OR)_3R'$ and $Si(OR)_2R'_2$, which are also known as organoalkoxysilanes where R' is also an alkyl group linked to Si through Si-C bond. Such unhydrolyzable organic groups are referred to as pendant organic groups. For organoalkoxysilanes, no three-dimensional network would be formed if there are more than one pendant organic group attached to each silicon atom. There are other forms of organic precursors in which unhydrolyzable organic groups bridge two silicon atoms. Such organic groups are referred to as bridge groups. Examples of such organoalkoxysilanes are given in Fig. 6.22.[202,203] Since metal-carbon bonds are very stable during sol-gel processing and unhydrolyzable, the organic group R' associated with the precursors will be incorporated into inorganic sol-gel network directly together with the metal cations. Typical hydrolysis and condensation reactions in the formation of such hybrids can be described as follows, taking silica-based hybrids as an example:

$$Si(OR)_4 + 4H_2O \Leftrightarrow Si(OH)_4 + 4\,HOR \qquad (6.2)$$
$$Si(OR)_3R' + 3H_2O \Leftrightarrow Si(OH)_3R' + 3HOR \qquad (6.3)$$
$$Si(OH)_4 + Si(OH)_3R' \Leftrightarrow (HO)_3Si-O-Si(OH)_2R' \qquad (6.4)$$
$$Si(OH)_3R' + Si(OH)_3R' \Leftrightarrow R'(HO)_2Si-O-Si(OH)_2R' \qquad (6.5)$$

It should be noted that although organoalkoxysilanes are the most useful and widely used family of organometallics for the synthesis of hybrid oxide-organic materials, other organometallics are also synthesized and used for

Special Nanomaterials

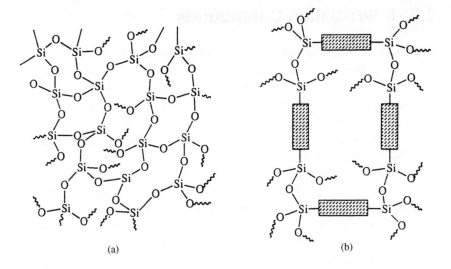

Fig. 6.22. Comparison of inorganic and hybrid structures: (a) silica network with some hydrolysable organic ligands and (b) an example of organoalkoxysilane with organic groups bridging two silicon atoms. In this structure, organic and inorganic components are chemically linked to form a single phase material. [K. Shea, D.A. Loy, and O. Webster, *J. Am. Chem. Soc.* **114**, 6700 (1992).]

the synthesis of organic–inorganic hybrids by co-condensation.[204] For example, butenyl chains were linked to Sn atom directly with C–Sn bonds.

The incorporation of organic components into inorganic matrix through either physical trapping or chemical bonding not only introduces and modifies various physical properties. The presence of organic components would also exert appreciable influences on the sol-gel processing and the resultant microstructures. Organic groups may have catalytic effects to promote hydrolysis and condensation reactions. Long-chained organic ligands may also introduce steric diffusion barrier or increase the viscosity of the sol, resulting in a diffusion-limited condensation or polymerization process. Depending on the nature and amount of organic components introduced into the systems, highly porous[205] or relatively dense hybrids[206,207] can be prepared without subjecting to heat-treatment at elevated temperatures. Some unique hierarchical microstructures can also be obtained by combining both highly porous and relatively dense structures with appropriately designed processing.[208] Although almost all the organic–inorganic hybrids are made through hydrolysis-condensation process, it has been demonstrated that non-hydrolytic sol-gel process is also capable of synthesizing hybrids.[209] Organic–inorganic hybrids with ordered nanostructures can be easily achieved by evaporation-induced self-assembly as demonstrated by Brinker and his coworkers.[210–213]

6.6. Intercalation Compounds

Intercalation compounds are a special family of materials. The intercalation refers to the reversible insertion of mobile guest species (atoms, molecules, or ions) into a crystalline host lattice that contains an interconnected system of empty lattice site of appropriate size, while the structural integrity of the host lattice is formally conserved.[214] The intercalation reactions typically occur around room temperature. A variety of host lattice structures have been found to undergo such low temperature reactions.[215] However, the intercalation reactions involving layered host lattices have been most extensively studied, partly due to the structural flexibility, and the ability to adapt to the geometry of the inserted guest species by free adjustment of the interlayer separation. In this section, only a brief summary on some layered intercalation compounds will be presented. For a more detailed discussion, the readers are referred to a comprehensive and excellent article on inorganic intercalation compounds.[214] In spite of the differences in chemical composition and lattice structure of the host sheets, all the layer hosts are characterized by strong intralayer covalent bonding and weak interlayer interactions. The weak interlayer interactions include van der Waals force or electrostatic attraction through oppositely charged species between two layers. Various host lattices can react with a variety of guest species to form intercalates. Examples of host lattices are metal dichalcogenides, metal oxyhalides, metal phosphorus trisulphides, metal oxides, metal phosphates, hydrogen phosphates, and phosphonates, graphite and layered clay minerals. Guest materials include metal ions, organic molecules and organometallic molecules. When guest species are incorporated into host lattices, various structural changes would take place. Figure 6.23 shows the principle geometrical transitions of layered host lattice matrices upon intercalation of guest species: (i) change in interlayer spacing, (ii) change in stacking mode of the layers, and (iii) forming of intermediate phases at low guest concentrations may exhibit staging.[216] Figure 6.24, as an example, shows the schematic structure and the interlayer spacing as a function of the organic chain length of intercalates of zirconium hydrogen phosphate.[217]

There are various synthesis methods for the formation of intercalation compounds.[214,218] The most commonly used and simplest method is the direct reaction of the guest species with the host lattice. The formation of $Li_xV_2O_5$ ($0 \leq x \leq 1$) from V_2O_5 and LiI is a typical example of such a synthesis method.[219] For direction reactions, intercalation reagent must be good reducing agents of the host crystals. Ion exchange is a method to replace the guest ion in an intercalation compound with another guest ion, which offers a useful route for intercalating large ions that do not directly intercalate.[220] Appropriate chosen solvents or electrolytes may assist the

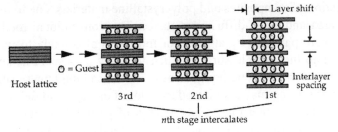

Fig. 6.23. Principle geometrical transitions of layered host lattice matrices upon intercalation of guest species: (1) change in interlayer spacing, (2) change in stacking mode of the layers, and (3) form intermediate phases at low guest concentrations may exhibit staging. [R. Schöllhorn, *NATO Ser.* **B172**, 149 (1987).]

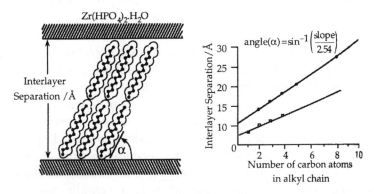

Fig. 6.24. The schematic structure and the interlayer spacing as a function of the organic chain length of intercalates of zirconium hydrogen phosphate. [U. Costantino, *J. Chem. Soc. Dalton Trans.* **402** (1979).]

ion exchange reactions by flocculating and reflocculating the host structure.[221] Electrointercalation is yet another method, in which the host lattice serves as the cathode of an electrochemical cell.[222]

6.7. Nanocomposites and Nanograined Materials

Nanocomposites and nanograined materials have been studied extensively mainly for improved physical properties.[223,224] Nanocomposites refer to materials consisting of at least two phases with one dispersed in another that is called matrix and forms a three-dimensional network, whereas nanograined materials are generally multi-grained single phase polycrystalline materials. A reduced particle size would definitely promote the densification of composites and polycrystalline materials, due to the large surface area and short diffusion distance. The conventional method for making small particles by attrition or milling is also likely to introduce impurity into the particle surface. Such impurity may serve as sintering aid, and may form eutectic liquid so as to introduce liquid phase sintering. Attrition or milling may also introduce a lot of damage and defects to the particle surface so that the surface energy of particles is high, which is again favorable to densification. Other methods of making nanosized powders, such as sol-gel processing and citrate combustion, would produce highly pure and less surface defect particles. As will be discussed briefly in Chapter 8, Hall–Petch relationship suggests that the mechanical properties increase inversely proportional to the square root of particle size at micrometer scale. However, the relationship between the mechanical properties and the particle size does not necessarily follow the Hall–Petch equation. More study is clearly needed to establish a better understanding on the size dependence in the nanometer scale. Obviously in nanocomposites and nanograined polycrystalline materials, surface or grain boundaries play a much more significant role in determining the mechanical properties than in large grained bulk materials.

Nanocomposites and nanograined materials are not necessarily limited to the bulk materials made by sintering nanosized powders. Deposition of a solid inside a porous substrate, by vapor chemical reactions, is one established technique, referred to as chemical vapor infiltration, for the synthesis of composite.[225–227] Ion implantation is a versatile and powerful technique for synthesizing nanometer-scale clusters and crystals embedded in the near-surface region of a variety of hosts. The principal features of this synthesis technique and various materials have been reviewed by Meldrum and co-workers.[228] Nanocomposites of polymers and metals or polymers and semiconductors are reviewed by Caseri.[229] Extensive research on various carbon nanotube composites were reviewed by Terrones.[20] A variety of nanostructured materials that have been discussed in previous chapters including this one can be perfectly grouped as nanocomposites or nanograined materials. For example, class I organic–inorganic hybrids can be considered as an organic–inorganic nanocomposite, anodic alumina membrane filled with metal nanowires is metal–ceramic composite.

6.8. Summary

In this chapter, we discussed various special nanomaterials that were not included in the previous three chapters, though they all possess characteristic dimension in the nanometer scale. Most of the nanomaterials discussed in this chapter do not exist in nature. Each of these materials brings unique physical properties and promises of potential and important applications. Such promises have made each of these materials an active research field. Although it is not known what types of new or artificial materials will be created in the near future, it is for sure that the members of the artificial material family will increase steadily, with more unknown physical properties.

References

1. M.S. Dresselhaus, *Ann. Rev. Mater. Sci.* **27**, 1 (1997).
2. M.S. Dresselhaus, G. Dresselhaus, and P.C. Eklund, *Science of Fullerenes and Carbon Nanotubes*, Academic Press, San Diego, CA, 1996.
3. H.W. Kroto, J.R. Heath, S.C. O'Brien, R.F. Curl, and R.E. Smalley, *Nature* **318**, 162 (1985).
4. F. Dierderich and R.L. Whetten, *Acc. Chem. Res.* **25**, 119 (1992).
5. K. Kikuchi, N. Nakahara, T. Wakabayashi, S. Suzuki, H. Shiromaru, Y. Miyake, K. Saito, I Ikemoto, M. Kainosho, and Y. Achiba, *Nature* **357**, 142 (1992).
6. D.E. Manolopoulos and P.W. Fowler, *Chem. Phys. Lett.* **187**, 1 (1991).
7. M.S. Dresselhaus and G. Gresselhaus, *Ann. Rev. Mater. Sci.* **25**, 487 (1995).
8. R.B. Fuller, in *The Artifacts of R. Buckminster Fuller: A Comprehensive Collection of His Designs and Drawings*, ed. W. Marlin, Garland, New York, 1984.
9. W.I.F. David, R.M. Ibberson, J.C. Matthewman, K. Prassides, T.J.S. Dennis, J.P. Hare, H.W. Kroto, R. Taylor, and D.R.M. Walton, *Nature* **353**, 147 (1991).
10. P.W. Stephens, L. Mihaly, P.L. Lee, R.L. Whetten, S.M. Huang, R. Kane, F. Deiderich, and K. Holczer, *Nature* **351**, 632 (1991).
11. J.E. Fischer, P.A. Heiney, A.R. McGhie, W.J. Romanow, A.M. Denenstein, J.P. McCauley, Jr., and A.B. Smith III, *Science* **252**, 1288 (1991).
12. J.E. Fischer, P.A. Heiney, and A.B. Smith III, *Acc. Chem. Res.* **25**, 112 (1992).
13. M.S. Dresselhaus, G. Dresselhaus, and P.C. Eklund, *J. Mater. Res.* **8**, 2054 (1993).
14. W. Krätschmer, L.D. Lamb, K. Fostiropoulos, and D.R. Huffman, *Nature* **347**, 354 (1990).
15. R. Taylor and D.R.M. Walton, *Nature* **363**, 685 (1993)
16. G.A. Olaf, I. Bucsi, R. Aniszfeld, and G.K. Surya Prakash, *Carbon* **30**, 1203 (1992).
17. A.R. Kortan, N. Kopylov, S. Glarum, E.M. Gyorgy, A.P. Ramirez, R.M. Fleming, F.A. Thiel, and R.C. Haddon, *Nature* **355**, 529 (1992).
18. R.E. Smalley and B.I. Yakobson, *Solid State Commun.* **107**, 597 (1998).
19. R.L. Meng, D. Ramirez, X. Jiang, P.C. Chow, C. Diaz, K. Matsuishi, S.C. Moss, P.H. Hor, and C.W. Chu, *Appl. Phys. Lett.* **59**, 3402 (1991).
20. M. Terrones, *Ann. Rev. Mater. Res.* **33**, 419 (2003).
21. P.J.F. Harris, *Carbon Nanotubes and Related Structures, New Materials for the Twenty-First Century*, Cambridge Univ. Press, Cambridge, 1999.
22. K. Tanaka, T. Yamabe, and K. Fukui, *The Science and Technology of Carbon Nanotubes*, Elsevier, Amsterdam, 1999.
23. R. Saito, G. Dresselhaus, and M.S. Dresselhaus, *Physical Properties of Carbon Nanotubes*, Imperial College Press, London, 1998.
24. M.S. Dresselhaus, G. Dresselhaus, and R. Saito, *Carbon* **33**, 883 (1995).

25. T.W. Ebbesen, *Ann. Rev. Mater. Sci.* **24**, 235 (1994).
26. T. Guo, P. Nikolaev, A. Thess, D.T. Colbert, and R.E. Smalley, *J. Phys. Chem.* **55**, 10694 (1995).
27. M. Endo, K. Takeuchi, S. Igarashi, K. Kobori, M. Shiraishi, and H.W. Kroto, *J. Phys. Chem. Solids* **54**, 1841 (1993).
28. O. Gröning, O.M. Kuttel, Ch. Emmenegger, P. Gröning, and L. Schlapbach, *J. Vac. Sci. Technol.* **B18**, 665 (2000).
29. W.K. Hsu, J.P. Hare, M. Terrones, H.W. Kroto, D.R.M. Walton, and P.J.F. Harris, *Nature* **377**, 687 (1995).
30. W.K. Hsu, M. Terrones, J.P. Hare, H. Terrones, H.W. Kroto, and D.R.M. Walton, *Chem. Phys. Lett.* **262**, 161 (1996).
31. S. Iijima, *Nature* **354**, 56 (1991).
32. S. Iijima and T. Ichihashi, *Nature* **363**, 603 (1993).
33. D.S. Bethune, C.H. Kiang, M.S. de Vries, G. Gorman, R. Savoy, J. Vazquez, and R. Beyers, *Nature* **363**, 605 (1993).
34. A. Thess, R. Lee, P. Nikolaev, H. Dai, P. Petit, J. Robert, C. Xu, Y.H. Lee, S.G. Kim, A.G. Rinzler, D.T. Colbert, G.E. Scuseria, D. Tomanek, J.E. Fischer, and R.E. Smalley, *Science* **273**, 483 (1996).
35. R.E. Smalley, *Mater. Sci. Engr.* **B19**, 1 (1993).
36. T.W. Ebbesen and P.M. Ajayan, *Nature* **358**, 220 (1992).
37. D.T. Colbert, J. Zhang, S.M. McClure, P. Nikolaev, Z. Chen, J.H. Hafner, D.W. Owens, P.G. Kotula, C.B. Carter, J.H. Weaver, A.G. Rinzler, and R.E. Smalley, *Science* **266**, 1218 (1994).
38. W.Z. Li, S.S. Xie, L.X. Qian, B.H. Chang, B.S. Zou, W.Y. Zhou, R.A. Zhano, and G. Wang, *Science* **274**, 1701 (1996).
39. Y.C. Choi, Y.M. Shin, Y.H. Lee, B.S. Lee, G.S. Park, W.B. Choi, N.S. Lee, and J.M. Kim, *Appl. Phys. Lett.* **76**, 2367 (2000).
40. C. Bower, W. Zhu, S. Jin, and O. Zhou, *Appl. Phys. Lett.* **77**, 830 (2000).
41. S.H. Tsai, C.W. Chao, C.L. Lee, and H.C. Shih, *Appl. Phys. Lett.* **74**, 3462 (1999).
42. R.T.K. Baker and P.S. Harris, *Chem. Phys. Carbon* **14**, 83 (1978).
43. J. Kong, H.T. Soh, A.M. Cassell, C.F. Quate, and H. Dai, *Nature* **395**, 878 (1998).
44. S. Fan, M.G. Chapline, N.R. Franklin, T.W. Tombler, A.M. Cassell, and H. Dai, *Science* **283**, 512 (1999).
45. W.Z. Li, S.S. Xie, L.X. Qian, B.H. Chang, B.S. Zou, W.Y. Zhou, R.A. Zhao, and G. Wang, *Science* **274**, 1701 (1996).
46. M. Terrones, N. Grobert, I. Olivares, I.P. Zhang, H. Terrones, K. Kordatos, W.K. Hsu, J.P. Hare, P.D. Townsend, K. Prassides, A.K. Cheetham, H.W. Kroto, and D.R.M. Walton, *Nature* **388**, 52 (1997).
47. C. Laurent, E. Flahaut, A. Peigney, and A. Rousset, *New J. Chem.* **22**, 1229 (1998).
48. X.Y. Liu, B.C. Huang, and N.J. Coville, *Carbon* **40**, 2791 (2002).
49. W. Qian, H. Yu, F. Wei, Q. Zhang, and Z. Wang, *Carbon* **40**, 2961 (2002).
50. M. Endo, K. Takeuchi, S. Igarashi, K. Kobori, M. Shiraishi, and H.W. Kroto, *J. Phys. Chem. Solids* **54**, 1841 (1993).
51. R.T.K. Baker, *Carbon* **27**, 315 (1989).
52. Z.F. Ren, Z.P. Huang, J.W. Xu, J.H. Wang, P. Bush, M.P. Siegal, and P.N. Provencio, *Science* **282**, 1105 (1998).
53. H. Murakami, M. Hirakawa, C. Tanaka, and H. Yamakawa, *Appl. Phys. Lett.* **76**, 1776 (2000).

54. Y. Chen, D.T. Shaw, and L. Guo, *Appl. Phys. Lett.* **76**, 2469 (2000).
55. D.C. Li, L. Dai, S. Huang, A.W.H. Mau, and Z.L. Wang, *Chem. Phys. Lett.* **316**, 349 (2000).
56. S. Fan, M.G. Chapline, N.R. Franklin, T.W. Tombler, A.M. Cassell, and H. Dai, *Science* **283**, 512 (1999).
57. C.J. Lee and J. Park, *Appl. Phys. Lett.* **77**, 3397 (2000).
58. Z.W. Pan, S.S. Xie, B.H. Chang, C.Y. Wang, L. Lu, W. Liu, W.Y. Zhou, and W.Z. Li, *Nature* **394**, 631 (1998).
59. Z.F. Ren, Z.P. Huang, J.W. Xu, J.H. Wang, P. Bush, M.P. Siegal, and P.N. Provencio, *Science* **282**, 1105 (1998).
60. M. Tanemura, K. Iwata, K. Takahashi, Y. Fujimoto, F. Okuyama, H. Sugie, and V. Filip, *J. Appl. Phys.* **90**, 1529 (2001).
61. M. Endo and H.W. Kroto, *J. Phys. Chem.* **96**, 6941 (1992).
62. T.W. Ebbessen, in *Carbon Nanotubes: Preparation and Properties*, ed. T.W. Ebbessen, CRC Press, Boca Raton, FL, p.139, 1997.
63. S.C. Tsang, P.J.F. Harris, and M.L.H. Green, *Nature* **362**, 520 (1993).
64. P.M. Ajayan, T.W. Ebbessen, T. Ichihashi, S. Iijima, K. Tanigaki, and H. Hiura, *Nature* **362**, 522 (1993).
65. H. Hiura, T.W. Ebbessen, and K. Tanigaki, *Adv. Mater.* **7**, 275 (1995).
66. F. Ikazaki, S. Oshima, K. Uchida, Y. Kuriki, H. Hayakawa, M. Yumura, K. Takahashi, and K. Tojima, *Carbon* **32**, 1539 (1994).
67. L. Langer, L. Stockman, J.P. Heremans, V. Bayot, C.H. Olk, C. Van Haesendonck, Y. Bruynseraede, and J.P. Issi, *J. Mater. Res.* **9**, 927 (1994).
68. W.A. de Heer, A. Chatelain, and D. Ugarte, *Science* **270**, 1179 (1995).
69. Y. Nakayama, S. Akita, and Y. Shimada, *Jpn. J. Appl. Phys.* **34**, L10 (1995).
70. M. Terrones, W.K. Hsu, A. Schilder, H. Terrones, N. Grobert, J.P. Hare, Y.Q. Zhu, M. Schwoerer, K. Prassides, H.W. Kroto, and D.R.M. Walton, *Appl. Phys.* **A66**, 307 (1998).
71. P.M. Ajayan, O. Stephan, C. Colliex, and D. Trauth, *Science* **265**, 1212 (1994).
72. R.S. Ruoff and D.C. Lorents, *Carbon* **33**, 925 (1995).
73. P. Kim, L. Shi, A. Majumdar, and P.L. McEuen, *Phys. Rev. Lett.* **87**, 215502 (2001).
74. A.C. Dillon, *Nature* **386**, 377 (1997).
75. G.E. Gadd, *Science* **277**, 933 (1997).
76. P.J. Briffo, K.S.M. Santhanam, and P.M. Ajayan, *Nature* **406**, 586 (2000).
77. S. Frank, P. Poncharal, Z.L. Wang, and W.A. de Heer, *Science* **280**, 1744 (1998).
78. P.G. Collins, A. Zettl, H. Bando, A. Thess, and R.E. Smalley, *Science* **278**, 100 (1997).
79. W.A. de Heer, A. Chetalain, and D. Ugarte, *Science* **270**, 1179 (1996).
80. H.J. Dai, J.H. Halfner, A.G. Rinzler, D.T. Colbert, and R.E. Smalley, *Nature* **384**, 147 (1996).
81. S. Ghosh, A.K. Sood, and N. Kumar, *Science* **299**, 1042 (2003).
82. H.D. Wagner, O. Lourie, Y. Feldman, and R. Tenne, *Appl. Phys. Lett.* **72**, 188 (1998).
83. K.S.W. Sing, D.H.W. Everett, R.A. Haul, L. Moscou, J. Pierotti, J. Rouquerol, and T. Siemieniewska, *Pure Appl. Chem.* **57**, 603 (1985).
84. G.J. de A.A. Soler-ILLia, C. Sanchez, B. Lebeau, and J. Patarin, *Chem. Rev.* **102**, 4093 (2002).
85. A. Galarneau, F. Di Renzo, F. Fajula, and J. Vedrine, eds., *Zeolites and Mesoporous Materials at the Dawn of the 21st Century*, Elsevier, Amsterdam, 2001.
86. A. Berthod, *J. Chim Phys.* (Fr.) **80**, 407 (1983).

87. K.L. Mittal and E.J. Fendler, eds., *Solution Behavior of Surfactants*, Plenum Press, New York, 1982.
88. A. Corma, *Chem. Rev.* **97**, 2373 (1997).
89. C.T. Kresge, M.E. Leonowicz, W.J. Roth, J.C. Vartulli, and J.S. Beck, *Nature* **359**, 710 (1992).
90. J. S. Beck, J.C. Vartuli, W.J. Roth, M.E. Leonowicz, C.T. Kresge, K.D. Schmitt, C.T.W. Chu, D.H. Olson, E.W. Sheppard, S.B. McCullen, J.B. Higgins, and J.L. Schlenker, *J. Am. Chem. Soc.* **114**, 10834 (1992).
91. P.T. Tanev and T.J. Pinnavaia, *Science* **267**, 865 (1995).
92. S. Forster and M. Antonietti, *Adv. Mater.* **10**, 195 (1998).
93. D. Zhao, J. Feng, Q. Huo, N. Melosh, G.H. Fredrickson, B.F. Chmelka, and G.D. Stucky, *Science* **279**, 548 (1998).
94. A. Sayari and S. Hamoudi, *Chem. Mater.* **13**, 3151 (2001).
95. P.D. Yang, D.Y. Zhao, D.I. Margoless, B.F. Chemelka, and G.D. Stucky, *Nature* **396**, 152 (1998).
96. D.M. Antonelli and J.Y. Ying, *Chem. Mater.* **8**, 874 (1996).
97. Z.R. Tian, W. Tong, J.Y. Wang, N.G. Duan, V.V. Krishnan, and S.L. Suib, *Science* **276**, 926 (1997).
98. A. Sayari and P. Liu, *Microporous Mater.* **12**, 149 (1997).
99. P.V. Braun, P. Oscar, and S.I. Stupp, *Nature* **380**, 325 (1996).
100. N. Ulagappan and C.N.R. Rao, *Chem. Commun.* 1685 (1996).
101. F. Schüth, *Chem. Mater.* **13**, 3184 (2001).
102. M. Mamak, N. Coombs, and G. Ozin, *Adv. Mater.* **12**, 198 (2000).
103. U. Ciesla, S. Schacht, G.D. Stucky, K.K. Unger, and F. Schuth, *Angew. Chem. Int. Ed. Engl.* **35**, 541 (1996).
104. T.T. Emons, J. Li, and L.F. Nazar, *J. Am. Chem. Soc.* **124**, 8516 (2002).
105. T. Asefa, C. Yoshina-Ishii, M.J. MacLachlan, and G.A. Ozin, *J. Mater. Chem.* **10**, 1751 (2000).
106. A. Stein, B.J. Melde, and R.C. Schroden, *Adv. Mater.* **12**, 1403 (2000).
107. S.H. Tolbert, T.E. Schaeffer, J. Feng, P.K. Hansma, and G.D. Stucky, *Chem. Mater.* **9**, 1962 (1997).
108. M. Templin, A. Franck, A. Du Chesne, H. Leist, Y. Zhang, R. Ulrich, V. Schädler, and U. Wiesner, *Science* **278**, 1795 (1997).
109. J. Liu, Y. Shin, Z. Nie, J.H. Chang, L.-Q. Wang, G.E. Fryxell, W.D. Samuels, and G.J. Exarhos, *J. Phys. Chem.* **A104**, 8328 (2000).
110. X. Feng, G.E. Fryxell, L.Q. Wang, A.Y. Kim, and J. Liu, *Science* **276**, 923 (1997).
111. J. Liu, X. Feng, G.E. Fryxell, L.Q. Wang, A.Y. Kim, and M. Gong, *Adv. Mater.* **10**, 161 (1998).
112. I.A. Aksay, M. Trau, S. Manne, I. Honma, N. Yao, L. Zhou, P. Fenter, P.M. Eisenberger, and S.M. Gruner, *Science* **273**, 892 (1996).
113. A.S. Brown, S.A. Holt, T. Dam, M. Trau, and J.W. White, *Langmuir* **13**, 6363 (1997).
114. Y. Lu, R. Ganguli, C.A. Drewien, M.T. Anderson, C.J. Brinker, W. Gong, Y. Guo, H. Soyez, B. Dunn, M.H. Huang, and J.I. Zink, *Nature* **389**, 364 (1997).
115. J.E. Martin, M.T. Anderson, J. Odinek, and P. Newcomer, *Langmuir* **13**, 4133 (1997).
116. H. Yang, N. Coombs, I. Sokolov, and G.A. Ozin, *Nature* **381**, 589 (1996).
117. H. Yang, A. Kuperman, N. Coombs, S. Mamiche-Afara, and G.A. Ozin, *Nature* **379**, 703 (1996).
118. A. Firouzi, D.J. Schaefer, S.H. Tolbert, G.D. Stucky, and B.F. Chmelka, *J. Am. Chem. Soc.* **119**, 9466 (1997).

119. H.W. Hillhouse, T. Okubo, J.W. van Egmond, and M. Tsapatsis, *Chem. Mater.* **9**, 1505 (1997).
120. H.P. Lin, S.B. Liu, C.Y. Mou, and C.Y. Tang, *Chem. Commun.* **583** (1999).
121. S.J. Limmer, T.L. Hubler, and G.Z. Cao, *J. Sol-Gel Sci. Technol.* **26**, 577 (2003).
122. H. Tanaka, *J. Non-Cryst. Solid* **65**, 301 (1984).
123. M.P. Thomas, R.R. Landham, E.P. Butler, D.R. Cowieseon, E. Burlow, and P. Kilmartin, *J. Membrane Sci.* **61**, 215 (1991).
124. R.L. Fleisher, P.B. Price, and R.M. Walker, *Nuclear Tracks in Solids*, University of California Press, Berkeley, CA, 1975.
125. S.S. Prakash, C.J. Brinker, and A.J. Hurd, *J. Non-Cryst. Solids* **188**, 46 (1995).
126. F.A.L. Dullien, Porous Media, *Fluid Transport and Pore Structure*, Academic Press, New York, 1979.
127. A.J. Burggraaf, K. Keizer, and B.A. van Hassel, *Solid State Ionics* **32/33**, 771 (1989).
128. S.S. Kistler, *Nature* **127**, 741 (1931).
129. J. Fricke, ed., *Aerogels*, Springer, Berlin, 1986.
130. R.W. Pekala and L.W. Hrubesh (guest editors), *J. Non-Cryst. Solids* **186** (1995)
131. A.C. Pierre and G.M. Pajonk, *Chem. Rev.* **102**, 4243 (2002).
132. D.W. Matson and R.D. Smith, *J. Am. Ceram. Soc.* **72**, 871 (1989).
133. G. Dagan and M. Tomkiewicz, *J. Phys. Chem.* **97**, 12651 (1993).
134. T. Osaki, T. Horiuchi, T. Sugiyama, K. Susuki, and T. Mori, *J. Non-Cryst. Solids* **225**, 111 (1998).
135. A.E. Gash, T.M. Tillotson, J.H. Satcher Jr., L.W. Hrubesh, and R.L. Simpson, *J. Non-Cryst. Solids* **285**, 22 (2001).
136. C. Hernandez and A.C. Pierre, *J. Sol-Gel Sci. Technol.* **20**, 227 (2001).
137. D.L. Ou and P.M. Chevalier, *J. Sol-Gel Sci. Technol.* **26**, 657 (2003).
138. G.Z. Cao and H. Tian, *J. Sol-Gel Sci. Technol.* **13**, 305 (1998).
139. S.S. Prakash, C.J. Brinker, A.J. Hurd, and S.M. Rao, *Nature* **374**, 439 (1995).
140. R.W. Pekala, *J. Mater. Sci.* **24**, 3221 (1989).
141. K. Barral, *J. Non-Cryst. Solids* **225**, 46 (1998).
142. H. Tamon, H. Ishizaka, T. Yamamoto, and T. Suzuki, *Carbon* **37**, 2049 (1999).
143. J.L. Schlenker and G.H. Kuhl, *Proc. Ninth Int. Zeolite Conf.*, ed. R. von Ballmoos, J.B. Higgins, and M.M. Treacy, Butterworth-Heinemann, Boston, MA, p. 3, 1993.
144. M.M. Helmkamp and M.E. Davis, *Ann. Rev. Mater. Sci.* **25**, 161 (1995).
145. J.V. Smith, *Chem. Rev.* **88**, 149 (1988).
146. J.M. Newsam, *Science* **231**, 1093 (1986).
147. H. van Bekkum, E.M. Flanigen, P.A. Jacobs, and J.C. Jansen, (eds.) *Introduction to Zeolite Science and Practice*, 2nd edition, Elsevier, Amsterdam, 2001.
148. Ch. Baerlocher, W.M. Meier, and D.H. Olson, (eds.), *Atlas of Zeolite Framework Types*, Elsevier, Amsterdam, 2001.
149. W.M. Meier and D.H. Olson, *Atlas of Zeolite Structure Types*, Butterworth-Heinemann, Boston, MA, 1992.
150. L.B. McCusker and C. Baerlocher, in *Introduction to Zeolite Science and Practice*, (2nd edition), eds. H. van Bekkum, E.M. Flanigen, P.A. Jacobs, and J.C. Jansen, Elsevier, Amsterdam, p. 37, 2001.
151. M.E. Davis, *Ind. Eng. Chem. Res.* **30**, 1675 (1991).
152. C.S. Cundy and P.A. Cox, *Chem. Rev.* **103**, 663 (2003).
153. J.C. Jansen, in *Introduction to Zeolite Science and Practice* (2nd edition), eds. H. van Bekkum, E.M. Flanigen, P.A. Jacobs, and J.C. Jansen, Elsevier, Amsterdam, p. 175, 2001.

154. S.L. Burkett and M.E. Davis, *J. Phys. Chem.* **98**, 4647 (1994).
155. C.E.A. Kirschhock, R. Ravishankar, L. Van Looveren, P.A. Jacobs, and J.A. Martens, *J. Phys. Chem.* **B103**, 4972 (1999).
156. T.V. Harris and S.I. Zones, *Stud. Surf. Sci. Catal.* **94**, 29 (1994).
157. S.I. Zones, M.M. Olmstead, and D.S. Santilli, *J. Am. Chem. Soc.* **114**, 4195 (1992).
158. S.I. Zones and D.S. Santilli, in *Proc. Ninth Int. Zeolite Conf.*, eds. R. von Ballmoos, J.B. Higgins, and M.M.J. Treacy, Butterworth-Heinemann, Boston, MA, p.171, 1993.
159. R.F. Lobo, M. Pan, I. Chan, S.I. Zones, P.A. Crozier, and M.E. Davis, *J. Phys. Chem.* **98**, 12040 (1994).
160. S.L. Lawton and W.J. Rohrbaugh, *Science* **247**, 1319 (1990).
161. K.D. Schmitt and G.J. Kennedy, *Zeolites* **14**, 635 (1994).
162. R. Szostak, *Handbook of Molecular Sieves*, Van Nostrand Reinhold, New York, 1992.
163. C.A. Fyfe, H. Gies, G.T. Kokotailo, B. Marler, and D.E. Cox, *J. Phys. Chem.* **94**, 3718 (1990).
164. M. Goepper, H.X. Li, and M.E. Davis, *J. Chem. Soc. Chem. Commun.* **22**, 1665 (1992).
165. P.M. Dove and D.A. Crerar, *Geochim. Cosmochim. Acta* **54**, 955 (1990).
166. P. Brady and J.V. Walther, *Chem. Geol.* **82**, 253 (1990).
167. J.B. Higgins, in *Reviews in Mineralogy: Silica Polymorphs*, Vol. 29, ed., P.H. Ribbe, Mineral. Soc. Am., Washington, DC, 1994.
168. S.I. Zones, *Microporous Mater.* **2**, 281 (1994).
169. K. Yamamoto, Y. Sakata, Y. Nohara, Y. Takahashi, and T. Tatsumi, *Science* **300**, 470 (2003).
170. C.W. Jones, K. Tsuji, and M.E. Davis, *Nature* **393**, 52 (1998).
171. L.M. Liz-Marzán, M. Giersig, and P. Mulvaney, *J. Chem. Soc. Chem. Commun.* **731** (1996).
172. L.M. Liz-Marzán, M. Giersig, and P. Mulvaney, *Langmuir* **12**, 4329 (1996).
173. R.J. Puddephatt, *The Chemistry of Gold*, Elsevier, Amsterdam, 1978.
174. E.P. Plueddermann, *Silane Coupling Agents*, 2nd edition, Plenum, New York, 1991.
175. B.V. Enüstün and J. Turkevich, *J. Am. Chem. Soc.* **85**, 3317 (1963).
176. R.K. Iler, *The Chemistry of Silica: Solubility, Polymerization, Colloid and Surface Properties, and Biochemistry*, John Wiley & Sons, New York, 1979.
177. C.J. Brinker and G.W. Scherer, *Sol-Gel Science: The Physics and Chemistry of Sol-Gel Processing*, Academic Press, San Diego, CA, 1990.
178. W. Stober, A. Fink, and E. Bohn, *J. Colloid Interfac. Sci.* **26**, 62 (1968).
179. W.D. Hergeth, U.J. Steinau, H.J. Bittrich, K. Schmutzler, and S. Wartewig, *Prog. Colloid Polym. Sci.* **85**, 82 (1991).
180. L. Quaroni and G. Chumanov, *J. Am. Chem. Soc.* **121**, 10642 (1999).
181. S.M. Marinakos, L.C. Brousseau, A. Jones, and D.L. Feldheim, *Chem. Mater.* **10**, 1214 (1998).
182. S.M. Marinakos, J.P. Novak, L.C. Brousseau, A.B. House, E.M. Edeki, J.C. Feldhaus, and D.L. Feldheim, *J. Am. Chem. Soc.* **121**, 8518 (1999).
183. S.M. Marinakos, D.A. Shultz, and D.L. Feldheim, *Adv. Mater.* **11**, 34 (1999).
184. F. Caruso, *Adv. Mater.* **13**, 11 (2001).
185. C.H.M. Hofman-Caris, *New J. Chem.* **18**, 1087 (1994).
186. R. Partch, S.G. Gangolli, E. Matijevic, W. Cai, and S. Arajs, *J. Colloid Interf. Sci.* **144**, 27 (1991).
187. H.T. Oyama, R. Sprycha, Y. Xie, R.E. Partch, and E. Matijevic, *J. Colloid Interf. Sci.* **160**, 298 (1993).

188. R. Sprycha, H.T. Oyama, A. Zelenzev, and E. Matijevic, *Colloid Polym. Sci.* **273**, 693 (1995).
189. C.L. Huang and E. Metijevic, *J. Mater. Res.* **10**, 1327 (1995).
190. A. Ulman, *An Introduction of Ultrathin Organic Films: From Langmuir–Blodgett to Self-Assembly*, Academic Press, San Diego, CA, 1991.
191. J.H. Fendler, *Nanoparticles and Nanostructured Films: Preparation, Characterization and Application*, Wiley-VCH, Weinhein, 1998.
192. D.C. Blackley, *Polymer Lattices: Science and Technology*, 2nd edition, Vol. 1, Chapman and Hall, London, 1997.
193. D. Avnir, D. Levy, and R. Reisfeld, *J. Phys. Chem.* **88**, 5956 (1984).
194. D. Levy, S. Einhorn, and D.J. Avnir, *J. Non-Cryst. Solids* **113**, 137 (1989).
195. M. Toki, T.Y. Chow, T. Ohnaka, H. Samura, and T. Saegusa, *Polym. Bull.* **29**, 653 (1992).
196. B.E. Yodas, *J. Mater. Sci.* **14**, 1843 (1979).
197. B.M. Novak and C. Davies, *Macromolecules* **24**, 5481 (1991).
198. F. Nishida, B. Dunn, E.T. Knobbe, P.D. Fuqua, R.B. Kaner, and B.R. Mattes, *Mater. Res. Soc. Symp. Proc.* **180**, 747 (1990).
199. R. Reisfeld, D. Brusilovsky, M. Eyal, E. Miron, Z. Bursheim, and J. Ivri, *Chem. Phys. Lett.* **160**, 43 (1989).
200. E. Ruiz-Hitchky, *Adv. Mater.* **5**, 334 (1993).
201. N. Gharbi, C. Sanchez, J. Livage, J. Lemerle, L. Nejem, and J. Lefebvre, *Inorg. Chem.* **21**, 2758 (1982).
202. K. Shea, D.A. Loy, and O. Webster, *J. Am. Chem. Soc.* **114**, 6700 (1992).
203. R.J.P. Corriu, J.J.E. Moreau, P. Thepot, and C.M. Wong, *Chem. Mater.* **4**, 1217 (1992).
204. C. Bonhomme, M. Henry, and J. Livage, *J. Non-Cryst. Solids* **159**, 22 (1993).
205. W.G. Fahrenholtz, D.M. Smoth, and D.W. Hua, *J. Non-Cryst. Solids* **144**, 45 (1992).
206. B. Yoldas, *J. Sol-Gel Sci. Technol.* **13**, 147 (1998).
207. C.M. Chan, G.Z. Cao, H. Fong, M. Sarikaya, T. Robinson, and L. Nelson, *J. Mater. Res.* **15**, 148 (2000).
208. S. Seraji, Y. Wu, M.J. Forbess, S.J. Limmer, T.P. Chou, and G.Z. Cao, *Adv. Mater.* **12**, 1695 (2000).
209. J.N. Hay and H.M. Raval, *Chem. Mater.* **13**, 3396 (2001).
210. C.J. Brinker, Y.F. Lu, A. Sellinger, and H.Y. Fan, *Adv. Mater.* **11**, 579 (1999).
211. Y. Lu, R. Ganguli, C. Drewien, M. Anderson, C.J. Brinker, W. Gong, Y. Guo, H. Soyez, B. Dunn, M. Huang, and J. Zink, *Nature* **389**, 364 (1997).
212. A. Sellinger, P.M. Weiss, A. Nguyen, Y. Lu, R.A. Assink, W. Gong, and C.J. Brinker, *Nature* **394**, 256 (1998).
213. Y. Lu, H. Fan, A. Stump, T.L. Ward, T. Rieker, and C.J. Brinker, *Nature* **398**, 223 (1999).
214. D. O'Hare, in *Inorganic Materials*, eds. D.W. Bruce and D. O'Hare, John Wiley & Sons, New York, p.165, 1991.
215. R. Schöllhorn, *Angew Chem. Int. Ed. Engl.* **19**, 983 (1980).
216. R. Schöllhorn, in Chemical Physics of Intercalation, eds. A.P. Legrand and S. Flandrois, Plenum, New York, *NATO Ser.* **B172**, 149 (1987).
217. U. Costantino, *J. Chem. Soc. Dalton Trans.* 402 (1979).
218. D.W. Murphy, S.A. Sunshine, and S.M. Zahurak, in *Chemical Physics of Intercalation*, eds. A.P. Legrand and S. Flandrois, Plenum, New York, *NATO Ser.* **B172**, 173 (1987).

219. D.W. Murphy, P.A. Christian, F.J. Disalvo, and J.V. Waszczak, *Inorg. Chem.* **24**, 1782 (1985).
220. R. Clement, *J. Am. Chem. Soc.* **103**, 6998 (1981).
221. L.F. Nazar and A.J. Jacobson, *J. Chem. Soc. Chem. Commun.* 570 (1986).
222. R. Schöllhorn, *Physics of Intercalation Compounds*, Springer-Verlag, Berlin, 1981.
223. R.W. Siegel, S.K. Chang, B.J. Ash, J. Stone, P.M. Ajayan, R.W. Doremus, and L. Schadler, *Scripta Mater.* **44**, 2063 (2001).
224. J.P. Tu, N.Y. Wang, Y.Z. Yang, W.X. Qi, F. Liu, X.B. Zhang, H.M. Lu, and M.S. Liu, *Mater. Lett.* **52**, 452 (2002).
225. W.V. Kotlensky, *Chem. Phys. Carbon* **9**, 173 (1973).
226. S. Vaidyaraman, W.J. Lackey, G.B. Freeman, P.K. Agrawal, and M.D. Langman, *J. Mater. Res.* **10**, 1469 (1995).
227. P. Dupel, X. Bourrat, and R. Pailler, *Carbon* **33**, 1193 (1995).
228. A. Meldrum, R.F. Haglund, Jr., L.A. Boatner, and C.W. White, *Adv. Mater.* **13**, 1431 (2001).
229. W. Caseri, *Macromol. Rapid Commun.* **21**, 705 (2000).

Chapter 7

Nanostructures Fabricated by Physical Techniques

7.1. Introduction

In the previous chapters, we have discussed various routes for the synthesis and fabrication of a variety of nanomaterials; however, the synthesis routes applied have been focused mainly on the chemical methods approaches. In this chapter, we will discuss a different approach: fabrication of nanoscale structures with various physical techniques. Compared to the general chemical fabrication and processing methods, physical fabrication techniques for producing nanostructures are derived mainly from the techniques applied for the fabrication of microstructures in semiconductor industry. Particularly the fundamentals and basic approaches are mostly based on microfabrications. In this chapter, the following techniques for the fabrication of nanostructures and nanopatterns are discussed:

(1) Lithographic techniques
  (a) Photolithography
  (b) Phase shifting optical lithography
  (c) Electron beam lithography
  (d) X-ray lithography
  (e) Focused ion beam lithography
  (f) Neutral atomic beam lithography

(2) Nanomanipulation and nanolithography
   (a) Scanning tunneling microscopy (STM)
   (b) Atomic force microscopy (AFM)
   (c) Near-field scanning optical microscopy (NSOM)
   (d) Nanomanipulation
   (e) Nanolithography
(3) Soft lithography
   (a) Microcontact printing
   (b) Molding
   (c) Nanoimprint
   (d) Dip-pen nanolithography
(4) Self-assembly of nanoparticles or nanowires
   (a) Capillary force induced assembly
   (b) Dispersion interaction assisted assembly
   (c) Shear force assisted assembly
   (d) Electric-field assisted assembly
   (e) Covalently linked assembly
   (f) Gravitational field assisted assembly
   (g) Template assisted assembly
(5) Other methods for microfabrication
   (a) LIGA
   (b) Laser direct writing
   (c) Excimer laser micromachining

Although all the above-mentioned processes are discussed in this chapter, not all methods have the same capability in fabricating nanoscaled structures. In addition, the fundamentals of various fabrication processes differ significantly from each other. Each method offers some advantages over other techniques, but suffers from other limitations and drawbacks. No attempt has been made to exhaustively list all the methods developed in the literature or the technical details for fabricating nanostructured devices. Similar to previous chapters, the attention has been focused mainly on the fundamental concepts and general technical approaches. However, more detailed discussion has been devoted to SPM based nanomanipulation and nanolithography, not only because the processes are relatively new, but also because they are truly capable of fabricating nanometer scaled structures and devices.

7.2. Lithography

Lithography is also often referred to as photoengraving, and is the process of transferring a pattern into a reactive polymer film, termed as resist, which will subsequently be used to replicate that pattern into an underlying thin film or substrate.[1-5] Many techniques of lithography have been developed in the last half a century with various lens systems and exposure radiation sources including photons, X-rays, electrons, ions and neutral atoms. In spite of different exposure radiation sources used in various lithographic methods and instrumental details, they all share the same general technical approaches and are based on similar fundamentals. Photolithography is the most widely used technique in microelectronic fabrication, particularly for mass production of integrated circuit.[2]

7.2.1. Photolithography

Typical photolithographic process consists of producing a mask carrying the requisite pattern information and subsequently transferring that pattern, using some optical technique into a photoactive polymer or photoresist (or simply resist). There are two basic photolithographic approaches: (i) shadow printing, which can be further divided into contact printing (or contact-mode printing) and proximity printing, and (ii) projection printing. The terms "printing" and "photolithography" are used interchangeably in the literature.

Figure 7.1 outlines the basic steps of the photolithographic process, in which the resist material is applied as a thin coating over some base and subsequently exposed in an image-wise fashion through a mask, such that light strikes selected areas of the resist material. The exposed resist is then subjected to a development step. Depending on the chemical nature of the resist material, the exposed areas may be rendered more soluble in some developing solvent than the unexposed areas, thereby producing a positive tone image of the mask. Conversely, the exposed areas may be rendered less soluble, producing a negative tone image of the mask. The effect of this process is to produce a three-dimensional relief image in the resist material that is a replication of the opaque and transparent areas of the mask. The areas of resist that remain following the imaging and developing processes are used to mask the underlying substrate for subsequent etching or other image transfer steps. The resist material resists the etchant and prevents it from attacking the underlying substrate in those areas where it remains in place after development. Following the etching process, the resist is removed by stripping to produce a positive or negative tone relief image in the underlying substrate.

Diffraction sets the limit of the maximum resolution or the minimum size of the individual elements by photolithography, which can be obtained.

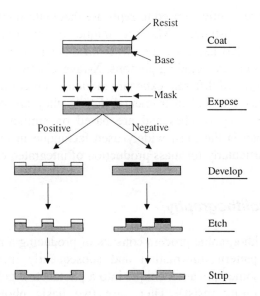

Fig. 7.1. Schematic representation of the photolithographic process sequences, in which images in the mask are transferred to the underlying substrate surface.

Diffraction refers to the apparent deviation of light from rectilinear propagation as it passes an obstacle such as an opaque edge and the phenomenon of diffraction can be understood qualitatively as follows. According to geometrical optics, if an opaque object is placed between a point light source and a screen, the edge of the object will cast a sharp shadow on the screen. No light will reach the screen at points within the geometrical shadow, whereas outside the shadow the screen will be uniformly illuminated. In reality, the shadow cast by the edge is diffuse, consisting of alternate bright and dark bands that extend into the geometrical shadow. This apparent bending of light around the edge is referred to as diffraction, and the resulting intensity distribution is called a diffraction pattern. Obviously diffraction causes the image of a perfectly delineated edge to become blurred or diffused at the resist surface. The theoretical resolution capability of shadow photolithography with a mask consisting of equal lines and spaces of width b is given by:

$$2b_{min} = 3\sqrt{\lambda\left(s+\frac{d}{2}\right)} \qquad (7.1)$$

where $2b$ is the grating period ($1/2b$ is the fundamental spatial frequency v), s the gap width maintained between the mask and the photoresist surface, λ the wavelength of the exposing radiation and d the photoresist thickness. For hard contact printing, s is equal to 0, and from the equation, the maximum resolution for 400 nm wavelength light and a 1 μm thick resist film will be slightly less than 1 μm.

In contact-mode photolithography, the mask and wafer are in intimate contact, and thus this method can transfer a mask pattern into a photoresist with almost 100% accuracy and provides the highest resolution. Other photolithographic techniques can approach but not exceed its resolution capabilities. However, the maximum resolution is seldom achieved because of dust on substrates and non-uniformity of the thickness of the photoresist and the substrate. Such problems can be avoided in proximity printing, in which, a gap between the mask and the wafer is introduced. However, increasing the gap degrades the resolution by expanding the penumbral region caused by diffraction. The difficulties in proximity printing include the control of a small and very constant space between the mask and wafer, which can be achieved only with extremely flat wafers and masks.

Projection printing differs from shadow printing. In projection printing techniques, lens elements are used to focus the mask image onto a wafer substrate, which is separated from the mask by many centimeters. Because of lens imperfections and diffraction considerations, projection techniques generally have lower resolution capability than that provided by shadow printing. The resolution limit in conventional projection photolithography is determined largely by the well-known Rayleigh's equation. The resolution, i.e. the minimum resolvable feature, R, and the corresponding depth of focus (DOF) are given by the following[6]:

$$R = \frac{k_1 \lambda}{NA} \quad (7.2)$$

$$\mathrm{DOF} = \frac{k_2 \lambda}{NA^2} \quad (7.3)$$

Here λ is the exposure wavelength, k_1 and k_2 are constants that depend on the specific resist material, process technology and image-formation techniques used, and NA is the numerical aperture of the optical system and is defined as

$$NA = n \sin \theta \quad (7.4)$$

where n is the index of refraction in image space and is usually equal to 1 (air or vacuum), and θ is the maximum cone angle of the exposure light beam. The diffraction limit is a very basic law of physics directly related to Heisenberg's uncertainty relation. It restricts any conventional imaging process to a resolution of approximately $\lambda/2$. Conventional photolithography is capable of fabricating features of 200 nm and above.[7]

To obtain higher resolutions, shorter wavelength light and lens systems with larger numerical apertures should be used. In general, the minimum feature size that can be obtained is almost the same as or slightly smaller than the wavelength of light used for the exposure, when a relatively larger numerical aperture (typically >0.5) is used. In such high NA lens systems, the depth of focus becomes very small and so the exposure process becomes sensitive to slight variations in the thickness and absolute position of the resist layer.[8]

Deep Ultra-Violet lithography (DUV) based on exposure at wavelengths below 300 nm, presents far more difficult technical challenges. Classical UV sources have lower output power in the DUV. Excimer lasers can provide 10 to 20 watts of power at any one of several wavelengths in the DUV. Of particular interest are the KrCl and KrF excimer lasers, which have outputs at 222 and 249 nm, respectively. High intensity, microwave powered emission sources provide substantially higher DUV output than classical electrode discharge mercury lamps.[9] Light sources with shorter wavelengths exploited for optical lithography include: KrF excimer laser with a wavelength of 249 nm, ArF excimer laser of 193 nm, F_2 excimer laser of 157 nm. With DUV, Optical lithography allows one to obtain patterns with a minimal size of ~100 nm.[10,11] Extreme UV (EUV) lithography with wavelengths in the range of 11–13 nm has also been explored for fabricating features with even smaller dimensions and is a strong candidate for achieving dimensions of 70 nm and below.[12,13] However, EUV lithography meets other problems. The adsorption of light in this wavelength regime is very strong, and therefore, refractive lens systems cannot be used. The reflectivity from reflective mirrors is rather low and, thus, the number of reflective mirrors should be kept as low as possible, not more than six. In addition, an extremely high precision metrology system is required to make this technique practically viable.[12]

Experimentally it is found that when the width of the slit is narrower than the wavelength, the radiation spreads out or is diffracted. Two edges very close together constitute a slit from which very distinct diffraction patterns are produced when illuminated with monochromatic light. The particular intensity distribution observed depends on the distance between the slit and the screen. For a short distance, the diffraction is Fresnel diffraction, which is the case in shadow printing. For a large distance such as in projection printing, the diffraction is Fraunhofer diffraction.

In addition to conventional photoresist polymers, Langmuir–Blodgett films and self-assembled monolayers have been used as resist in photolithography.[14,15] In such applications, photochemical oxidation, cross-linking, or generation of reactive groups are used to transfer patterns from the mask to the monolayers.[16,17]

7.2.2. Phase-shifting photolithography

Phase-shifting photolithography was first developed by Levenson et al.[18] In this method, a transparent mask induces abrupt changes of the phase of the light used for exposure, and cause optical attenuation at desired locations. These phase masks, also known as phase shifters, have produced futures of ~100 nm in photoresist.[19,20] Figure 7.2 schematically illustrates the principles of phase-shifting lithography. A clear film, i.e. a phase shifter or a phase mask, whose thickness is $\lambda/2(n-1)$ is placed on a photoresist with conformal contact, the phase angle of the exposure light passing through the film shifts by the amount of π to the incident light arriving at the surface of the photoresist. Here λ is the wavelength of the exposure light and n is the index of refraction of the phase mask. Because the light phase angle between the phase shifter and the photoresist is inverted, the electric field at the phase shifter edge is 0. So the intensity of the exposure light at the surface of the photoresist would be zero. An image having zero intensity can be formed about the edge of the phase shifter. Phase masks can be used in both projection and contact-mode photolithographic techniques. For a phase-shifting contact-mode photolithography, there are two possible approaches to increase the resolution: (i) reducing the wavelength of the source of exposure light and (ii) increasing the index of refraction of

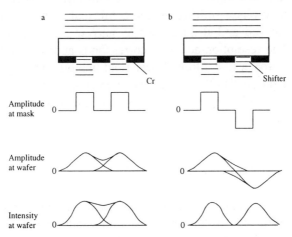

Fig. 7.2. Schematic illustrating the principles of phase shifting lithography, which utilizes the optical phase change at the phase shifter edge.

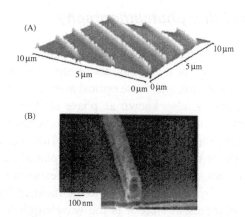

Fig. 7.3. Parallel lines formed in photoresist using near field contact-mode photolithography have widths on the order of 100 nm and are ~300 nm in height as imaged by (A) AFM and (B) SEM. [J.A. Rogers, K.E. Paul, R.J. Jackman, and G.M. Whitesides, *J. Vac. Sci. Technol.* **B16**, 59 (1998).]

the photoresist. The achievable photolithographic resolution is roughly of $\sim\lambda/4n$, where λ is the wavelength of the exposure light and n is the refractive index of the photoresist. Although contact-mode photolithography with a phase-shifting mask has a higher resolution, conformal contact between the phase-shifting mask and the photoresist on wafer is difficult to achieve, due to the presence of dust, non-uniformities in the thickness of the photoresist, and bowing of the mask or the substrate. However, by introducing elastomeric phase-shifting masks, conformal contact can be relatively easily achieved and feature lines as narrow as 50 nm have been generated.[21,22] The resolution achieved corresponds approximately to $\lambda/5$. An improved approach to conformal near field photolithography is to use masks constructed from "soft" organic elastomeric polymers.[23–25] Figure 7.3 shows a pattern created using such a contact-mode phase-shifting photolithographic process.[25]

7.2.3. Electron beam lithography

A finely focused beam of electrons can be deflected accurately and precisely over a surface. When the surface is coated with a radiation sensitive polymeric material, the electron beam can be used to write patterns of very high resolution.[26–29]. The first experimental electron beam writing systems were designed to take advantage of the high resolution capabilities in the late sixties.[30] Electron beams can be focused to a few nanometers in diameter and rapidly deflected either electromagnetically or electrostatically. Electrons possess both particle and wave properties; however, their wavelength is on the order of a few tenths of angstrom, and therefore their resolution is not limited by diffraction considerations. Resolution of electron beam lithography is, however, limited by forward scattering of the electrons in the resist layer and back scattering from the underlying substrate. Nevertheless, electron beam lithography is the most powerful tool for the fabrication of feathers as small as 3–5 nm.[31,32]

When an electron beam enters a polymer film or any solid material, it loses energy via elastic and inelastic collisions known collectively as electron scattering. Elastic collisions result only in a change of direction of the electrons, whereas inelastic collisions lead to energy loss. These scattering processes lead to a broadening of the beam, i.e. the electrons spread out as they penetrate the solid producing a transverse or lateral electron flux normal to the incident beam direction, and cause exposure of the resist at points remote from the point of initial electron incidence, which in turn results in developed resist images wider than expected. The magnitude of electron scattering depends on the atomic number and density of both the resist and substrate as well as the velocity of the electrons or the accelerating voltage.

Exposure of the resist by the forward and backscattered electrons depends on the beam energy, film thickness and substrate atomic number. As the beam energy increases, the energy loss per unit path length and scattering cross-sections decreases. Thus the lateral transport of the forward scattered electrons and the energy dissipated per electron decrease while the lateral extent of the backscattered electrons increases due to the increased electron range. As the resist film thickness increases, the cumulative effect of the small angle collisions by the forward scattered electrons increases. Thus the area exposed by the scattered electrons at the resist-substrate interface is larger in thick films than in thin films. Proper exposure requires that the electron range in the polymer film be greater than the film thickness in order to ensure exposure of the resist at the interface. As the substrate atomic number increases, the electron reflection coefficient increases which in turn increases the backscattered contribution.

Electron beam systems can be conveniently considered in two broad categories: those using scanned, focused electron beams which expose the wafer in serial fashion, and those projecting an entire pattern simultaneously onto a wafer. Scanning beam systems can be further divided into Gausian or round beam systems and shaped beam systems. All scanning beam systems have four typical subsystems: (i) electron source (gun),

(ii) electron column (beam forming system), (iii) mechanical stage and (iv) control computer which is used to control the various machine subsystems and transfer pattern data to the beam deflection systems.

Electron sources applicable to electron beam lithography are the same as those used in conventional electron microscopes. These sources can be divided into two groups: thermionic and field emission. Thermionic guns rely on the emission of electrons from a material that is heated above a critical temperature beyond which electrons are emitted from the surface. These sources are prepared from materials such as tungsten, thoriated tungsten, or lanthanum hexaboride. Field emission sources use a high electric field surrounding a very sharp point of tungsten. The electric field extracts electrons at the tip of the source, forming a Gaussian spot of only a few tens of angstroms in diameter.

It is impossible to deflect an electron beam to cover a large area, in a typical electron beam lithography system, mechanical stages are required to move the substrate through the deflection field of the electron beam column. Stages can be operated in a stepping mode in which the stage is stopped, an area of the pattern written and then the stage moved to a new location where an adjacent pattern area is exposed. Alternatively, stages can be operated in a continuous mode where the pattern is written on the substrate while the stage is moving. Figure 7.4 shows SEM images of a 40 nm pitch pillar grating after nickel lift-off when developing with ultrasonic agitation.[31]

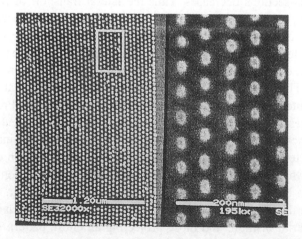

Fig. 7.4. SEM images of a 40 nm pitch pillar grating after nickel lift-off when developing with ultrasonic agitation. [C. Vieu, F. Carcenac, A. Pepin, Y. Chen, M. Mejias, A. Lebib, L. Manin-Ferlazzo, L. Couraud, and H. Lunois, *Appl. Surf. Sci.* **164**, 111 (2000).]

7.2.4. X-ray lithography

X-rays with wavelengths in the range of 0.04 to 0.5 nm represent another alternative radiation source with potential for high-resolution pattern replication into polymeric resist materials.[33] X-ray lithography was first demonstrated that to obtain high-resolution patterns using X-ray proximity printing by Spears and Smith.[34] The essential ingredients in X-ray lithography include:

(1) A mask consisting of a pattern made with an X-ray absorbing material on a thin X-ray transparent membrane,
(2) An X-ray source of sufficient brightness in the wavelength range of interest to expose the resist through the mask, and
(3) An X-ray sensitive resist material.

There are two X-ray radiation sources: (i) electron impact and (ii) synchrotron sources. Conventional electron impact sources produce a broad spectrum of X-rays, centered about a characteristic line of the material, which are generated by bombardment of a suitable target material by a high energy electron beam. The synchrotron or storage ring produces a broad spectrum of radiation stemming from energy loss of electrons in motion at relativistic energies. This radiation is characterized by an intense, continuous spectral distribution from the infrared to the long wavelength X-ray region. It is highly collimated and confined near the orbital plane of the circulating electrons, thereby requiring spreading in the vertical direction of moving the mask and wafer combination with constant speed through the fan of synchrotron radiation. Synchrotrons offer the advantage of high power output.

Absorption of an X-ray photon results in the formation of a photoelectron which undergoes elastic and inelastic collisions within the absorbing material producing secondary electrons which are responsible for the chemical reactions in the resist film. The range of the primary photoelectrons is on the order of 100–200 nm. A major limitation is that of penumbral shadowing, since the X-ray source is finite in size and separated from the mask and the edge of the mask does not cast a sharp shadow. Low mask contrast is another factor that degrades the pattern resolution. It is very important to keep the radiation source in a small area in order to minimize penumbral shadowing and with a maximum intensity of X-rays to minimize exposure time. X-ray proximity lithography is known to provide a one to one replica of the features patterned on the mask, and the resolution limit of the X-ray lithography is ~25 nm.[35,36] Figure 7.5 shows the SEM micrographs of 35 nm wide Au lines and 20 nm wide W dots

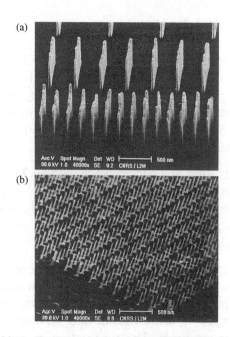

Fig. 7.5. (a) 35 nm wide Au lines grown by electroplating using a template fabricated by X-ray lithography. The mean thickness is about 450 nm, which corresponds to an aspect ratio close to 13. (b) 20 nm wide W dots obtained after reactive ion etching of 1250 nm thick W layer. [G. Simon, A.M. Haghiri-Gosnet, J. Bourneix, D. Decanini, Y. Chen, F. Rousseaux, H. Launios, and B. Vidal, *J. Vac. Sci. Technol.* **B15**, 2489 (1997).]

fabricated by electroplating and reactive ion etching in combination with X-ray lithography.[35]

7.2.5. Focused ion beam (FIB) lithography

Since the development of liquid metal ion (LMI) source in 1975,[37] focused ion beam has been rapidly developed into a very attractive tool in lithography, etching, deposition, and doping.[38] Since scattering of ions in the MeV range is several orders of magnitude less than that for electrons, ion beam lithography has long been recognized to offer improved resolution.[39,40] The commonly used FIBs are Ga and Au-Si-Be alloys LMI sources due to their long lifetime and high stability.[41,42] FIB lithography is capable of producing electronic devices with submicrometer dimensions.[43] The advantages of FIB lithography include its high resist exposure sensitivity, which is two or more orders of magnitude higher than that of electron beam lithography, and its negligible ion scattering in the resist and low back scattering from the substrate.[44] However, FIB lithography suffers from some drawbacks such as lower throughput and extensive substrate damage. Therefore, FIB lithography is more likely to find applications in fabricating devices where substrate damage is not critical.

FIB etching includes physical sputtering etching and chemical assisted etching. Physical sputtering etching is straightforward and is to use the highly energetic ion beams to bombard the area to be etched and to erode material from the sample. The advantages of this process are simple, capable of self-alignment, and applicable to any sample material. Chemical etching is based on chemical reactions between the substrate surface and gas molecules adsorbed on the substrate. Chemical etching offers several advantages: an increased etching rate, the absence of redeposition and little residual damage. Particularly, the chemical assisted etching rate ranges 10 to 100 folds for various combinations of materials and etchant gases, and the absence of redeposition permits very high aspect ratios.[44]

FIB can also be used for depositing. Similar to etching, there are direct deposition and chemical assisted deposition. Direct deposition uses low energy ions, whereas chemical assisted deposition relies on chemical reactions between the substrate surface and molecules adsorbed on the substrate. For example, a regular array of 36 gold pillars as shown in Fig. 7.6, each corresponding to an individual ion beam spot has been created using chemical assisted FIB deposition.[45]

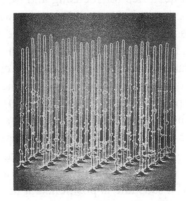

Fig. 7.6. SEM image showing a regular array of 36 gold pillars in each corresponding to an individual ion beam spot created using chemical assisted FIB deposition. [A. Wargner, J.P. Levin, J.L. Mauer, P.G. Blauner, S.J. Kirch, and P. Longo, *J. Vac. Sci. Technol.* **B8**, 1557 (1990).]

FIB lithography offers several advantages for the fabrication and processing of magnetic nanostructures in comparison with electron beam lithography. Ions are substantially heavier than electrons, and thus the FIB is much less influenced by magnetic properties of the material. Another advantage is its ability of achieving direct etching and/or deposition without using extra patterning steps. Magnetic nanostructures have been fabricated using FIB etching and deposition.[46] A ring-type nanomagnetic head was fabricated via FIB etching followed by FIB deposition of nonmagnetic tungsten into the etched trenches. Magnetic pole tips each with a cross-section as narrow as $140 \times 60\,nm^2$ and with a length as tall as 500 nm were protected and supported from all the sides, and had the desired magnetic properties. FIB doping can be considered essentially the same as that of conventional ion implantation.

7.2.6. Neutral atomic beam lithography

In neutral atomic beams, no space charge effects make the beam divergent; therefore, high kinetic particle energies are not required. Diffraction is no severe limit for the resolution because the de Broglie wavelength of thermal atoms is less than 1 angstrom. These atomic beam techniques rely either on direct patterning using light forces on atoms that stick on the surface,[47–50] or on patterning of a special resist.[51–53]

Interaction between neutral atoms and laser light has been explored for various applications, such as reduction of the kinetic-energy spread into the nanokelvin regime, trapping atoms in small regions of space or manipulation of atomic trajectories for focusing and imaging.[54–56]

Basic principle of atomic beam lithography with light forces can be understood in a classical model as follows.[57] The induced electric dipole moment of an atom in an electromagnetic wave can be resonantly enhanced by tuning the oscillation frequency of the light ω_L close to an atomic dipole transition with frequency ω_A. Depending on the sign of the detuning $\delta = \omega_L - \omega_A$, the dipole moment is in phase ($\delta < 0$) or out of phase ($\delta > 0$). In an intensity gradient, this induced dipole feels a force towards the local minimum ($\delta < 0$) or maximum ($\delta > 0$) of the spatial light intensity distribution. Therefore, a standing light wave acts as a periodic conservative potential for the motion of the atoms and forms the analogue of an array of cylindrical lenses. If a substrate is positioned at the focal plane of this lens array, a periodic structure is written onto the surface. Figure 7.7 schematically illustrats the basic principles of neutral atom lithography with light forces and Fig. 7.8 shows the resulting chromium nanowires of 64 nm on silicon substrate grown by neutral atomic beam deposition with laser forces.[57]

Nanostructures Fabricated by Physical Techniques

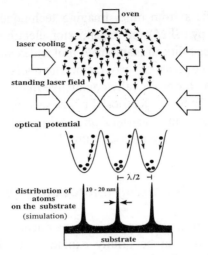

Fig. 7.7. Schematic illustrating the basic principles of neutral atom lithography with light forces. [B. Brezger, Th. Schulze, U. Drodofsky, J. Stuhler, S. Nowak, T. Pfau, and J. Mlynek, *J. Vac. Sci. Technol.* **B15**, 2905 (1997).]

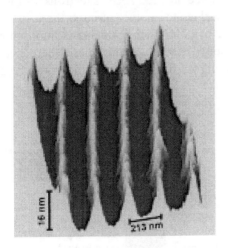

Fig. 7.8. SEM image showing chromium nanowires of 64 nm on silicon substrate grown by neutral atomic beam deposition with laser forces. [B. Brezger, Th. Schulze, U. Drodofsky, J. Stuhler, S. Nowak, T. Pfau, and J. Mlynek, *J. Vac. Sci. Technol.* **B15**, 2905 (1997).]

7.3. Nanomanipulation and Nanolithography

Nanomanipulation and nanolithography are based on scanning probe microscopy (SPM). So before discussing the details of nanomanipulation and nanolithography, an introduction to SPM will be presented first in this section. SPM differs from other imaging techniques such as scanning electron microscopy (SEM) and transmission electron microscopy (TEM) and offers the possibility to manipulate molecules and nanostructures on a surface. SPM consists of two major members: scanning tunneling microscopy (STM) for electrically conductive materials and atomic force microscopy (AFM) for dielectrics. In this section, near-field scanning optical microscopy and near-field photolithography are also included, since they share a lot of similarities with SPM.

7.3.1. Scanning tunneling microscopy (STM)

STM relies on electron tunneling, which is a phenomenon based on quantum mechanics, and can be briefly explained as follows.[58] For more detailed discussion on the fundamentals, the readers are referred to excellent books.[59,60] Let us first consider a situation where two flat surfaces of a metal or semiconductor are separated by an insulator or a vacuum as schematically illustrated in Fig. 7.9.[61] Electrons in the material cannot transfer from one surface to another through the insulator, since there is an energy barrier. However, when a voltage is imposed between the two, the shape of the energy barrier is changed and there is a driving force for electrons to move across the barrier by tunneling, resulting in a small

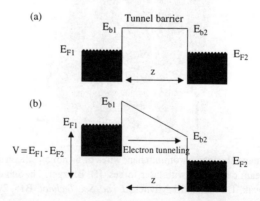

Fig. 7.9. The energy levels in two solids (metals or semiconductors) separated by an insulating or vacuum barrier (a) with no electric field applied between the solids and (b) with an applied electric field. Energies of the electrons in the solids are indicated by the shaded areas up to E_{F1} and E_{F2}, which are the Fermi levels of the respective solids. The applied bias V is $E_{F1} - E_{F2}$, and z is the distance between the two solids. [D.A. Bonnell and B.D. Huey, in *Scanning Probe Microscopy and Spectroscopy*, ed., D. Bonnell, Wiley-VCH, New York, p. 7, 2001.]

current when the distance is sufficiently small so that the electron wave functions extended from the two surfaces overlap. The tunneling current, I, is given by:

$$I \propto e^{-2kz} \tag{7.5}$$

where z is the distance between the two metals or the thickness of the insulator and k is given by:

$$k = \frac{\sqrt{2m(V-E)}}{h} \tag{7.6}$$

where m is the mass of an electron, h is Planck's constant, E is the energy of electron, and V is the potential in the insulator. Similar discussion is applicable to a tip-planar surface geometry, the configuration of a STM. However, the tunneling current is then given by:

$$I = C\rho_t \rho_s e^{-zk^{\frac{1}{2}}} \tag{7.7}$$

where z is the distance between the tip and the planar surface or sample, ρ_t is the tip electronic structure, ρ_s is the sample electronic structure, and C is a constant dependent on the voltage applied between the tip and the sample surface. The tunneling current decays exponentially with the tip-sample distance. For example, a 0.1 nm decrease in the distance will increase the tunneling current by one order of magnitude. Such a quantum mechanical property has been utilized in the STM.

In a typical STM, a conductive tip is positioned above the surface of a sample. When the tip moves back and forth across the sample surface at very small intervals, the height of the tip is continually adjusted to keep the tunneling current constant. The tip positions are used to construct a topographic map of the surface. Figure 7.10 schematically depicts a STM structure. An extremely sharp tip usually made of metals or metal alloys, such as tungsten or PtIr alloy is mounted on to a three-dimensional positioning stage made of an array of piezoelectrics. Such a tip would move above the sample surface in three dimensions accurately controlled by the piezoelectric arrays. Typically the distance between the tip and the sample surface falls between 0.2 and 0.6 nm, thus a tunneling current in the scale of 0.1–10 nA is commonly generated. The scanning resolution is about 0.01 nm in XY direction and 0.002 nm in Z direction, offering true atomic resolution three-dimensional image.

STM can be operated in two modes. In constant current imaging, a feedback mechanism is enabled that a constant current is maintained while a constant bias is applied between the sample and tip. As the tip scans over the sample, the vertical position of the tip is altered to maintain the constant separation. An alternating imaging mode is the constant height operation in

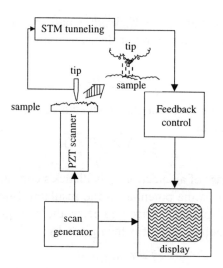

Fig. 7.10. Schematic of scanning tunneling microscope. Two operation modes are commonly used. (i) Constant current mode, in which the distance between the tip and sample surface is kept constant. (ii) Constant voltage mode, in which the tip position is held unchanged during scanning the sample surface.

which constant height and bias are simultaneously maintained. A variation in current results as the tip scans the sample surface because a topographic structure varies the tip-sample separation. The constant current mode produces a contrast directly related to electron charge density profiles, whereas the constant height mode permits faster scan rates. STM was first developed by Binnig and Rohrer in 1982,[62] and STM was first to demonstrate its atomic scale resolution in an image of silicon 7 × 7 restructured (111) surface.[63]

7.3.2. Atomic force microscopy (AFM)

In spite of atomic resolution and other advantages, STM is limited to an electrically conductive surface since it is dependent on monitoring the tunneling current between the sample surface and the tip. AFM was developed as a modification of STM for dielectric materials.[64] A variety of tip–sample interactions may be measured by an AFM, depending on the separation. At short distances, the van der Waals interactions are predominant. Van der Waals force consists of interactions of three components: permanent dipoles, induced dipoles and electronic polarization. A more detailed discussion on van der Waals force was presented in Chapter 2. Long-range forces act in addition to short-range forces between the tip and sample, and become significant when the tip–sample distance increases such that the van der Waals forces become negligible. Examples of such forces include electrostatic attraction or repulsion, current-induced or static-magnetic interactions, and capillary forces due to the condensation of water between the sample and tip. Readers are referred to an excellent book by Israelachevili[65] for details and insight of interactions of surfaces and molecules.

In AFM, the motion of a cantilever beam with an ultra small mass is measured, and the force required to move this beam through measurable distance (10^{-4} Å) can be as small as 10^{-18} N. Figure 7.11 is a schematic drawing, which shows how the AFM works. The instrument consists of a cantilever with a nanoscale tip, a laser pointing at the end of a cantilever, a mirror and a photodiode collecting the reflected laser beam, and a three-dimensional positioning sample stage which is made of an array of piezo-electrics. Similar to STM, the images are also generated by scanning the tip across the surface. However, instead of adjusting the height of the tip to maintain a constant distance between the tip and the surface, and thus a constant tunneling current as in STM, the AFM measures the minute upward and downward deflections of the tip cantilever while maintaining a constant force of contact.

A combination of STM and AFM is also commonly referred to as Scanning Probe Microscope (SPM). There are other variations of microscopes using various tip–surface forces. For example, the magnetic force microscope, the scanning capacitance microscope, and the scanning

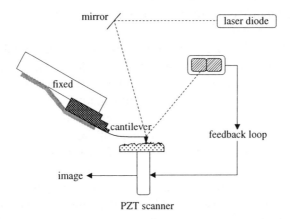

Fig. 7.11. Operating principle of an atomic force microscope. The sample is mounted on a scanner, and the cantilever and tip are positioned near the surface with a macroscopic positioning device. Cantilever deflected with a photo diode that records the position of a laser beam that has been reflected off the top of the cantilever.

acoustic microscope are also members of SPM.[66] SPM has proved its suitability in various fields of applications. First, SPM is capable of imaging the surface of all kinds of solids virtually under any kind of environment. Secondly, with various modifications of tips and operating conditions, SPM can be used to measure local chemical and physical properties of sample surface. Thirdly, SPM has been explored as a useful tool in nano-manipulation and nanolithography in fabrication and processing of nanostructures. Fourthly, SPM has also been investigated as various nanodevices, such as nanosensors and nanotwizers. In this chapter, our discussion will be focused only on the applications of SPM on nano-manipulation and fabrication of nanostructures as well as the surface chemistry modification. Imaging of surface topography and surface property measurements will be discussed in Chapter 8, whereas the nanodevices derived from SPM will be one of the subjects of discussion in Chapter 9.

7.3.3. Near-field scanning optical microscopy (NSOM)

The concept of near-field scanning for imaging purposes using 3 cm microwaves was first studied in 1972[67] and near-field optical microscopy was first developed early in 1980s.[68,69] In NSOM, a resolution of ~30 nm is achievable.[70] NSOM has been used as a tool for photolithography on the submicron length scale. In this application, the fiber optical probe is used as a light source to expose a photoresist, and patterns are generated by scanning the probe over the resist surface. Patterns on conventional polymer resists, amorphous silicon photoresists, and ferroelectric surfaces have been demonstrated.[71–74]

In NSOM, the incident radiation is forced through a subwavelength aperture. In terms of wave propagation theory, this is only possible by formation of wavelets with wavelengths similar to the aperture diameter. The latter is known as evanescent waves and cannot propagate in free space. However, they can wind themselves around the aperture and therefore, transmit radiative energy to the other side of the screen. Such a radiation on the other side of the screen varies significantly with the distance from the screen, and three zones of energy density can be distinguished as shown in Fig. 7.12.[75] Next to the aperture, within a proximity sheet of 2–5 nm, the intensity stays nearly constant and has a relatively large value of the order of 10^{-3}–10^{-4}. This is the evanescent wave regime and an absorbing object in this zone strongly influences the radiation from the aperture. A little away from the aperture (5–500 nm), the intensity decays approximately proportional to $s^{-3.7}$, where s is the distance from the aperture. This is the

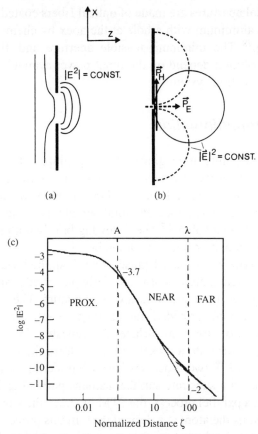

Fig. 7.12. (a) Schematic showing lines of equal electrical-energy density near a small aperture. (b) Same for far field and equivalent dipoles. (c) Calculated on axis electrical-energy density versus distance from aperture, magnetic excitation. [U. Dürig, D.W. Pohl, and F. Rohner, *J. Appl. Phys.* **59**, 3318 (1986).]

zone where the evanescent waves evanesce, and is the better known part of the near-field regime. In this regime, the energy density is already very small and varies between 10^{-4} and 10^{-10}. Absorbing objects in this zone have much less influence per unit volume on the evolving radiative field than in the proximity zone. Further away from the aperture with a distance larger than the wavelength, the radiation from the aperture enters the far-field regime and the energy density decreases by s^{-2}. The control of the distance between tip and sample in the nanometer range, i.e. in the near-field region, is crucial, so that the intensity of the evanescent wave remains sufficient for the detection system.

The NSOM setup is similar to AFM. The ideal aperture is a transparent hole in a thin perfectly conducting metal film at optical frequencies. In practice, typical apertures are made of optical fibers coated with a layer of metal such as aluminum with a hole at the apex by chemical etching[76,77] or by pulling.[78] The minimum feasible aperture and, thus, maximum achievable resolution depend on the input power available and the sensitivity of the detection system.

7.3.4. Nanomanipulation

In addition to the ability of imaging surface topography at atomic resolution, the interactions or forces between the tip and the sample surface offer a means to carry out precise and controlled manipulation of atoms, molecules and nanostructures on a surface. Photolithography is capable of fabricating features of 200 nm and above.[7] The following briefly outlines some examples of nanomanipulation and fabrication by STM.

Eigler and coworkers[79] used pulse voltage applied via a STM tip to move and place xenon atoms onto an orderly patterned structure. They did this in ultrahigh vacuum and ultra low temperature. The low temperature and ultrahigh vacuum provided the stability, cleanliness and absence of thermal diffusion of atoms on surface. The tungsten tip was used to position 35 xenon atoms onto a nickel surface to form three letters "IBM" as shown in Fig. 7.13.[79] Two processes have been identified for the manipulation of atoms on a substrate surface, namely, parallel and perpendicular processes.[80] In a parallel process, the STM tip drags the atom along the surface and positions the atom at a desired spot. In this process, the motion of the manipulated adsorbed atom or molecule is parallel to the surface, and the bond between the manipulated atom or molecule and the underlying

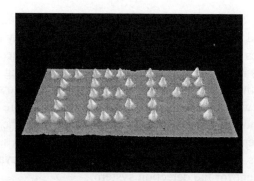

Fig. 7.13. The tungsten tip was used to position 35 xenon atoms onto a nickel surface to form three letters "IBM". [D.M. Eigler and E.K. Schweizer, *Nature* **344**, 524 (1990).]

surface is never broken. The relevant energy barrier for such a process is the energy required for diffusion across the surface, or the corrugation energy, which is typically in the range of 1/10 to 1/3 of the adsorption energy and thus varies from tens of millivolts for a weakly bound physisorbed atom on a closely packed metal surface up to about 0.1 to 1.0 eV for a strongly bound chemisorbed atom.[80] Parallel process can be further divided into two groups: field-assisted diffusion and sliding process. The field-assisted diffusion is based on the presence of the intense and inhomogeneous electric field between the STM probe tip and the surface, which interacts with the dipole moment of an adsorbed atom[81,82] and results in a directional diffusion of the adatom.[80] Electric field assisted directional diffusion in the STM has been demonstrated with Cs atoms on GaAs and InSb surfaces.[83] The sliding process is based on the force between the STM and the adatom, such as interatomic potential or chemical binding force and the directional motion of the adatom is achieved by adjusting the position of the tip, so that the force between the tip and the adatom will pull the adatom across the surface with the tip.[80] The ability of manipulation of surface atoms by this sliding process has been demonstrated in several systems including Xe on Ni (100) surface,[79] CO on Pt (111) surface[84] and Pt on Pt (111) surface.[85]

In a perpendicular process, the STM tip first lifts the adsorbed atom or molecule from the surface, and hover the atom above the substrate surface to a desired position where the atom is dropped from the tip.[80] The energy barrier for such a process is that lift-off of the adatom from the surface with the STM tip varies from the adsorption energy in the limit of large tip–surface separation and zero when the tip is close enough to the adatom. Depending on the mechanisms how an atom is transferred between the tip and surface, several methods have been developed. Transfer on- or near-contact relies on the stronger attraction force between the tip and the adatom than that between the adatom and the surface, when the STM tip is brought in contact with the adatom during the lift step, and the opposite during the drop-off step. Obviously, such a process requires transfer adatoms between different surfaces. Field evaporation is another method to transfer an atom between the tip and surface by the application of a voltage pulse. Field evaporation is described as thermally activated evaporation ions over the "Schottky" barrier formed by the lowering of the potential energy outside the conductor by the application of an electric field.[86] The ability of transfer atoms reversibly has been demonstrated between a silicon surface and a tungsten tip of a STM in UHV at room temperature.[87] Electromigration is yet another phenomenon explored in nano manipulation by STM[88] and the ability to reversibly transfer Xe atoms between a Ni (100) surface and the tip of an STM at 4 K by application of voltage pulses has been demonstrated.[89]

STM has also been explored for chemical manipulation and the ability of single molecule dissociation and construction has been demonstrated.[90] In the study, the STM tip was positioned above the iodobenzene (C_6H_5I) molecules adsorbed at a Cu (111) step-edge and then the 1.5 eV tunneling electrons were injected into the molecules, resulting in the break of C-I bond, while the C_6H_5 radicals remained intact. In the second step, two C_6H_5 radicals on the same Cu (111) step-edge were brought together through lateral manipulation. Finally, the tunneling electrons were injected to provide the two radicals the required energy for the formation of a biphenyl, $C_{12}H_{10}$. Figure 7.14 schematically illustrated the process.[90] Cu also served as a catalyst in this dissociation and construction process. Electric field and tunneling electron beams may also have significant applications in surface modification. For example, the electrical conductivity of a polyimide Langmuir–Blodgett film on Au electrode was found to increase significantly when exposed to an electric field induced by STM tip.[91]

Similarly, AFM has been explored for nano manipulation and fabrication, though the interaction force between AFM tip and substrate surface or object on a surface is different from that between STM tip and substrates surface. The AFM tip or needle is literally dragged across the sample surface, and therefore can be used to manipulate the surface atoms and

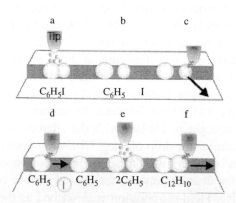

Fig. 7.14. Schematic illustration of the STM tip-induced synthesis steps of a biphenyl molecule. (a and b) Electron-induced selective abstraction of iodine from iodobenzene. (c) Removal of the iodine atom to a terrace site by lateral manipulation. (d) Bringing together two phenyls by lateral manipulation. (e) Electron-induced chemical association of the phenyl couple to biphenyl. (f) Pulling the synthesized molecule by its front end with the STM tip to confirm the association. [S.W. Hla, L. Bartels, G. Meyer, and K.H. Rieder, *Phys. Rev. Lett.* **85**, 2777 (2000).]

Nanostructures Fabricated by Physical Techniques 301

molecules. Depending on the nature of the interactions between the tip and adatoms, three basic manipulation modes have been explored: pushing, pulling and sliding.[92] The ability of manipulation by AFM has been demonstrated with gold nanoparticles on a mica surface. An AFM tip was used to mechanically push the gold nanoparticles along a mica surface by the repulsive forces between tip and the particle, as shown in Fig. 7.15.[92] Figure 7.16 demonstrated that the patterns of Au nanoparticles can be accurately and reliably positioned using these pushing protocols.[92] Such a

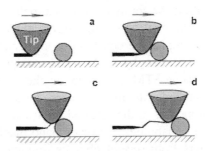

Fig. 7.15. Schematic diagram of the relative motion of the tip and nanoparticles during manipulation. The full heavy line is the path of the tip apex, and the line thickness indicates the tip vibration amplitude. [C. Baur, A. Bugacov, B.E. Koel, A. Madhukar, N. Montoya, T.R. Ramachandran, A.A.G. Requicha, R. Resch, and P. Will, *Nanotechnology* **9**, 360 (1998).]

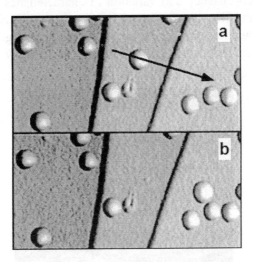

Fig. 7.16. A 30 nm Au particle before (a) and after (b) being pushed over a 10 nm high step along the direction indicated by the arrow. Image sizes are both $1 \times 0.5\,\mu m$. [C. Baur, A. Bugacov, B.E. Koel, A. Madhukar, N. Montoya, T.R. Ramachandran, A.A.G. Requicha, R. Resch, and P. Will, *Nanotechnology* **9**, 360 (1998).]

mechanical pushing is a very versatile process and applicable to a wide range of environments and weakly coupled particle/substrate systems.

In comparison with other nanofabrication processes, manipulation and fabrication by SPM offer a promising alternative with some distinct advantages. First, the SPM tip has a nano-scale sharp point and is the best nano manipulation tool, which offers extremely fine positional control in all three dimensions. SPM manipulation allows the tip to be brought within a few atomic diameters, i.e. approximately a nanometer of the surface of the sample. It therefore promises the prospect of manipulating a single atom. Secondly, the SPM offers the ability of both manipulation and characterization *in situ*. With a SPM, one can see the structure while under every construction step. For example, the construction of a quantum corral of Fe atoms on Cu (111) surface by STM nano-manipulation was *in situ* monitored using the same STM instrument as shown in Fig. 7.17.[93–95] *In situ* characterization includes the measurement of various physical, chemical and biological properties of the material or structure, when the SPM is functioned with these capabilities. However, nano manipulation and fabrication by SPM suffers from several obvious limitations. First of all, the scanning area is very small, typically less than $250 \times 250\,\mu m$ and the scanning speed is very slow. Only one nanostructure can be fabricated at each time with one SPM instrument. Second, SPM tips must possess the high quality and invariant size and shape for nano manipulation and fabrication. Any inconsistency and variation in characteristics of the tip may cause a great deviation in the resultant nanostructure. In addition, SPM tips for nano manipulation and fabrication can be easily damaged and contaminated. Thirdly, substrate surface is required to be extremely flat and smooth

Fig. 7.17. Construction of a quantum corral of Fe atoms on Cu (111) surface by STM nano manipulation was *in situ* monitored using the same STM instrument. [M.F. Crommie, C.P. Lutz, D.M. Eigler, E.J. Heller, *Surf. Rev. Lett.* **2**, 127 (1995).]

and no contamination is allowed, otherwise, the tips may be damaged and resolution may be lost. Lastly, in general, a well-controlled fabrication environment is required. UHV and extremely low temperatures are commonly used. Moisture and dust are big hazards in SPM nanofabrication.

7.3.5. Nanolithography

SPM-based nanolithography has been exploited for local oxidation and passivation,[96] localized chemical vapor deposition,[97] electrodeposition,[98] mechanical contact of the tip with the surface,[99] and deformation of the surface by electrical pulses.[100] There are direct anodic oxidation of the sample surface[101–103] and exposure of electron resist.[104] Patterns with a minimal size of 10–20 nm[105] or to 1 nm in UHV[106] have been demonstrated.

Nanometer holes can be formed using low energy electrons from a STM tip when a pulsed electric voltage is applied at the presence of sufficient gas molecules between the substrate and the tip. For example, holes of 7 nm deep and 6 nm wide on HOPG substrate were formed in nitrogen at a pressure of 25 bar by applying a -7 V pulse to the tip for 130 ms with the distance between the tip and the substrate being 0.6–1 nm.[107] A possible mechanism is that the electric field induces the ionization of gas molecules near the STM tip, and accelerates the ions towards the substrate. Ions bombard the substrate and consequently nanometer-sized holes are created. A certain electric field is required to generate field emitted electrons.[108] The diameter of electron beam ejected from a STM tip is dependent on the applied bias voltage and the diameter of the tip. At low bias (<12 V), the diameter of the ejected electron beam remains almost constant; however, the beam diameter changes significantly with bias voltage and the diameter of the tip.[109]

Nanostructures can be created using field evaporation by applying bias pulses to the STM tip–sample tunneling junction. For example, nano-dots, lines, and corrals of gold on a clean stepped Si(111) surface were fabricated by applying a series of bias pulses (<10 V and $\sim 30\,\mu$s) to a STM gold tip at UHV (a base pressure of $\sim 10^{-10}$ mbar).[110] Nano-dots with diameter as small as a few nanometers can be realized. By decreasing the distance between adjacent nano-dots, it was possible to create continuous nano-lines of a few nanometer wide and over a few hundred nanometer long. A nano-corral of about 40 nm in diameter formed by many Au nano-dots for a few nanometer in diameter each was also created on the silicon (111) surface.

Field Evaporation, or field desorption, is a basic physical process in field-ion microscopy (FIM).[111] The theory has also been developed for the

STM configuration and is briefly summarized below.[112] Considering the distance between a STM tip and the sample, d, is comparatively large, the tip atom interaction potential energy curve U_{at} and that of the atom–sample interaction U_{as} do not overlap significantly as shown in Fig. 7.18a. The binding energy of the atom with the tip, Λ_t, is too strong for the chemisorbed adatom to be thermally activated to the potential well of the tip–atom interaction. However, when the tip–sample distance d is shortened, U_{at} and U_{as} start to overlap and the total potential-energy curve of the atom interacting with both the tip and the sample, $U_a = U_{at} + U_{as}$, shows a hump of height Q_o from the tip side and $Q_o' = Q_o + (\Lambda_s - \Lambda_t)$ from the sample (Fig. 7.18b). At room temperature, the rate of transfer of the atom from the tip side to the sample side $k = \nu \exp(-Q_o/kT)$ becomes 1 s^{-1}, if Q_o is reduced to about 0.772 eV when ν is taken to be $\sim 10^{13}$ s^{-1}. An atom at the sample side can also be thermally activated over to the tip surface, although with a lower rate of $k' = \nu \exp(-Q_o'/kT)$. This explains that one can deposit atoms from the tip to the surface in a controlled manner using the STM, provided that Λ_t is smaller than Λ_s, or to remove an atom from the sample surface to the tip if Λ_s is smaller than Λ_t. It should be noted that the above discussion was on

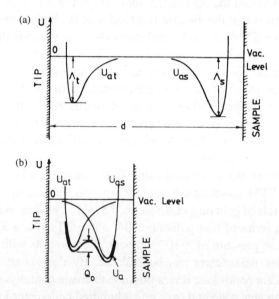

Fig. 7.18. (a) When the tip to same distance d is large, the atom-tip and atom-sample interactions U_{at} and U_{as} do not overlap. (b) When d is small, the two start to overlap and U_a, the sum of U_{at} and U_{as}, exhibits a double-well structure having a small activation barrier. The atom can either be transferred from the tip to the sample or from the sample to the tip. [T.T. Tsong, *Phys. Rev.* **B44**, 13703 (1991).]

the tip–atom–sample interaction in the absence of electric field. Therefore, the theory is also applicable to AFM.

When an electric field is applied between the tip and the sample, two theoretical models, known as the charge-exchange model[113] and the image-hump model,[114] have been developed and well accepted, which can be directly applied in the STM configuration. Similar discussion to the preceding paragraph can be applied here, except that an applied electric field, E, is present and the interaction between the electric field and charged species needs to be considered.[112] When the tip–sample distance is shortened, both the atomic potential curve and the ionic potential curve are changed. In the absence of an electric field, the atomic and ionic curves are simply the sum of U_{as} and U_{at} and that of U_{is} and U_{it}, respectively. When a positive electric field is applied to the tip, an externally applied electric potential, $-neEz$, is added to the ionic potential and these potential curves are modified to $U_I = U_I(0) - neEz$. Here n is the charge state of ions and z is the distance from the tip. As a result, the potential barrier a tip atom has to activate over to reach the sample is greatly reduced, in the case of field evaporation of positive ions. On the other hand, the potential barrier an atom at the sample surface has to activate over to reach the tip surface is greatly increased. Thus, atom transfer between the tip and the sample surface can occur only from the positive electrode to the negative electrode, not the other way around, in the case of field evaporation of positive ions. It should be noted that field evaporation of negative ions is a more complicated issue, since field electron emission starts at an electric field of ~0.3 V/Å. When the electric field increases to 0.6 V/Å, the field emission current density will be large enough to melt tips of most metals by a resistive heating.[112]

Field-gradient induced surface diffusion. Figure 7.19 explains the basic mechanism of field-gradient-induced surface diffusion.[81,82,112,115,116] In the absence of a voltage pulse, the field at the sample surface produced by the probing tip is too small to have an effect. Thus, the adatom sees a horizontal but periodic surface potential, assuming the sample has a periodic structure such as single crystal. There will be no net diffusion. However, when a voltage pulse is applied to either the tip or the sample, a field with a large gradient will be created at the sample surface around the tip due to the asymmetry of the tip–sample configuration. As a result, the polarization energy is position dependent and given by $E_p(r) = -\mu E - 1/2 \alpha E^2$. When this energy is added to the periodic surface potential, the potential–energy curve becomes inclined toward the center where the field is the strongest. Therefore, the surface diffusion becomes directional and the adatoms always move from the

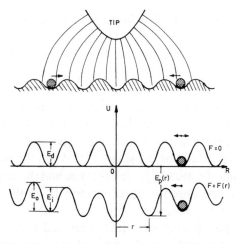

Fig. 7.19. Diagrams explaining why adsorbed atoms will migrate toward the tip by applying voltage to the tip or to the sample. Regardless of the polarity of the voltage pulses, the adsorbed atoms will always migrate toward the center where the field is highest. [T.T. Tsong, *Phys. Rev.* **B44**, 13703 (1991).]

outer region toward the position directly below the tip. Although surface diffusion is a thermally activated process, the activation energy is relatively low and reduced by the field gradient. Further, when a voltage pulse is applied, the tunneling current is greatly increased with the addition of the field emission current that will heat the sample surface slightly and thus promote the surface diffusion. It should be noted that field gradient induced surface diffusion can occur for either polarity of the voltage pulse.

The field-gradient-induced surface diffusion can not only be used to sharpen the tip and create a cusp-shaped cone; it can also be used to attract adsorbed atoms on the sample surface to the position directly under the probing tip as demonstrated by Whitman et al.[83] They manipulated adsorbed atoms and molecules to diffuse toward the probing tip by applying voltage pulses to the tip. When a voltage pulse is applied, a high electric field regardless of whether it is a positive or negative field, the tunneling current from the tip is suddenly increased greatly by the onset of field electron emission which produces a heat pulse to the tip by a joule heating. If the temperature reaches near the melting point, a cusp-shaped liquid metal cone will be formed by either field gradient induced surface diffusion or by a hydrodynamic flow of atoms, as illustrated in Fig. 7.20.[112] STM has also been used to impress or deposit molecules and mounts of atoms on a surface[117,118] and to remove molecules from the surface by applying voltage pulses.[119,120]

AFM based nanolithography. Direct contacting, writing, or scratching is referred to as a mechanical action of the AFM tip that is used as a sharply

Nanostructures Fabricated by Physical Techniques   307

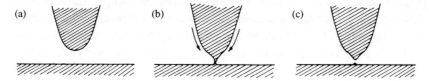

Fig. 7.20. Diagrams illustrating how piles of metal atoms can be deposited on a sample surface by applying either negative or positive voltage pulses to either the sample or the tip. When a high voltage pulse is applied, field electrons are emitted either from the tip or the sample according to the polarity of the pulse. This electron current will heat up or even melt the tip. Because the field gradient exists at the tip surface, atoms will migrate from the tip shank to the tip apex either by a directional surface diffusion or by a hydrodynamic flow of atoms, resulting in the formation of a liquid like cone, which will touch the sample. When the pulse is over and the liquid like metal cone cools down, the neck is broken by the surface tension leaving a mount of tip atoms on the sample surface. [T.T. Tsong, *Phys. Rev.* **B44**, 13703 (1991).]

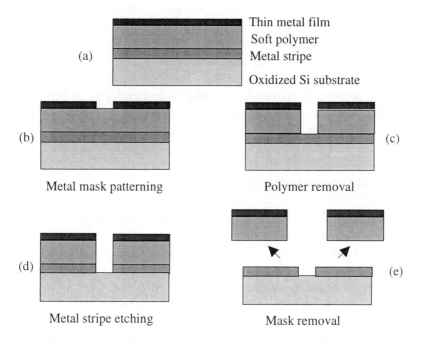

Fig. 7.21. Layout of the sample and the process steps: (a) sample multilayer structure, (b) thin mask patterning by AFM lithography, (c) polymer removal in plasma oxygen, (d) titanium stripe etching, and (e) resulting electrodes after sacrificial layers removal. [A. Notargiacomo, V. Foglietti, E. Cianci, G. Capellini, M. Adami, P. Faraci, F. Evangelisti, and C. Nicolini, *Nanotechnology* **10**, 458 (1999).]

pointed tool in order to produce fine grooves on sample surfaces.[121–125] Although direct scratching creates grooves with high precision, low quality results are often obtained due to tip wear during the process. An alternative approach is to combine scratching on a soft resist polymer layer, such as PMMA or polycarbonate, as a mask for the etching process and subsequent etching to transfer the pattern to the sample surface. This method ensures reduced tip damage, but also precludes an accurate alignment to the structures underneath. A two-layer mask has been investigated as a further improvement. For example, a mask coating consisting of a thin layer of polycarbonate of 50–100 nm and a film of easy-to-deform and fusible metal such as indium or tin was used to create 50 nm wide structures.[125] Figure 7.21 is the typical layout of the sample and the process steps with AFM lithography.[124]

7.4. Soft Lithography

Soft lithography is a general term describing a set of non-photolithographic techniques for microfabrication that are based on the printing of SAMs and molding of liquid precursors. Soft lithography techniques include contact printing, micromolding in capillaries, microtransfer molding and replica molding. Soft lithography has been developed as an alternative to photolithography and a replication technology for both micro- and nanofabrication. The techniques of soft lithography were developed at Whitesides' group and have been summarized in excellent review articles.[126–128] In this section, only a brief introduction to the method will be presented.

7.4.1. Microcontact printing

Microcontact printing is a technique that uses an elastomeric stamp with relief on its surface to generate patterned SAMs on the surface of both planar and curved substrate.[129,130] The procedure of microcontact printing is experimentally simple and inherently parallel. The elastomeric stamp is fabricated by casting and polymerizing PDMS monomer in a master mold, which can be prepared by photolithography or other relevant techniques. The stamp with a desired pattern is brought in contact with "ink", a solution to form a SAM on the surface of the stamp. The inked stamp then contacts a substrate and transfers the SAM onto the substrate surface with patterned structure. A very important advantage that the microcontact printing offers over other patterning techniques is the capability to fabricate a patterned structure on a curved surface.[131,132] The success of microcontact printing relies (i) on the conformal contact between the stamp and the surface of the substrate, (ii) on the rapid formation of highly ordered monolayers as a result of self-assembly,[133] and (iii) on the autophobicity of the SAM, which effectively blocks the reactive spreading of the ink across the surface.[134]

Microcontact printing has been used with a number of systems including SAMs of alkanethiolates on gold, silver and copper, and SAMs of alkylsiloxanes on HO-terminated surfaces.[126] Microcontact printing can routinely form patterns of alkanethiolate SAMs on gold and silver with in-pane dimensions at the scale of ~500 nm. But smaller futures, such trenches in gold as ~35 nm wide and separated by ~350 nm can be fabricated with a combination of microcontact printing of alkanethiolate SAMs and wet etching.[135] Figure 7.22 is the schematic showing of the

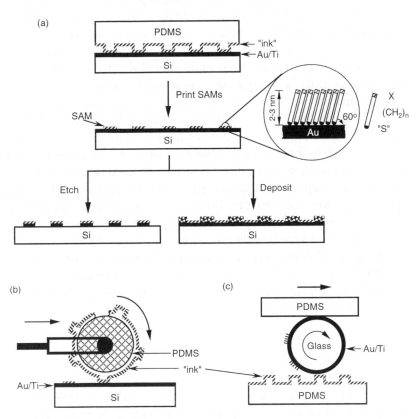

Fig. 7.22. Schematic showing the principal procedures of a typical microcontact printing: (a) printing on a planar substrate with a planar PDMS stamp, (b) printing on a planar substrate with a rolling stamp, and (c) printing on a curved substrate with a planar stamp. [D. Qin, Y.N. Xia, J.A. Rogers, R.J. Jackman, X.M. Zhao, and G.M. Whitesides, *Top. Curr. Chem.* **194**, 1 (1998).]

principal procedures of a typical microcontact printing: (a) printing on a planar substrate with a planar PDMS stamp, (b) printing on a planar substrate with a rolling stamp, and (c) printing on a curved substrate with a planar stamp.[7]

7.4.2. Molding

A number of molding techniques have been developed for the fabrication of microstructures, but are also capable of fabricating nanostructures. These techniques include micromolding in capillaries,[136] microtransfer molding,[137] and replica molding.[138] An elastomeric (PDMS) stamp with relief on its surface is central to each of these procedures. In micromolding in capillaries, a liquid precursor wicks spontaneously by capillary action into the network of channels formed by conformal contact between an elastomeric stamp and a substrate. In microtransfer molding, the recessed regions of an elastomeric mold are filled with a liquid precursor, and the filled mold is brought into contact with a substrate. After solidifying, the mold is removed, leaving a micro- or nanostructure on the substrate. Micromolding in capillaries can only be used to fabricate interconnected structures, whereas microtransfer molding is capable of generating both isolated and interconnected structures. In replica molding, micro- or nanostructures are directly formed by casting and solidifying a liquid precursor against an elastomeric mold. This method is effective for replicating feature sizes ranging from ~30 nm to several centimeters and Fig. 7.23 shows AFM images of such prepared structures.[139] Replica molding also offers a convenient route to fabricating structures with high aspect ratios. Molding has been used to fabricate microstructures and nanostructures of a wide range of materials, including polymers,[140] inorganic and organic salts, sol-gels,[141] polymer beads, and precursor polymers to ceramics and carbon.[142]

7.4.3. Nanoimprint

Nanoimprint lithography was developed in the middle of 1990's and is a conceptually straightforward method in fabrication of patterned nanostructures.[143] Nanoimprint lithography has demonstrated both high resolution and high throughput for making nanometer scale structures.[144,145] Figure 7.24 schematically illustrates the principal steps of a typical nanoimprint process.[146] First a stamp with the desired features is fabricated, for example,

Nanostructures Fabricated by Physical Techniques

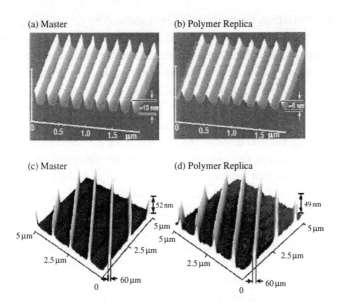

Fig. 7.23. (a and b) AFM images of chromium nanostructures on a master, and a polyurethane replica prepared from a PDMS mold cast from this master. (c and d) AFM images of Gold nanostructures on another master, and a polyurethane replica produced from a different PDMS mold cast from this master. [Y.N. Xia, J.J. McClelland, R. Gupta, D. Qin, X.M. Zhao, L.L. Sohn, R.J. Celotta, and G.M. Whitesides, *Adv. Mater.* **9**, 147 (1997).]

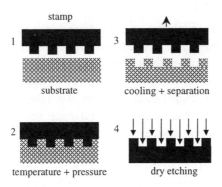

Fig. 7.24. Principal steps of a typical nanoimprint process. A stamp with the desired features is pressed on the polymer layer with the temperature raised above the glass transition point for a certain period of time to allow the plastic to deform. The stamp is separated from the polymer after cooling and the patterned polymer left on the substrate are used for further processing, such as drying etching or lift off, or for use directly as a device component. [S. Zankovych, T. Hoffmann, J. Seekamp, J.U. Bruch, and C.M. Sotomayor Torres, *Nanotechnology* **12**, 91 (2001).]

by optical or electron beam lithography followed by dry etching or reactive ion etching. The material to be printed, typically a thermoplastic polymer, is spun onto a substrate where the nanostructures are to be fabricated. The second step is to press the stamp on the polymer layer with the temperature raised above the glass transition point for a certain period of time to allow the plastic to deform. In the third step, the stamp is separated from the polymer after cooling. The patterned polymer left on the substrate is used for further processing, such as dry etching or lift-off, or for use directly as a device component. Although the process is technically straightforward, there are several key issues that require special attention to make the process competitive as a nanofabrication technology as briefly summarized below.[146]

The first challenge for the nanoimprint lithography is the multilevel capability or the ability of exact alignment of multilayers. Various approaches have been explored to achieve exact alignment including use of commercially available stepper and aligner.[147–149] Stamp size should be controlled, since large stamp size may introduce potential drawback, such as the parallelity of the substrate and stamp and thermal gradients in printing.[150] The flow of the displaced polymer could set a limit to the feature density that imprinting stamps can achieve. Imprint of 50 nm features separated by 50 nm spaces within an area of 200×200 μm^2 has been demonstrated.[146] Sticking is another challenge to the nanoimprint lithography. Ideally, there should be no sticking at all between the polymer layers to be imprinted and the stamp. The choice of printing temperature, the viscoelastic properties of the polymer and the interfacial energy are among the key factors.[151]

Processing control includes printing temperature and pressure and curing time. In principle, both temperature and pressure should be chosen as low as possible in view of the time needed for temperature and pressure cycling. Pressure is less important, since its application takes only a minute or so; more significant may be the rate at which pressure is increased, again with respect to the mechanical recovery of the polymer. Various nanostructured devices have been demonstrated by nanoimprint lithography. For example, InP/GaInAs two-dimensional electron gas three-terminal ballistic junction devices were fabricated using NIL.[152] SiO_2/Si stamps fabricated by electron beam lithography and reactive ion etching were used to transfer sub-100 nm features into a high-mobility InP-based two-dimensional electron gas material. After NIL, the resist residues are removed in oxygen plasma and followed by wet etching of InP/GaInAs to create the desired three-terminal junctions devices. Figure 7.25 gives an example of nanostructures created by means of nanoimprint lithography.[152]

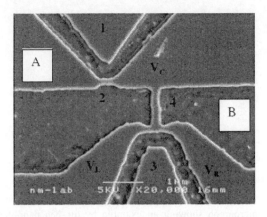

Fig. 7.25. SEM image of a device structure in InP/GaInAs with electron waveguide (A) and three-terminal ballistic junction (B) areas fabricated by nanoimprint lithography. Electrode (1) is a side gate used to control electron waveguide (2); gate (3) controls the TBJ (4) device. The voltages on the TBJ electrodes are denoted V_L, V_R, and V_C. [I. Maximov, P. Carlberg, D. Wallin, I. Shorubalko, W. Seifert, H.Q. Xu, L. Montelius, and L. Samuelson, *Nanotechnology* **13**, 666 (2002).]

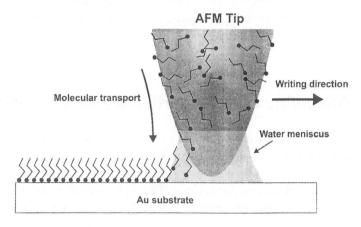

Fig. 7.26. Schematic illustrating the concepts of dip-pen nanolithography. Chemisorption acts as a driving force for moving the molecules from the AFM tip to the substrate via the water filled capillary, when the tip is scanned across a substrate. [R.D. Piner, J. Zhu, F. Xu, S. Hong, and C.A. Mirkin, *Science* **283**, 661 (1999).]

7.4.4. Dip-pen nanolithography

Dip-pen nanolithography is a direct-write method based upon an AFM and works under ambient conditions.[153,154] Figure 7.26 illustrates the concepts of dip-pen nanolithography.[153] Chemisorption is acted as a driving force for moving the molecules from the AFM tip to the substrate via the

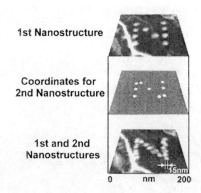

Fig. 7.27. SEM micrograph showing an example by dip-pen nanolithography: fifteen nanometer dots spaced ~5 nm apart in the form of an "N" on an Au (111) substrate. [C.A. Mirkin, *Inorg. Chem.* **39**, 2258 (2000).]

water filled capillary, when the tip is scanned across a substrate. Various nanostructures have been demonstrated and multicomponent nanostructures can readily be created.[155] Figure 7.27 shows an example by DPN: fifteen nanometer dots spaced ~5 nm apart in the form of an "N" on an Au (111) substrate.[155]

7.5. Assembly of Nanoparticles and Nanowires

Self-assembly as a processing technique for the deposition of thin films using molecules as building blocks has been discussed in Chapter 5; however, this deceivingly simple technique also offers a versatile approach to the fabrication of nanoscale devices. The key idea of a self-assembly process is that the final structure or assembly is close to or at a thermodynamic equilibrium, it thus tends to form spontaneously and to reject defects. Self-assembly usually provides routes to structures of greater order than that can be reached in non-self-assembly structures. A variety of interactions between the substrate and building blocks and between building blocks have been exploited as the driving forces for the formation of self-assembled structures. For example, molecular self-assembly, in general, involves noncovalent interactions including van der Waals, electrostatic and hydrophobic interactions as well as hydrogen and coordination bonds. In the self-assembly of meso- and macroscopic objects with nanostructures as building blocks, other forces may play a significant role.[156,157] Examples include gravitation, electromagnetic field, shear force, capillary and entropy.[158] Although the fabrication of functional nanoscale devices is still a subject of intensive research and many new techniques are being discovered, the following discussion will provide a general picture about the most commonly used approaches in the self-assembly of nanoclusters and nanocrystal[159–161] and nanorods.[162]

7.5.1. Capillary forces

One of the commonly used strategies of self-assembly of nanoparticles into ordered 2D arrays is based on the lateral capillary interactions. The origin of the lateral capillary forces is the deformation of the liquid surface, which is supposed to be flat in the absence of particles. The magnitude of the capillary interaction between two colloidal particles is directly proportional to the interfacial deformation created by the particles. The capillary interactions between two adjacent particles either floating on the liquid–air surface or partially immersed into a liquid film on a substrate are briefly introduced below.[163] Figure 7.28 schematically shows two typical

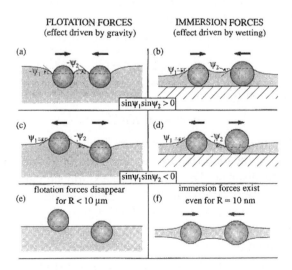

Fig. 7.28. Schematic showing two typical approaches of self-assembly of colloidal particles by capillary force. Flotation (a,c,e) and immersion (b,d,f) lateral capillary forces between two particles attached to fluid interface: (a and b) two similar particles; (c) a light and a heavy particle; (d) a hydrophilic and a hydrophobic particle; (e) small floating particles that do not deform the interface; (f) small particles captured in a thin liquid film deforming the interfaces due to the wetting effects. [P.A. Kralchevsky, K.D. Danov, and N.D. Denkov, in *Handbook of Surface and Colloid Chemistry*, ed. K.S. Birdi, CRC Press, Boca Raton, FL, p. 333, 1997.]

approaches of self-assembly of colloidal particles by capillary force.[163] In the first method, solid particles are partially immersed in a liquid after they have been spread onto the surface of the air–liquid interface through a spreading agent. The arrays assembled on the surface of liquid are then transferred onto solid substrates. The quality of the 2D arrays formed using this method can be fine-tuned by controlling the particle size, the number of particles, the surface properties and charge density of the particles, and the properties of the underlying liquid.[164–166] In the second method, particles are partially immersed into a liquid and are directly in contact with the substrate. Deformation of liquid surface is related to the wetting properties of particles. A complete wetting of liquid or colloidal dispersion on the substrate and an electrostatic repulsion between the colloidal particles and the substrate are critical in obtaining a uniform monolayer. The wetting can be improved by adding surfactants to the colloidal dispersion or simply by precoating the substrate with a thin layer of surfactants.[167] By using the two approaches described above, spherical colloids have been organized into hexagonal closely packed 2D arrays either on solid substrates or in thin films of liquids.[168–171] It should be noted that lateral attractive capillary forces could be directly used to form 3D structures as well.[172]

When two particles have equal size and are not in contact, the capillary force can be simplified as follows, for floating force[163]:

$$F \propto \left(\frac{R^6}{\sigma}\right) K_1(L) \qquad (7.8)$$

And for immersion force:

$$F \propto \sigma R^2 K_1(L) \qquad (7.9)$$

where σ is the interfacial tension between air and liquid, R is the radius of particles, $K_1(L)$ is the modified Bessel function of the first order, and L is the inter-particle distance. These two forces exhibit similar dependence on the inter-particle separation, but very different dependence on the size of particles and the interfacial tension. The flotation force decreases, while the immersion force increases, when the interfacial tension increases. In addition, the flotation force decreases much faster with the decrease of radius than the immersion force. Figure 7.29 is SEM images of 3D structures of nanospheres self-assembled using capillary force.[172]

7.5.2. Dispersion interactions

Ohara and coworkers[173] reported that for the unique cases of metallic nanoparticles, size dependent interparticle dispersion attractions were

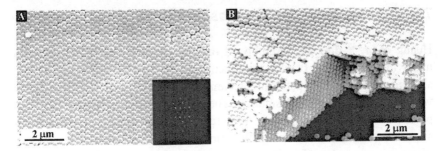

Fig. 7.29. (A) Typical scanning electron micrograph (SEM) of a sample (top view) showing spheres of 298.6 nm diameter. The inset shows a Fourier transform of a 40 × 40 μm² region. (B) Typical SEM side view of the same sample at the same magnification (×12 000), showing a perspective view of the cleaved face and the underlying substrate. [P. Jiang, J.F. Bertone, K.S. Hwang, and V.L. Colvin, *Chem. Mater.* **11**, 2132 (1999).]

sufficient to drive size segregation and assembly of nanoparticles. According to Hamaker theory, the dispersion interaction force between two finite-volume spheres is a function of separation between the spheres.[174] At large interparticle separations, D, the attractive potential, V, is proportional to D^{-6}, whereas at small interparticle separation distances, V is inversely proportional to D. So for the self-assembly, V must be comparable to kT. If $V \ll kT$, there is no driving force for assembly. On the other hand, if $V \gg kT$, the particles will aggregate. In order to form long-range ordered arrays, interparticle attractions must be strong enough to drive nanoparticles to assemble into ordered structure, yet weak enough to allow annealing.[175] A narrow size distribution is another critical requirement for the formation of long-range order.[176] However, other factors, such as the surface chemistry and the capping material, and the nature of the nanoparticle, all play important roles.[177] For example, semiconductors are characterized by substantially weaker interparticle dispersion attractions than are metal nanoparticles, and thus a very narrow size distribution is necessary to form ordered arrays of semiconductor nanoparticles.

Nanoparticle superlattices with nanoparticles as building blocks have been synthesized from colloidal suspensions, and have been summarized by Collier *et al.*[177] Both the 2-D and 3-D superlattices can be readily formed, with structures of body-centered cubic (bcc), face-centered cubic (fcc) and hexagonally closest packed (hcp).[178] The spontaneous formation of ordered 3-D arrays of iron oxide particles occurs when a drop of colloidal solution containing the particles is placed on a TEM grid and the solvent is allowed to evaporate.[179,180] The nanoparticle packing arrangements were found to be dependent on the shape and faceted morphology of the particles and the organic capping molecules tethered to various facets.

7.5.3. Shear force assisted assembly

Huang et al.[162] directed assembly of one-dimensional nanostructures with nanowires as building blocks into functional networks by combining fluidic alignment with surface-patterning techniques. Nanowires of GaP, InP and Si were suspended in ethanol solution. The suspensions of nanowires are then passed through fluidic channel structures formed between a poly(dimethylsiloxane) (PDMS) mold and a flat substrate. Parallel and cross arrays of nanowires can be readily achieved with single and sequential crossed flows, respectively. All the nanowires are aligned along the flow direction over hundreds of micrometers. It was also found that the degree of alignment could be controlled by the flow rates. Such a self-assembly can be explained by means of shear flow.[181,182] Specifically, the channel flow near the substrate surface resembles a shear flow and aligns the nanowires in the flow direction before they are immobilized on the substrate. Higher flow rates produce larger shear forces and hence lead to better alignment. Extended deposition time would result in a reduced separation space between assembled nanowire arrays. Furthermore, the deposition rate and hence the average separation versus time depend strongly on the surface chemical functionality.

7.5.4. Electric-field assisted assembly

Electric field has also been explored to assist the assembly of rod-shaped metallic nanoparticles, carbon nanotubes and metallic nanowires.[183–185] A non-uniform alternating electric field ranging from 1×10^4 to 14×10^4 V/cm has been demonstrated to be able to precisely align metallic nanowires of 70–350 nm in diameter from a colloidal suspension (isopropyl alcohol used as solvent) between two lithographically defined metal fingers.[185] The alignment of the nanowires between the electrodes is due to forces that direct the nanowires toward regions of high field strength. The metallic nanowires polarize readily in the alternating electric field due to charge separation at the surface of the nanowires. Because the nanowires are more polarizable than the dielectric medium, they will experience a dielectrophoretic force that produces net movement in the direction of increasing field strength.[186] As the nanowires approach the electrodes, the electric field strength between the electrodes and nanowire tips increases inversely proportional to the distance, and such a strong near field strength connects the metallic nanowires with the electrodes in addition to the alignment. Ordered hexagonal monolayers of 14-nm nanoparticles can be formed using electrophoretic deposition.[187,188]

7.5.5. Covalently linked assembly

Another approach for the rational construction of complex assemblies of nanopaprticles and nanowires is to use traditional organic synthetic methods to covalently bind the nanostructures. Judicious choice of the functionality of the coordinating ligands allows chemical reactivity to direct the assembly, in potentially very specific ways. Covalently assembled particles or rods form irreversible and much more stable cross-linkages; however, long-range order is difficult to achieve. Typically, such covalently linked assembly is used to build devices requiring short-range order, such as single-electron tunnel junctions, nanoelectrodes or surface enhanced Raman spectroscopy (SERS) substrates. For example, gold nanoparticles capped with dithiols can be self-assembled into 3-D networks.[189] Anchoring metal nanoparticles onto polymers of functionalized alkoxysilanes is a convenient way to attach metal nanoparticles onto solid substrates, and has been used to form colloidal gold and silver films attached to solid supports.[190,191] Substrate surface can also be organically derivatized and terminated with specific active groups or ligands so as to promote substrate-selective assembly.[192,193] Further, when the substrate surface is patterned, a spatially patterned self-assembly can be formed.[194,195]

7.5.6. Gravitational field assisted assembly

Sedimentation in a gravitational field is another method used in self-assembly of nanoparticles[196] and used in the growth of colloidal crystals.[197] A number of parameters have to be carefully chosen to grow highly ordered colloidal crystals, which include the size and density of the particles and the rate of sedimentation. The sedimentation process has to be slow enough so that the colloids concentrated at the bottom of the container will undergo a hard-sphere disorder-to-order phase transition to form a three-dimensionally ordered structure.[198,199] The major disadvantage of the sedimentation method is that it has very little control over the morphology of the top surface and the number of layers. It also takes relatively long time to complete the sedimentation.

7.5.7. Template-assisted assembly

Template-assisted assembly is to introduce surface or spatial confinement to self-assembly. Various approaches have been explored. For example, the surface confinement provided by liquid droplets has been used to assemble colloidal particles or microfabricated building blocks into spherical objects.[200,201] Patterned arrays of relieves on solid substrates were used to grow colloidal crystals.[202,203] Patterned monolayers have been explored to direct the deposition of colloidal particles onto designated regions on a solid substrate.[204] A fluidic cell has also been investigated for self-assembly of colloidal crystals.[205,206]

Other forces may also play important roles in self-assembly process. For example, sonication has been used in the self-assembly of spherical particles into closely packed structures. Magnetic field, similar to electric field, would be another force applicable in directing self-assembly of magnetic nanostructures. Figure 7.30 shows SEM micrographs of various structures fabricated by template-assisted assembly.[207]

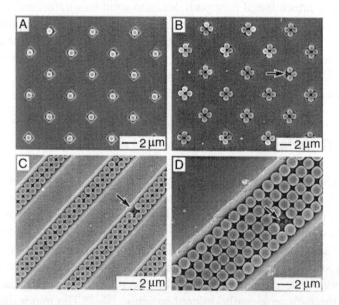

Fig. 7.30. The SEM images of 2D arrays of colloidal aggregates that were assembled under the confinement of templates etched in the surfaces of Si(100) substrates: (A) 800-nm PS beads in square pyramidal cavities 1.2 μm wide at the base; (B) 1.0 μm silica colloids in square pyramidal cavities 2.2 μm wide at the base; (C) 0.8 μm PS beads in V-shaped grooves 2.5 μm wide at the top; and (D) 1.6 μm PS beads in V-shaped grooves 10 μm wide at the top. Note that the use of V-shaped grooves as the templates also allowed one to control the orientation of the colloidal crystals. In parts C and D, the face-center-cubic structures have a (100) orientation rather than (111), the one that is most commonly observed when spherical colloids are crystallized into three-dimensional lattices. The arrows indicate defects, where one can also see the colloidal beads underneath the first layer of the structure. [Y. Yin, Y. Lu, B. Gates, and Y. Xia, *J. Am. Chem. Soc.* **123**, 8718 (2001).]

7.6. Other Methods for Microfabrication

In this section, we will briefly summarize some of the important fabrication methods for patterns with sizes in the micrometer range.

Laser direct writing is a technique that combines laser assisted deposition and a high resolution transformational stage to fabricate patterned microstructures from a wide range of materials.[208–211] For example, laser-assisted deposition can be used for generating micropatterns of seeding materials for electroless plating.[212] Laser-assisted polymerization enables the fabrication of patterned microstructures of polymers.[213] Stereolithography, based on laser-assisted processing, can be used to fabricate three-dimensional microstructures.[214,215]

There are two basic techniques for deposition, which use gas-phase reagents, are pyrolytical (or thermochemical) deposition and photolytical (or photochemical) deposition. In the former, the substrate is heated to decompose the gases on the surface.[216,217] In the latter, molecules in the gas or weakly bound on the substrate or film are directly dissociated on the substrate by relying on an electronic transition through photon absorption.[218–220] The very different laser chemical interactions in these two approaches result in different advantages and disadvantages for use in the writing process. For example, pyrolytic deposition is more substrate sensitive, but yields a deposit with better microstructure and properties. Photolytic deposition is substrate insensitive and permits substantial reaction selectivity, since the chemistry is efficiently driven by non-equilibrium process.

LIGA (Lithography, Electroforming and Molding) is a technique that combines X-ray (or synchrotron) lithography, electroplating, and molding for fabricating microstructures with high aspect ratios and relatively large feature sizes.[221,222] Although the standard equipment for UV exposure can be adapted for this application, special optics and alignment systems are needed for structures thicker than 200 μm.

Excimer laser micromachining is a technique based on laser ablation.[223,224] All types of materials can be ablated routinely, including polymers, glasses, ceramics and metals. The minimum size of the features that this method can produce is limited by diffraction and by heat and mass transport.

7.7. Summary

Various physical techniques for fabrication of micro- and nanostructures have been discussed in this chapter. In some of these physical techniques, chemical reaction or process plays an dispensable role. However, physical processes determine the size and shape of the resultant nanostructures. Although lithographic techniques are not new, with continuous improvement, they are capable of mass production of nanostructures. Feature size less than 100 nm can now be routinely achieved by these methods, and it is expected that the feature size will keep decreasing further. Limitation of these physical approaches will be the damage on the surfaces of nanostructures fabricated. Such surface damages can have significant impact on the physical properties, and thus the performance of the resultant nanostructures and nanodevices.

SPM based nanomanipulation and nanolithography are relatively new and promise the capability of fabricating structures with atoms and molecules as building blocks; however, the processes are very slow and not capable of mass production. Most of the work done so far remains as proof of concepts. Soft lithography is another relatively new technique and would find their roles in the fabrication of nanostructures and nanodevices. There is no doubt that the self-assembly will play a crucial role in the fabrication of macroscale structures and devices using molecules, nanoparticles and nanowires as fundamental building blocks. There are a lot of things to be learned.

References

1. L.F. Thompson, in *Introduction to Microlithography*, eds., L.F. Thompson, C.G. Willson, and M.J. Bowden, The American Chemical Society, Washington, DC, p.1, 1983.
2. W.M. Moreau, *Semiconductor Lithography: Principles and Materials*, Plenum, New York, 1988.
3. K. Suzuki, S. Matsui, and Y. Ochiai, *Sub-Half-Micron Lithography for ULSIs*, Cambridge University Press, Cambridge, 2000.
4. M. Gentili, C. Giovannella, and S. Selci, *Nanolithography: A Borderland between STM, EB, IB, and X-Ray Lithographies*, Kluwer, Dordrecht, The Netherlands, 1993.
5. D. Brambley, B. Martin, and P.D. Prewett, *Adv. Mater. Opt. Electron.* **4**, 55 (1994).
6. M.V. Klein, *Optics*, Wiley, New York, 1970.
7. D. Qin, Y.N. Xia, J.A. Rogers, R.J. Jackman, X.M. Zhao, and G.M. Whitesides, *Top. Curr. Chem.* **194**, 1 (1998).
8. S. Okazaki, *J. Vac. Sci. Technol.* **B9**, 2829 (1991).
9. C.G. Willson, in *Introduction to Microlithography*, eds., L.F. Thompson, C.G. Willson, and M.J. Bowden, The American Chemical Society, Washington, DC, p. 87, 1983.
10. C.C. Davis, W.A. Atia, A. Gungor, D.L. Mazzoni, S. Pilevar, and I.I. Smolyaninov, *Laser Phys.* **7**, 243 (1997).
11. M.K. Herndon, R.T. Collins, R.E. Hollinsworth, P.R. Larson, and M.B. Johnson, *Appl. Phys. Lett.* **74**, 141 (1999).
12. T. Ito and S. Okazaki, *Nature* **406**, 1027 (2000).

13. J.E. Bjorkholm, J. Bokor, L. Lichner, R.R. Freeman, J. Gregus, T.E. Jewell, W.M. Mansfield, A.A. MacDowell, E.L. Raab, W.T. Silfvast, L.H. Szeto, D.M. Tennant, W.K. Waskiewicz, D.L. White, D.L. Windt, O.R. Wood II, and J.H. Bruning, *J. Vac. Sci. Technol.* **B8**, 1509 (1990).
14. A. Kumar, N.A. Abbot, E. Kim, H.A. Biebuyck, and G.M. Whitesides, *Acc. Chem. Res.* **28**, 219 (1995).
15. A. Ulman, *An Introduction to Ultrathin Organic Films: From Langmuir–Blodgett to Self-Assembly*, Academic Press, San Diego, CA, 1991.
16. J. Huang, D.A. Dahlgren, and J.C. Hemminger, *Langmuir* **10**, 626 (1994).
17. K.C. Chan, T. Kim, J.K. Schoer, and R.M. Crooks, *J. Am. Chem. Soc.* **117**, 5875 (1995).
18. M.D. Levnson, N.S. Viswanathan, and R.A. Simpson, *IEEE Trans. Electron Devices* **ED-29**, 1828 (1982).
19. T. Tananka, S. Uchino, N. Hasegawa, T. mYamanaka, T. Terasawa, and S. Okazaki, *Jpn. J. Appl. Phys. Part 1*, **30**, 1131 (1991).
20. J.A. Rogers, K.E. Paul, R.J. Jackman, and G.W. Whitesides, *J. Vac. Sci. Technol.* **B16**, 59 (1998).
21. J. Aizenberg, J.A. Rogers, K.E. Paul, and G.M. Whitesides, *Appl. Opt.* **37**, 2145 (1998).
22. J. Aizenberg, J.A. Rogers, K.E. Paul, and G.M. Whitesides, *Appl. Phys. Lett.* **71**, 3773 (1997).
23. J.A. Rogers, K.E. Paul, R.J. Jackman, and G.M. Whitesides, *Appl. Phys. Lett.* **70**, 2658 (1997).
24. J.L. Wilbur, E. Kim, Y. Xia, and G.M. Whitesides, *Adv. Mater.* **7**, 649 (1995).
25. J.A. Rogers, K.E. Paul, R.J. Jackman, and G.M. Whitesides, *J. Vac. Sci. Technol.* **B16**, 59 (1998).
26. G.R. Brewer, *Electron-Beam Technology in Microelectronic Fabrication*, Academic Press, New York, 1980.
27. W. Chen and H. Ahmed, *Appl. Phys. Lett.* **62**, 1499 (1993).
28. H.G. Craighead, R.E. Howard, L.D. Jackel, and P.M. Mankievich, *Appl. Phys. Lett.* **42**, 38 (1983).
29. S.Y. Chou, *Proc. IEEE* **85**, 652 (1997).
30. T.H.P. Chang and W.C. Nixon, *J. Sci. Instrum.* **44**, 230 (1967).
31. C. Vieu, F. Carcenac, A. Pepin, Y. Chen, M. Mejias, A. Lebib, L. Manin-Ferlazzo, L. Couraud, and H. Lunois, *Appl. Surf. Sci.* **164**, 111 (2000).
32. S. Yesin, D.G. Hasko, and H. Ahmed, *Appl. Phys. Lett.* **78**, 2760 (2001).
33. L.F. Thompson and M.J. Bowden, in *Introduction to Microlithography*, eds. L.F. Thompson, C.G. Willson, and M.J. Bowden, The American Chemical Society, Washington, DC, p. 15, 1983.
34. D.L. Spears and H.I. Smith, *Solid State Technol.* **15**, 21 (1972).
35. G. Simon, A.M. Haghiri-Gosnet, J. Bourneix, D. Decanini, Y. Chen, F. Rousseaux, H. Launios, and B. Vidal, *J. Vac. Sci. Technol.* **B15**, 2489 (1997).
36. T. Kitayama, K. Itoga, Y. Watanabe, and S. Uzawa, *J. Vac. Sci. Technol.* **B18**, 2950 (2000).
37. V.E. Krohn and G.R. Ringo, *Appl. Phys. Lett.* **27**, 479 (1975).
38. P.D. Prewett and G.L.R. Mair, eds., *Focused Ion Beams from Liquid Metal Ion Sources*, Wiley, New York, 1991.
39. T.M. Hall, A. Wagner, and L.F. Thompson, *J. Vac. Sci. Technol.* **16**, 1889 (1979).
40. R.L. Seliger, R.L. Kubena, R.D. Olney, J.W. Ward, and V. Wang, *J. Vac. Sci. Technol.* **16**, 1610 (1979).
41. L.W. Swanson, G.A. Schwind, and A.E. Bell, *J. Appl. Phys.* **51**, 3453 (1980).
42. E. Miyauchi, H. Arimoto, H. Hashimoto, T. Furuya, and T. Utsumi, *Jpn. J. Appl. Phys.* **22**, L287 (1983).

43. S. Matsui, Y. Kojima, Y. Ochiai, and T. Honda, *J. Vac. Sci. Technol.* **B9**, 2622 (1991).
44. S. Matsui and Y. Ochiai, *Nanotechnology* **7**, 247 (1996).
45. A. Wargner, J.P. Levin, J.L. Mauer, P.G. Blauner, S.J. Kirch, and P. Longo, *J. Vac. Sci. Technol.* **B8**, 1557 (1990).
46. S. Khizroev, J.A. Bain, and D. Litvinov, *Nanotechnology* **13**, 619 (2002).
47. G. Timp, R.E. Behringer, D.M. Tennant, J.E. Cunningham, M. Prentiss, and K.K. Berggren, *Phys. Rev. Lett.* **69**, 1636 (1992).
48. J.J. McClelland, R.E. Scholten, E.C. Palm, and R.J. Celotta, *Science* **262**, 877 (1993).
49. R.W. MaGowan, D.M. Giltner, and S.A. Lee, *Opt. Lett.* **20**, 2535 (1995).
50. U. Drodofsky, J. Stuhler, B. Brezger, Th. Schulze, M. Drewsen, T. Pfau, and J. Mlynek, *Microelectron. Eng.* **35**, 285 (1997).
51. K.K. Berggren, A. Bard, J.L. Wilbur, J.D. Gillaspy, A.G. Helg, J.J. McClelland, S.L. Rolston, W.D. Phillips, M. Prentiss, and G.M. Whitesides, *Science* **269**, 1255 (1995).
52. S. Nowak, T. Pfau, and J. Mlynek, *Appl. Phys. B: Lasers Opt.* **63**, 3 (1996).
53. M. Kreis, F. Lison, D. Haubrich, D. Meschede, S. Nowak, T. Pfau, and J. Mlynek, *Appl. Phys. B: Lasers Opt.* **63**, 649 (1996).
54. C.S. Adams, M. Sigel, and J. Mlynek, *Phys. Rep.* **240**, 143 (1994).
55. J. Dalibard and C. Cohen-Tannoudji, *J. Opt. Soc. Am.* **B2**, 1701 (1985).
56. H. Metcalf and P. van der Straten, *Phys. Rep.* **244**, 203 (1994).
57. B. Brezger, Th. Schulze, U. Drodofsky, J. Stuhler, S. Nowak, T. Pfau, and J. Mlynek, *J. Vac. Sci. Technol.* **B15**, 2905 (1997).
58. P.K. Hansma and J. Tersoff, *J. Appl. Phys.* **61**, R1 (1987).
59. J.D. Jackson, *Classical Electrodynamics*, John Wiley & Sons, New York, 1998.
60. A. Zangwill, *Physics at Surfaces*, Cambridge University Press, Cambridge, 1988.
61. D.A. Bonnell and B.D. Huey, in *Scanning Probe Microscopy and Spectroscopy*, ed., D. Bonnell, Wiley-VCH, New York, p. 7, 2001.
62. G. Binnig, H. Rohrer, Ch. Gerber, and E. Weibel, *Phys. Rev. Lett.* **49**, 57 (1982).
63. G. Binnig, H. Rohrer, Ch. Gerber, and E. Weibel, *Phys. Rev. Lett.* **50**, 120 (1983).
64. G. Binnig, C.F. Quate, and Ch. Gerber, *Phys. Rev. Lett.* **56**, 930 (1986).
65. J.N. Israelachevili, *Intermolecular and Surface Forces*, Academic Press, San Diego, CA, 1992.
66. H.K. Wickramsinghe, *Scientific American*, October, p. 98, 1989.
67. E.A. Ash and G. Nichols, *Nature* **237**, 510 (1972).
68. U.Ch. Fischer, *J. Vac. Sci. Technol.* **B3**, 386 (1985).
69. A. Lewis, M. Isaacson, A. Murray, and A. Harootunian, *Biophys. J.* **41**, 405a (1983).
70. G.A. Massey, *Appl. Opt.* **23**, 658 (1984).
71. J. Massanell, N. Garcia, and A. Zlatkin, *Opt. Lett.* **21**, 12 (1996).
72. S. Davy and M. Spajer, *Appl. Phys. Lett.* **69**, 3306 (1996).
73. I.I. Smolyaninov, D.L. Mazzoni, and C.C. Davis, *Appl. Phys. Lett.* **67**, 3859 (1995).
74. M.K. Herndon, R.T. Collins, R.E. Hollingsworth, R.R. Larson, and M.B. Johnson, *Appl. Phys. Lett.* **74**, 141 (1999).
75. U. Dürig, D.W. Pohl, and F. Rohner, *J. Appl. Phys.* **59**, 3318 (1986).
76. P. Hoffmann, B. Dutoit, and R.P. Salathe, *Ultramicroscopy* **61**, 165 (1995).
77. T. Saiki, S. Mononobe, M. Ohtsu, N. Saito, and J. Kusano, *Appl. Phys. Lett.* **68**, 2612 (1996).
78. G.A. Valaskovic, M. Holton, and G.H. Morrison, *Appl. Opt.* **34**, 1215 (1995).
79. D.M. Eigler and E.K. Schweizer, *Nature* **344**, 524 (1990).
80. J.A. Stroscio and D.M. Eigler, *Science* **254**, 1319 (1991).

81. T.T. Tsong and G.L. Kellogg, *Phys. Rev.* **B12**, 1343 (1975).
82. S.C. Wang and T.T. Tsong, *Phys. Rev.* **B26**, 6470 (1982).
83. L.J. Whitman, J.A. Stroscio, R.A. Dragoset, and R.J. Celotta, *Science* **251**, 1206 (1991).
84. *New Scientist* **129**, p. 20 (23 February 1991).
85. P.F. Schewe (ed.), *Physics News* in 1990, The American Institute of Physics, New York, p.73 and cover, 1990.
86. R. Gomer, *IBM J. Res. Dev.* **30**, 428 (1986).
87. I.-W. Lyo and P. Avouris, *Science* **253**, 173 (1991).
88. K.S. Ralls, D.C. Ralph, and R.A. Buhrman, *Phys. Rev.* **B40**, 11561 (1989).
89. D.M. Eigler, C.P. Lutz, and W.E. Rudge, *Nature* **352**, 600 (1991).
90. S.W. Hla, L. Bartels, G. Meyer, and K.H. Rieder, *Phys. Rev. Lett.* **85**, 2777 (2000).
91. K. Takimoto, H. Kawade, E. Kishi, K. Yano, K. Sakai, K. Hatanaka, K. Eguchi, and T. Nakagiri, *Appl. Phys. Lett.* **61**, 3032 (1992).
92. C. Baur, A. Bugacov, B.E. Koel, A. Madhukar, N. Montoya, T.R. Ramachandran, A.A.G. Requicha, R. Resch, and P. Will, *Nanotechnology* **9**, 360 (1998).
93. M.F. Crommie, C.P. Lutz, and D.M. Eigler, *Physica D: Nonlinear Phenomena* **83**, 98 (1995).
94. M.F. Crommie, C.P. Lutz, D.M. Eigler, and E.J. Heller. *Surf. Rev. Lett.* **2**, 127 (1995).
95. M.F. Crommie, C.P. Lutz, and D.M. Eigler, *Science* **262**, 218 (1993).
96. A.E. Gordon, R.T. Fayfield, D.D. Litfin, and T.K. Higman, *J. Vac. Sci. Technol.* **B13**, 2805 (1995).
97. E.E. Ehrichs, S. Yoon, and A.L. de Lozanne, *Appl. Phys. Lett.* **53**, 2287 (1988).
98. F.R.F. Fan and A.J. Bard, *J. Electrochem. Soc.* **136**, 3216 (1989).
99. R.C. Jaklevic and L. Ellie, *Phys. Rev. Lett.* **60**, 120 (1988).
100. J.P. Rabe and S. Buchholz, *Appl. Phys. Lett.* **58**, 702 (1991).
101. P.A. Fontaine, E. Dubois, and D. Stievenard, *J. Appl. Phys.* **84**, 1776 (1998).
102. Y. Okada, S. Amano, M. Kawabe, and J.S. Harris, *J. Appl. Phys.* **83**, 7998 (1998).
103. B. Legrand and D. Stievenard, *Appl. Phys. Lett.* **74**, 4049 (1999).
104. K. Wilderm, C. Quate, D. Adderton, R. Bernstein, and V. Elings, *Appl. Phys. Lett.* **73**, 2527 (1998).
105. K. Matsumoto, M. Ishii, K. Segawa, Y. Oka, B.J. Vartanian, and J.S. Harris, *Appl. Phys. Lett.* **68**, 34 (1996).
106. J.W. Lyding, T.C. Shen, J.S. Tucher, and G.C. Abeln, *Appl. Phys. Lett.* **64**, 2010 (1994).
107. B.L. Weeks, A. Vollmer, M.E. Welland, and T. Rayment, *Nanotechnolgy* **13**, 38 (2002).
108. C. Wang, C. Bai, X. Li, G. Shang, I. Lee, X. Wang, X. Qiu, and F. Tian, *Appl. Phys. Lett.* **69**, 348 (1996).
109. T.M. Mayer, D.P. Adams, and B.M. Marder, *J. Vac. Sci. Technol.* **B14**, 2438 (1996).
110. X. Hu, D. Sarid, and P. von Blanckenhagen, *Nanotechnology* **10**, 209 (1999).
111. T.T. Tsong, *Atom-Probe Field Ion Microscopy*, Cambridge University Press, Cambridge, 1990.
112. T.T. Tsong, *Phys. Rev.* **B44**, 13703 (1991).
113. G. Gomer and L.W. Swanson, *J. Chem. Phys.* **38**, 1613 (1963).
114. E.W. Muller, *Phys. Rev.* **102**, 618 (1956).
115. E.V. Kimenko and A.G. Naumovets, *Sov. Phys. Solid State* **13**, 25 (1971).
116. E.V. Kimenko and A.G. Naumovets, *Sov. Phys. Solid State* **15**, 2181 (1973).
117. H.J. Mamin, P.H. Geuthner, and D. Rugar, *Phys. Rev. Lett.* **65**, 2418 (1990).
118. J.S. Foster, J.E. Frommer, and P.C. Arnett, *Nature* **331**, 324 (1988).

119. R. Emch, J. Nagami, M.M. Dovek, C.A. Lang, and C.F. Quate, *J. Microsc.* **152**, 129 (1988).
120. Y.Z. Li, R. Vazquez, R. Pinter, R.P. Andres, and R. Reifenberger, *Appl. Phys. Lett.* **54**, 1424 (1989).
121. H. Bruckl, R. Ranh, H. Vinzelberg, I. Monch, L. Kretz, and G. Reiss, *Surf. Interf. Anal.* **25**, 611 (1997).
122. S. Hu, S. Altmeyer, A. Hamidi, B. Spangenberg, and H. Kurz, *J. Vac. Sci. Technol.* **B16**, 1983 (1998).
123. S. Hu, S.A. Hamidi, Altmeyer, T. Koster, B. Spangenberg, and H. Kurz, *J. Vac. Sci. Technol.* **B16**, 2822 (1998).
124. A. Notargiacomo, V. Foglietti, E. Cianci, G. Capellini, M. Adami, P. Faraci, F. Evangelisti, and C. Nicolini, *Nanotechnology* **10**, 458 (1999).
125. V.F. Dryakhlushin, A. Yu Klimov, V.V. Rogov, V.I. Shashkin, L.V. Sukhodoev, D.G. Volgunov, and N.V. Vostokov, *Nanotechnology* **11**, 188 (2000).
126. Y. Xia, J.A. Rogers, K.E. Paul, and G.M. Whitesides, *Chem. Rev.* **99**, 1823 (1999).
127. Y. Xia and G.M. Whitesides, *Angew. Chem. Int. Ed. Engl.* **37**, 550 (1998).
128. Y. Xia and G.M. Whitesides, *Annu. Rev. Mater. Sci.* **28**, 153 (1998).
129. R. Jackman, R. Wilbur, and G.M. Whitesides, *Science* **269**, 664 (1995).
130. Y. Xia and G.M. Whitesides, *Langmuir* **13**, 2059 (1997).
131. T.P. Moffat and H.J. Yang, *J. Electrochem. Soc.* **142**, L220 (1995).
132. Y. Xia, E. Kim, and G.M. Whitesides, *J. Electrochem. Soc.* **143**, 1070 (1996).
133. N.B. Larsen, H. Biebuyck, E. Delamarche, and B. Michel, *J. Am. Chem. Soc.* **119**, 3017 (1997).
134. H.A. Biebuyck and G.M. Whitesides, *Langmuir* **10**, 4581 (1994).
135. H.A. Biebuyck, N.B. Larsen, E. Delamarche, and B. Michel, *IBM J. Res. Dev.* **41**, 159 (1997).
136. E. Kim, Y. Xia, and G.M. Whitesides, *Nature* **376**, 581 (1995).
137. X.M. Zhao, Y. Xia, and G.M. Whitesides, *Adv. Mater.* **8**, 837 (1996).
138. Y. Xia, E. Kim, X.M. Zhao, J.A. Rogers, M. Prentiss, and G.M. Whitesides, *Science* **273**, 347 (1996).
139. Y. Xia, J.J. McClelland, R. Gupta, D. Qin, X.M. Zhao, L.L. Sohn, R.J. Celotta, and G.M. Whitesides, *Adv. Mater.* **9**, 147 (1997).
140. X.M. Zhao, A. Stoddart, S.P. Smith, E. Kim, Y. Xia, M. Pretiss, and G.M. Whitesides, *Adv. Mater.* **8**, 420 (1996).
141. S. Seraji, N.E. Jewell-Larsen, Y. Wu, M.J. Forbess, S.J. Limmer, T.P. Chou, and G.Z. Cao, *Adv. Mater.* **12**, 1421 (2000).
142. E. Kim, Y. Xia, and G.M. Whitesides, *Adv. Mater.* **8**, 245 (1996).
143. S.Y. Chou, P.R. Krauss, and P.J. Renstrom, *Appl. Phys. Lett.* **76**, 3114 (1995).
144. B. Heidari, I. Maximov, E.L. Sarwe, and L. Montelius, *J. Vac. Sci. Technol.* **B17**, 2961 (1999).
145. S.Y. Chou, P.R. Krauss, and P.J. Renstrom, *J. Vac. Sci. Technol.* **B14**, 4129 (1996).
146. S. Zankovych, T. Hoffmann, J. Seekamp, J.U. Bruch, and C.M. Sotomayor Torres, *Nanotechnology* **12**, 91 (2001).
147. T. Haatainen, J. Ahopelto, G. Gruetzner, M. Fink, and K. Pfeiffer, *Proc. SPIE* **3997**, 874 (2000).
148. X. Sun, L. Zhuang, W. Zhang, and S.Y. Chou, *J. Vac. Sci. Technol.* **B16**, 3922 (1998).
149. D.L. White and O.R. Wood II, *J. Vac. Sci. Technol.* **B18**, 3552 (2000).
150. B. Heidari, I. Maximov, E.L. Sarwe, and L. Montelius, *J. Vac. Sci. Technol.* **B18**, 3557 (2000).

151. H. Schift, C. David, J. Gobrecht, A.D. Amore, D. Simoneta, W. Kaiser, and M. Gabriel, *J. Vac. Sci. Technol.* **B18**, 3564 (2000).
152. I. Maximov, P. Carlberg, D. Wallin, I. Shorubalko, W. Seifert, H.Q. Xu, L. Montelius, and L. Samuelson, *Nanotechnology* **13**, 666 (2002).
153. R.D. Piner, J. Zhu, F. Xu, S. Hong, and C.A. Mirkin, *Science* **283**, 661 (1999).
154. S. Hong, J. Zhu, and C.A. Mirkin, *Science* **286**, 523 (1999).
155. C.A. Mirkin, *Inorg. Chem.* **39**, 2258 (2000).
156. Z.L. Wang (ed.), *Characterization of Nanophase Materials*, Wiley-VCH, New York, 2000.
157. J.Z. Zhang, J. Liu, Z.L. Wang, S.W. Chen, and G.Y. Liu, *Chemistry of Self-Assembled Nanostructures*, Kluwer, New York, 2002.
158. D.N. Reinhoudt, *Supermolecular Technology*, John Wiley & Sons, New York, 1999.
159. Y. Lin, H. Skaff, T. Emrick, A.D. Dinsmore, and T.P. Russell, *Science* **299**, 226 (2003).
160. W.R. Bowen and A.O. Sharif, *Nature* **393**, 663 (1998).
161. Z.L. Wang, *J. Phys. Chem.* **B104**, 1153 (2000).
162. Y. Huang, X. Duan, Q. Wei, and C.M. Lieber, *Science* **291**, 630 (2001).
163. P.A. Kralchevsky, K.D. Danov, and N.D. Denkov, in *Handbook of Surface and Colloid Chemistry*, ed., K.S. Birdi, CRC Press, Boca Raton, FL, p. 333, 1997.
164. A.J. Hurd and D.W. Schaefer, *Phys. Rev. Lett.* **54**, 1043 (1985).
165. H.H. Wickman and J.N. Korley, *Nature* **393**, 445 (1998).
166. P.A. Kralchevsky and K. Nagayama, *Langmuir* **10**, 23 (1994).
167. J.C. Hulteen, D.A. Treichel, M.T. Smith, M.L. Duval, T.R. Jensen, and R.P.V. Duyne, *J. Phys. Chem.* **B103**, 3854 (1999).
168. P.A. Kralchevsky and N.D. Denkov, *Curr. Opin. Colloid Interf. Sci.* **6**, 383 (2001).
169. C.A. Murray and D.H.V. Winkle, *Phys. Rev. Lett.* **58**, 1200 (1987).
170. A.T. Skjeltorp and P. Meakin, *Nature* **335**, 424 (1988).
171. N.D. Denkov, O.D. Velev, P.A. Kralchevsky, I.B. Ivanov, H. Yoshimura, and K. Nagayama, *Nature* **361**, 26 (1993).
172. P. Jiang, J.F. Bertone, K.S. Hwang, and V.L. Colvin, *Chem. Mater.* **11**, 2132 (1999).
173. P.C. Ohara, D.V. Leff, J.R. Heath, and W.M. Gelbart, *Phys. Rev. Lett.* **75**, 3466 (1995).
174. H.C. Hamaker, *Physica* **4**, 1058 (1937).
175. P.C. Ohara, J.R. Heath, and W.M. Gelbart, *Angew. Chem. Int. Ed. Engl.* **36**, 1078 (1997).
176. S. Murthy, Z.L. Wang, and R.L. Whetten, *Phil. Mag.* **L75**, 321 (1997).
177. C.P. Collier, T. Vossmeyer, and J.R. Heath, *Ann. Rev. Phys. Chem.* **49**, 371 (1998).
178. S.A. Harfenist, Z.L. Wang, R.L. Whetten, I. Vezmar, and M.M. Alvarez, *Adv. Mater.* **9**, 817 (1997).
179. M.D. Bentzon, J. van Wonterghem, S. Morup, A. Thlen, and C.J. Koch, *Phil. Mag.* **B60**, 169 (1989).
180. M.D. Bentzon and A. Tholen, *Ultramicroscopy* **38**, 105 (1990).
181. C.A. Stover, D.L. Koch, and C. Cohen, *J. Fluid Mech.* **238**, 277 (1992).
182. D.L. Koch and E.S.G. Shaqfeh, *Phys. Fluids* **A2**, 2093 (1990).
183. B.M.I. van der Zande, G.J.M. Koper, and H.N.W. Lekkerkerker, *J. Phys. Chem.* **B103**, 5754 (1999).
184. J.S. Yamamoto, S. Akita, and Y. Nakayama, *J. Phys.* **D31**, L34 (1998).
185. P.A. Smith, C.D. Nordquist, T.N. Jackson, T.S. Mayer, B.R. Martin, J. Mbindyo, and T.E. Malloouk, *Appl. Phys. Lett.* **77**, 1399 (2000).
186. H.A. Pohl, *Dielectrophoresis*, Cambridge University Press, Cambridge, 1978.
187. M. Giersig and P. Mulvaney, *J. Phys. Chem.* **97**, 6334 (1993).
188. M. Giersig and P. Mulvaney, *Langmuir* **9**, 3408 (1993).

189. M. Brush, D. Bethell, D.J. Schiffrin, and C.J. Kiely, *Adv. Mater.* **7**, 795 (1995).
190. R.G. Freeman, K.C. Grabar, K.J. Allison, R.M. Bright, J.A. Davis, A.P. Guthrie, M.B. Hommer, M.A. Jackson, P.C. Smith, D.G. Walter, and M.J. Natan, *Science* **267**, 1629 (1995).
191. G. Chumanov, K. Sokolov, B.W. Gregory, and T.M. Cotton, *J. Phys. Chem.* **99**, 9466 (1995).
192. V.L. Colvin, A.N. Goldstein, and A.P. Alivisatos, *J. Am. Chem. Soc.* **114**, 5221 (1992).
193. S. Peschel and G. Schmid, *Angew. Chem. Int. Ed. Engl.* **34**, 1442 (1995).
194. P.C. Hidber, W. Helbig, E. Kim, and G.M. Whitesides, *Langmuir* **12**, 1375 (1996).
195. T. Vossmeyer, E. Delonno, and J.R. Heath, *Angew. Chem. Int. Ed. Engl.* **36**, 1080 (1997).
196. A.K. Arora and B.R.V. Tata, *Ordering and Phase Transitions in Colloidal Systems*, VCH, Weinheim, 1996.
197. J.V. Sanders, *Nature* **204**, 1151 (1964).
198. K.E. Davis, W.B. Russel, and W.J. Glantschnig, *Science* **245**, 507 (1989).
199. P.N. Pusey and W. van Megen, *Nature* **320**, 340 (1986).
200. W.T.S. Huck, J. Tien, and G.M. Whitesides, *J. Am. Chem. Soc.* **120**, 8267 (1998).
201. O.D. Velev, A.M. Lenhoff, and E.W. Kaler, *Science* **287**, 2240 (2000).
202. A. van Blaaderen, R. Ruel, and P. Wiltzius, *Nature* **385**, 321 (1997).
203. K.H. Lin, J.C. Crocker, V. Prasad, A. Schofield, D.A. Weitz, T.C. Lubensky, and A.G. Yodh, *Phys. Rev. Lett.* **85**, 1770 (2000).
204. J. Aizenberg, P.V. Braun, and P. Wiltzius, *Phys. Rev. Lett.* **84**, 2997 (2000).
205. S. Mazur, R. Beckerbauer, and J. Buckholz, *Langmuir* **13**, 4287 (1997).
206. Y. Lu, Y. Yin, B. Gates, and Y. Xia, *Langmuir* **17**, 6344 (2001).
207. Y. Yin, Y. Lu, B. Gates and Y. Xia, *J. Am. Chem. Soc.* **123**, 8718 (2001).
208. O. Lehmann and M. Stuke, *Appl. Phys. Lett.* **61**, 2027 (1992).
209. N. Kramer, M. Niesten, and C. Schonenberger, *Appl. Phys. Lett.* **67**, 2989 (1995).
210. T.W. Weidman and A.M. Joshi, *Appl. Phys. Lett.* **62**, 372 (1993).
211. R.M. Osgood and H.H. Gilgen, *Ann. Rev. Mater. Sci.* **15**, 549 (1985).
212. T.J. Hirsch, R.F. Miracky, and C. Lin, *Appl. Phys. Lett.* **57**, 1357 (1990).
213. A. Torres-Filho and D.C. Neckers, *Chem. Mater.* **7**, 744 (1995).
214. F.T. Wallenberger, *Science* **267**, 1274 (1995).
215. O. Lehmann and M. Stuke, *Science* **270**, 1644 (1995).
216. S.D. Allen, *J. Appl. Phys.* **52**, 6301 (1981).
217. C.P. Christensen and K.M. Larkin, *Appl. Phys. Lett.* **32**, 254 (1978).
218. D.J. Ehrlich, R.M. Osgood Jr., and T.F. Deutsch, *J. Vac. Sci. Technol.* **21**, 23 (1982).
219. I.J. Rigby, *J. Chem. Soc. Faraday Trans.* **65**, 2421 (1969).
220. Y. Rytz-Froidevaux, R.P. Salathé, and H.H. Gilgen, in *Laser Diagnostics and Photochemical Processing for Semiconductor Devices*, eds., R.M. Osgood, S.R.J. Brueck, and H. Schlossberg, Elsevier, Amsterdam, p. 29, 1983.
221. B. Lochel, A. Maciossek, H.J. Quenzer, and B. Wagner, *J. Electrochem. Soc.* **143**, 237 (1996).
222. V. White, R. Ghodssi, C. Herdey, D.D. Denton, and L. McCaughan, *Appl. Phys. Lett.* **66**, 2072 (1995).
223. R.S. Patel, T.F. Redmond, C. Tessler, D. Tudryn, and D. Pulaski, *Laser Focus World*, p. 71 (January, 1996).
224. T. Lizotte, O. Ohar, and T. O'Keefe, *Solid State Technol.* **39**, 120 (1996).

Chapter 8

Characterization and Properties of Nanomaterials

8.1. Introduction

Materials in the nanometer scale, such as colloidal dispersions and thin films, have been studied over many years and many physical properties related to the nanometer size, such as coloration of gold nanoparticles, have been known for centuries. One of the critical challenges faced currently by researchers in the nanotechnology and nanoscience fields is the inability and the lack of instruments to observe, measure and manipulate the materials at the nanometer level by manifesting at the macroscopic level. In the past, the studies have been focused mainly on the collective behaviors and properties of a large number of nanostructured materials. The properties and behaviors observed and measured are typically group characteristics. A better fundamental understanding and various potential applications increasingly demand the ability and instrumentation to observe, measure and manipulate the individual nanomaterials and nanostructures. Characterization and manipulation of individual nanostructures require not only extreme sensitivity and accuracy, but also atomic-level resolution. It therefore leads to various microscopy that will play a central role in characterization and measurements of nanostructured materials and nanostructures. Miniaturization of instruments is obviously not the only challenge; the new phenomena, physical properties and short-range forces, which do not play a noticeable role in macroscopic level characterization, may have significant impacts in the nanometer scale. The development of novel tools and instruments is one of the greatest challenges in nanotechnology.

In this chapter, various structural characterization methods that are most widely used in characterizing nanomaterials and nanostructures, are first discussed. These include: X-ray diffraction (XRD),[1,2] various electron microscopy (EM) including scanning electron microscopy (SEM) and transmission microscopy (TEM),[3-6] and scanning probe microscopy (SPM).[7] Then some typical chemical characterization techniques are discussed. Examples include optical and electron spectroscopy and ionic spectrometry. Then the relationships between the physical properties of nanomaterials and dimensions are briefly discussed. The physical properties discussed in this chapter include thermal, mechanical, optical, electrical and magnetic properties. The discussion in this chapter is focused mainly on the fundamentals and basic principles of the characterization methods and physical properties. Technical details, operation procedures and instrumentations are not the subjects of detailed discussion here. The intention of this chapter is to provide readers with the basic information on the fundamentals that the characterization methods are based on. For technique details, readers are recommended to relevant literature.[8,9]

8.2. Structural Characterization

Characterization of nanomaterials and nanostructures has been largely based on the surface analysis techniques and conventional characterization methods developed for bulk materials. For example, XRD has been widely used for the determination of crystallinity, crystal structures and lattice constants of nanoparticles, nanowires and thin films; SEM and TEM together with electron diffraction have been commonly used in characterization of nanoparticles; optical spectroscopy is used to determine the size of semiconductor quantum dots. SPM is a relatively new characterization technique and has found wide spread applications in nanotechnology. The two major members of the SPM family are scanning tunneling microscopy (STM) and atomic force microscopy (AFM). Although both STM and AFM are true surface image techniques that can produce topographic images of a surface with atomic resolution in all three dimensions, combining with appropriately designed attachments, the STM and AFM have found a much broadened range of applications, such as nanoindentation, nanolithography (as discussed in the previous chapter), and patterned self-assembly. Almost all solid surfaces, whether hard or soft, electrically conductive or not, can all be studied with STM and AFM. Surfaces can be studied in gas such as air or vacuum or in liquid. In the following, we will briefly discuss the aforementioned characterization techniques and their applications in nanotechnology.

8.2.1. X-ray diffraction (XRD)

XRD is a very important experimental technique that has long been used to address all issues related to the crystal structure of solids, including lattice constants and geometry, identification of unknown materials, orientation of single crystals, preferred orientation of polycrystals, defects, stresses, etc. In XRD, a collimated beam of X-rays, with a wavelength typically ranging from 0.7 to 2 Å, is incident on a specimen and is diffracted by the crystalline phases in the specimen according to Bragg's law:

$$\lambda = 2d \sin\theta \tag{8.1}$$

where d is the spacing between atomic planes in the crystalline phase and λ is the X-ray wavelength. The intensity of the diffracted X-rays is measured as a function of the diffraction angle 2θ and the specimen's orientation. This diffraction pattern is used to identify the specimen's crystalline phases and to measure its structural properties. XRD is nondestructive and does not require elaborate sample preparation, which partly explains the wide usage of XRD method in materials characterization. For more details, the readers are highly recommended to an excellent book by Cullity and Stock.[1]

Diffraction peak positions are accurately measured with XRD, which makes it the best method for characterizing homogeneous and inhomogeneous strains.[1,10] Homogeneous or uniform elastic strain shifts the diffraction peak positions. From the shift in peak positions, one can calculate the change in d-spacing, which is the result of the change of lattice constants under a strain. Inhomogeneous strains vary from crystallite to crystallite or within a single crystallite and this causes a broadening of the diffraction peaks that increase with $\sin\theta$. Peak broadening is also caused by the finite size of crystallites, but here the broadening is independent of $\sin\theta$. When both crystallite size and inhomogeneous strain contribute to the peak width, these can be separately determined by careful analysis of peak shapes.

If there is no inhomogeneous strain, the crystallite size, D, can be estimated from the peak width with the Scherrer's formula[11]:

$$D = \frac{K\lambda}{B \cos\theta_B} \tag{8.2}$$

where λ is the X-ray wavelength, B is the full width of height maximum (FWHM) of a diffraction peak, θ_B is the diffraction angle, and K is the Scherrer's constant of the order of unity for usual crystal. However, one should be alerted to the fact that nanoparticles often form twinned structures; therefore, Scherrer's formula may produce results different from the true particle sizes. In addition, X-ray diffraction only provides the collective information of the particle sizes and usually requires a sizable amount of powder. It should be noted that since the estimation would work only for very small particles, this technique is very useful in characterizing nanoparticles. Similarly, the film thickness of epitaxial and highly textured thin films can also be estimated with XRD.[12]

One of the disadvantages of XRD, compared to electron diffraction, is the low intensity of diffracted X-rays, particularly for low-Z materials. XRD is more sensitive to high-Z materials, and for low-Z materials, neutron or electron diffraction is more suitable. Typical intensities for electron diffraction are $\sim 10^8$ times larger than for XRD. Because of small diffraction intensities, XRD generally requires large specimens and the information acquired is an average over a large amount of material. Figure 8.1 shows the powder XRD spectra of a series of InP nanoparticles with different sizes.[13]

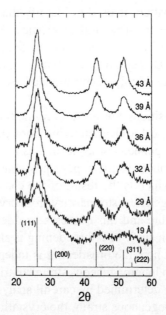

Fig. 8.1. Powder X-ray diffraction of a series of InP nanocrystal sizes. The stick spectrum gives the bulk reflections with relative intensities. [A.A. Guzelian, J.E.B. Katari, A.V. Kadavanich, U. Banin, K. Hamad, E. Juban, A.P. Alivisatos, R.H. Wolters, C.C. Arnold, and J.R. Heath, *J. Phys. Chem.* **100**, 7212 (1996).]

8.2.2. Small angle X-ray scattering (SAXS)

SAXS is another powerful tool in characterizing nanostructured materials. Strong diffraction peaks result from constructive interference of X-rays scattered from ordered arrays of atoms and molecules. A lot of information can be obtained from the angular distribution of scattered intensity at low angles. Fluctuations in electron density over lengths on the order of 10 nm or larger can be sufficient to produce an appreciable scattered X-ray intensities at angles $2\theta < 5°$. These variations can be from differences in density, from differences in composition or from both, and do not need to be periodic.[14,15] The amount and angular distribution of scattered intensity provides information, such as the size of very small particles or their surface area per unit volume, regardless of whether the sample or particles are crystalline or amorphous.

Let us consider a body with an inhomogeneous structure, assuming it consists of two phases separated by well-defined boundaries, such as nanoparticles dispersed in a homogeneous medium, the electron density of such a two-phase structure can be schematically described in Fig. 8.2. The variation of electron density should evidently be divided into two categories. The first type is the deviation resulting from the atomic structure of each of the phases, and the second type is due to the heterogeneity of the material.[16] SAXS is the scattering due to the existence of inhomogeneity regions of sizes of several nanometers to several tens nanometers, whereas XRD, is to determine the atomic structures.

The SAXS intensity, $I(q)$, scattered from a collection of a number, N, of noninteracting nanoparticles that have uniform electron density, ρ, in

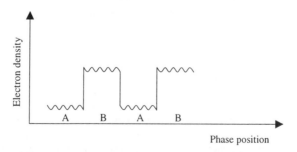

Fig. 8.2. Schematic representing the electron density of a two-phase structure. The variation of electron density can evidently be divided into two categories. The first type is the deviation resulting from the atomic structure of each of the phases, and the second type is due to the heterogeneity of the material.

a homogeneous medium of electron density, ρ_o is given by a simplified form[17,18]:

$$I(q) = I_0 N (\rho - \rho_0)^2 F^2(q) \qquad (8.3)$$

I_0 is the incident X-ray intensity and $F(q)$ is the form factor—the Fourier transform of the shape of the scattering object. For spheres of radius, R, the form factor is expressed by[14,15]:

$$F(q) = 4\pi R^3 \left\{ \frac{\sin(qR) - qR\cos(qR)}{(qR)^3} \right\} \qquad (8.4)$$

where $q = [4\pi \sin(\theta/2)]/\lambda$, λ is the X-ray wavelength, and θ is the angle between a primary and a scattering X-ray beam. SAXS has been widely used in the characterization of nanocrystals[17–19] and Fig. 8.3 shows simulated and measured SAXS spectra of CdSe nanocrystals with various sizes and shapes.[19] Figure 8.4 shows the scattering patterns and their corresponding laminate structures.[1] Small angle scattering has been widely used for

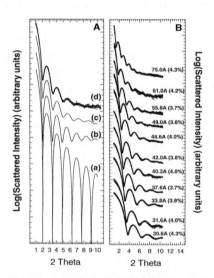

Fig. 8.3. (A) SAXS patterns for model structures having 4500 atoms, comparable to a 62 Å diameter CdSe nanocrystals. The curves are models for (a) 62 Å spheres of uniform electron density, (b) monodisperse, 4500 atom spherical fragments of the bulk CdSe lattice, (c) monodisperse, 4500 atom ellipsoidal fragments of the bulk CdSe lattice, having a 1.2 aspect ratio, and (d) fit to SAXS data (dots) assuming a Gaussian distribution of elliposoids (as in curve c), yielding the nanocrystal sample size and size distribution. (B) SAXS patterns for CdSe nanocrystal samples ranging from 30 to 75 Å in diameter (dots). Fits are used to devise the nanocrystal sample size, reported in equivalent diameters, and size distributions, ranging from 3.5 to 4.5% for the samples shown. [C.B. Murray, C.R. Kagan, and M.G. Bawendi, *Ann. Rev. Mater. Sci.* **30**, 545 (2000).]

the determination of size and ordering of mesoporous materials synthesized by organic-templated condensation, which were discussed in Chapter 6. Further information on the structure can be obtained by studying the asymptotic behavior of the intensity. For large enough q values and for spherical particles with uniform size, the following Porod's Law is observed[20,21]:

$$I(q) = \frac{8\pi(\rho - \rho_o)^2 N \pi R^2}{q^4} \qquad (8.5)$$

where N is the total number of spherical particles with a radius of R. Deviations from Porod's Law can be due to two reasons: (i) the presence of smeared transition boundaries between the phases and (ii) the existence of electron density fluctuations in inhomogeneity regions, over distances exceeding interatomic ones. Further information concerning the system geometry can be conveyed by the slope of the curve of the dependency of $\log I(q)$ on $\log q$ in the intermediate range of angles, where Porod's Law has not yet been observed. For example, with a fibrous or a foliage-shaped structure the curve has a smaller slope.[16]

Apparatus for measuring the distribution of small angle scattering generally employ the transmission geometry using a fine monochromatic radiation beam. SAXS permits measuring the size of inhomogeneity regions in the range from 1 to 100 nm. Applications of small angle scattering span

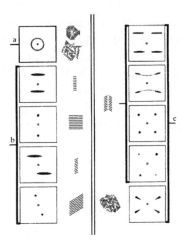

Fig. 8.4. Schematic of long-range ordered structures and corresponding diffraction patterns in the small angle region. (a) A ring pattern corresponds to spherically symmetric assemblies of crystallites or unoriented stacks of lamellar crystallites. (b) Two-point/line patterns reveal the oriented stacks of lamellar crystallites. (c) Four-point/line patterns indicate lamellar crystallites stacked in two different orientations. [B.D. Cullity and S.R. Stock, *Elements of X-Ray Diffraction*, 3rd edition, Prentice Hall, Upper Saddle River, NJ, 2001.]

fields from biological structures to porosity in coals to dispersoids in structural engineering materials. It should also be noted that the theory of visible light scattering[22] is almost identical to that of SAXS described above if the following condition is met:

$$\frac{8\pi R(n_1 - n_2)}{n_2 \lambda} \ll 1 \tag{8.6}$$

where n_1 and n_2 are the refractive indices of a particle and its environment, respectively. However, visible light scattering is limited to systems only when R is larger than approximately 80 nm.

8.2.3. Scanning electron microscopy (SEM)

SEM is one of the most widely used techniques used in characterization of nanomaterials and nanostructures. The resolution of the SEM approaches a few nanometers, and the instruments can operate at magnifications that are easily adjusted from ~10 to over 300,000. Not only does the SEM produce topographical information as optical microscopes do, it also provides the chemical composition information near the surface.

In a typical SEM, a source of electrons is focused into a beam, with a very fine spot size of ~5 nm and having energy ranging from a few hundred eV to 50 KeV, that is rastered over the surface of the specimen by deflection coils. As the electrons strike and penetrate the surface, a number of interactions occur that result in the emission of electrons and photons from the sample, and SEM images are produced by collecting the emitted electrons on a cathode ray tube (CRT). Various SEM techniques are differentiated on the basis of what is subsequently detected and imaged, and the principle images produced in the SEM are of three types: secondary electron images, backscattered electron images and elemental X-ray maps. When a high-energy primary electron interacts with an atom, it undergoes either inelastic scattering with atomic electrons or elastic scattering with the atomic nucleus. In an inelastic collision with an electron, the primary electron transfers part of its energy to the other electron. When the energy transferred is large enough, the other electron will emit from the sample. If the emitted electron has energy of less than 50 eV, it is referred to as a secondary electron. Backscattered electrons are the high-energy electrons that are elastically scattered and essentially possess the same energy as the incident or primary electrons. The probability of backscattering increases with the atomic number of the sample material. Although backscattering images cannot be used for elemental identification, useful contrast can develop between regions of the specimen that differ widely in atomic number, Z. An additional electron interaction in the

Characterization and Properties of Nanomaterials

SEM is that the primary electron collides with and ejects a core electron from an atom in the sample. The excited atom will decay to its ground state by emitting either a characteristic X-ray photon or an Auger electron, both of which have been used for chemical characterization and will be discussed later in this chapter. Combining with chemical analytical capabilities, SEM not only provides the image of the morphology and microstructures of bulk and nanostructured materials and devices, but can also provide detailed information of chemical composition and distribution.

The theoretical limit to an instrument's resolving power is determined by the wavelengths of the electron beam used and the numerical aperture of the system. The resolving power, R, of an instrument is defined as:

$$R = \frac{\lambda}{2NA} \tag{8.7}$$

where λ is the wavelength of electrons used and NA is the numerical aperture, which is engraved on each objective and condenser lens system, and a measure of the electron gathering ability of the objective, or the electron providing ability of the condenser. Figure 8.5 shows SEM pictures (A and B), together with TEM image with electron diffraction

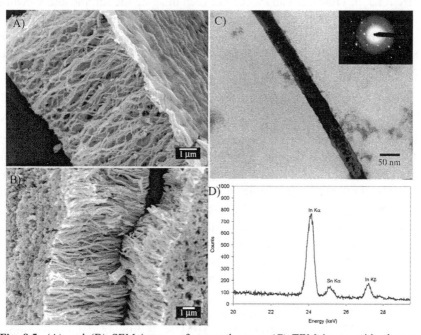

Fig. 8.5. (A) and (B) SEM images of nanorod arrays (C) TEM image with electron diffraction pattern, and (D) the EDS spectrum of indium doped tin oxide (ITO) grown by template-based sol-gel electrophoretic deposition. [S.J. Limmer, S. Vince Cruz, and G.Z. Cao, unpublished results (2003).] (only A and B)

pattern (C) and the spectrum of EDS (D) of nanorod arrays of indium doped tin oxide (ITO) grown by sol-gel electrophoretic deposition.[23]

8.2.4. Transmission electron microscopy (TEM)

In TEM, electrons are accelerated to 100 KeV or higher (up to 1 MeV), projected onto a thin specimen (less than 200 nm) by means of the condenser lens system, and penetrate the sample thickness either undeflected or deflected. The greatest advantages that TEM offers are the high magnification ranging from 50 to 10^6 and its ability to provide both image and diffraction information from a single sample.

The scattering processes experienced by electrons during their passage through the specimen determine the kind of information obtained. Elastic scattering involves no energy loss and gives rise to diffraction patterns. Inelastic interactions between primary electrons and sample electrons at heterogeneities such as grain boundaries, dislocations, second-phase particles, defects, density variations, etc., cause complex absorption and scattering effects, leading to a spatial variation in the intensity of the transmitted electrons. In TEM one can switch between imaging the sample and viewing its diffraction pattern by changing the strength of the intermediate lens.

The high magnification or resolution of all TEM is a result of the small effective electron wavelengths, λ, which is given by the de Broglie relationship:

$$\lambda = \frac{h}{\sqrt{2mqV}} \qquad (8.8)$$

where m and q are the electron mass and charge, h is Planck's constant, and V is the potential difference through which electrons are accelerated. For example, electrons of 100 KeV energy have wavelengths of 0.37 nm and are capable of effectively transmitting through ~0.6 μm of silicon. The higher the operating voltage of a TEM instrument, the greater its lateral spatial resolution. The theoretical instrumental point-to-point resolution is proportional[24] to $\lambda^{3/4}$. High-voltage TEM instruments (with e.g. 400 KV) have point-to-point resolutions better than 0.2 nm. High-voltage TEM instruments have the additional advantage of greater electron penetration, because high-energy electrons interact less strongly with matter than lower-energy electrons. So it is possible to work with thicker samples on a high-voltage TEM. One shortcoming of TEM is its limited depth resolution. Electron scattering information in a TEM image originates from a three-dimensional sample, but is projected onto a two-dimensional detector.

Therefore, structure information along the electron beam direction is superimposed at the image plane. Although the most difficult aspect of the TEM technique is the preparation of samples, it is less so for nanomaterials.

Selected-area diffraction (SAD) offers a unique capability to determine the crystal structure of individual nanomaterials, such as nanocrystals and nanorods, and the crystal structures of different parts of a sample. In SAD, the condenser lens is defocused to produce parallel illumination at the specimen and a selected-area aperture is used to limit the diffracting volume. SAD patterns are often used to determine the Bravais lattices and lattice parameters of crystalline materials by the same procedure used in XRD.[1] Although TEM has no inherent ability to distinguish atomic species, electron scattering is exceedingly sensitive to the target element and various spectroscopy are developed for the chemical composition analysis. Examples include Energy-dispersive X-ray Spectroscopy (EDS) and Electron Energy Loss Spectroscopy (EELS).

In addition to the capability of structural characterization and chemical analyses, TEM has been also explored for other applications in nanotechnology. Examples include the determination of melting points of nanocrystals, in which, an electron beam is used to heat up the nanocrystals and the melting points are determined by the disappearance of electron diffraction.[25] Another example is the measurement of mechanical and electrical properties of individual nanowires and nanotubes.[26–28] The

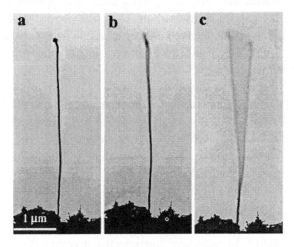

Fig. 8.6. A selected Si nanowires at (a) stationary, (b) the first harmonic resonance with the vibration plane parallel to the viewing direction, and (c) the resonance with the vibration plane perpendicular to the viewing direction. A slight difference in the resonance frequencies in (b) and (c) results from the anisotropic structure of the nanowire. [Z.L. Wang, *Adv. Mater.* **12**, 1295 (2000).]

technique allows a one-to-one correlation between the structure and properties of the nanowires. Figure 8.6 shows TEM micrographs of a silicon nanowire when stationary and vibrating at resonance frequencies,[26] from which the Young's modulus of this silicon nanowire is determined.

8.2.5. Scanning probe microscopy (SPM)

SPM is unique among imaging techniques in that it provides three-dimensional (3-D) real-space images and among other analysis techniques in that it allows spatially localized measurements of structure and properties. Under optimum conditions subatomic spatial resolution is achieved. SPM is a general term for a family of microscopes depending on the probing forces used. Two major members are STM and AFM. The principles of electron tunneling and atomic forces have been already discussed in Chapter 7. For more details, the readers are recommended to an excellent book[7] and references therein.

STM was first developed by Binnig and his coworkers in 1981,[29] and AFM was invented a few years later.[30] The limitation of STM, which is restricted to electrically conductive sample surface, is complemented by AFM, which does not require conductive sample surface. Therefore, almost any solid surface can be studied with SPM: insulators, semiconductors and conductors, magnetic, transparent and opaque materials. In addition, surface can be studied in air, in liquid, or in ultrahigh vacuum, with fields of view from atoms to greater than 250 μm × 250μm, and vertical ranges of about 15 μm.[31] In addition, sample preparation for SPM analysis is minimal.

STM was first used in the study of the Si (111) surface.[32] In ultrahigh vacuum (UHV), STM resolved 7 × 7 reconstruction on Si (111) surface in real space with atomic resolution shown in Fig. 8.7. The experimental procedures can be summarized below. After etching the oxide with an HF solution, the (111) silicon wafer was immediately transferred to the STM in UHV chamber. Repeated heating to 900°C in a vacuum not exceeding 3×10^{-8} Pa resulted in effective sublimation of the SiO layer grown during the transfer, resulting in a clean surface. The micrographs were taken at 2.9 V with tip positive. Only unidirectional scans were recorded to avoid nonlinear effects of the scanning piezoelectric drives.

As summarized by Lang et al.[33] in their excellent tutorial article, SPM has been developed to a wide spectrum of techniques using various probe and sample surface interactions, as shown in Fig. 8.8. The interaction force may be the interatomic forces between the atoms of the tip and those

Characterization and Properties of Nanomaterials

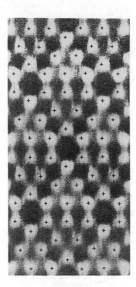

Fig. 8.7. (a) STM imagine of 7 × 7 reconstruction on Si (111) surface in real space with atomic resolution. (b) Modified adatom model. The underlying top-layer atom positions are shown by dots, and the remaining atoms with unsatisfied dangling bonds carry circles, whose thickness indicates the measured depth. The adatoms are represented by large dots with corresponding bonding arms. The empty potential adatom position is indicated by an empty circle in the triangle of adjacent rest atoms. The grid indicates the 7 × 7 unit cells. [G. Binnig, H. Rohrer, C. Gerber, and E. Weibel, *Phys. Rev. Lett.* **50**, 120 (1983).]

of a surface, short-range van der Waals forces, or long-range capillary forces, or stick-slip processes producing friction forces. Modifying the tip chemically allows various properties of the sample surface to be measured. Depending on the type of interactions between the tip and the sample surface used for the characterization, various types of SPM have been developed. Electrostatic force microscopy is based on local charges on the tip or surface, which lead to electrostatic forces between tip and sample, which allow a sample surface to be mapped, i.e. local differences in the distribution of electric charge on a surface to be visualized. In a similar way, magnetic forces can be imagined if the tip is coated with a magnetic material, e.g. iron, that has been magnetized along the tip axis, which is magnetic force microscopy.[34] The tip probes the stray field of the sample and allows the magnetic structure of the sample to be determined. When the tip is functionalized as a thermal couple, temperature distribution on the sample surface can be measured, which is scanning thermal microscopy.[35] The capacity change between tip and sample is evaluated in scanning capacitance microscopy,[36] whereas locally resolved measurement of the chemical potential is done by Kelvin probe microscopy.[37] The

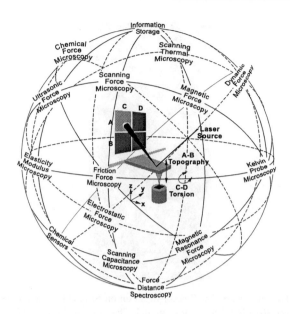

Fig. 8.8. SPM consists of a wide spectrum of techniques using various probe and sample surface interactions. [H.P. Lang, M. Hegner, E. Meyer, and Ch. Gerber, *Nanotechnology* **13**, R29 (2002).]

tip can be driven in an oscillating mode to probe the elastic properties of a surface, which is referred to as elastic modulus microscopy. At high oscillation frequencies (cantilevers with high resonance frequency), further information on inter-atomic forces between tip and sample can be obtained, which is referred to as dynamic force microscopy.

Near-Field Scanning Optical Microscopy (NSOM) can be considered as yet another member of SPM. The fundamentals of NSOM have been summarized in a previous chapter when the near-field optical lithography for the fabrication of nanostructures was discussed, and detailed information can be found in Refs. 38–41. NSOM breaks the diffraction limit ($\lambda/2$) to the resolution of ordinary microscopy by scanning an optical probe (source or detector) in close proximity to the sample. The resolution of a NSOM is dependent on the probe size and the probe and sample separation. When both dimensions are much smaller than the optical wavelength, the resolution in the NSOM experiment is also much smaller than the optical wavelength. In typical NSOM apparatus and experiments, the sample is irradiated through a sub-wavelength aperture in the probe, which is typically a tapered, metal-coated single mode optical fiber with an aperture of a few tens of nanometers at one end.[42] The probe-sample distance is regulated by scanning the lateral shear force interaction of the probe with the sample during the scanning process.[43] During the scanning process two simultaneous images are recorded: the scanning force microscopy topographic image and the near-field optical image. A resolution approaching 1 nm, by NSOM using apertureless NSOM probes coupled to far-field excitation has been achieved.[44]

8.2.6. Gas adsorption

Physical and chemical adsorption isotherm is a powerful technique in determining the surface area and characteristic sizes of particles and porous structures regardless of their chemical composition and crystal structures. When a gas comes in contact with a solid surface, under suitable temperature and pressure, gas molecules will adsorb onto the surface so as to reduce the imbalanced attractive force on surface atoms, and thus to reduce the surface energy. Adsorption may be either physical or chemical in nature.[45,46] Physically adsorbed gases can be removed readily from the solid surface by reducing the partial pressure, whereas chemisorbed gases are difficult to remove unless heated to higher temperatures. For physical adsorption, the amount of gas needed to form a monolayer or to fill pores in various sizes can be measured as a function of gas pressure; such a plot is referred to as gas adsorption isotherm.

Physical adsorption is particularly useful in the determination of specific surface area and pore volume in mesopores (2~50 nm) or micropores (<2 nm) materials. When a vapor of a condensable gas is brought in contact with porous media at constant temperature, several mechanisms of adsorption occur successively on the inner surface of the pore as the relative pressure increases from zero to unity. With increasing relative vapor pressure, first a monomolecular layer is formed on the inner surface of the pores. As the relative vapor pressure increases further, a multi-molecular layer starts to form. Pore volume is based on the assumption that all pores are filled up through capillary condensation. When the relative pressure continues to increase further, capillary condensation will occur on the inner surface of the pores in accordance with the Kelvin equation:

$$\ln\left(\frac{P}{P_0}\right) = \frac{-2\gamma V}{rR_g T} \tag{8.9}$$

This equation relates the equilibrium vapor pressure, P, of a curved surface, such as that of a liquid in a capillary or pore of radius r, to the equilibrium pressure, P_0, of the same liquid on a plane surface. The remaining terms, γ, V, θ, R_g and T, represent the surface tension, molar volume, contact angle of the adsorbate, the gas constant and absolute temperature, respectively. According to this equation, vapor will condense into pores of

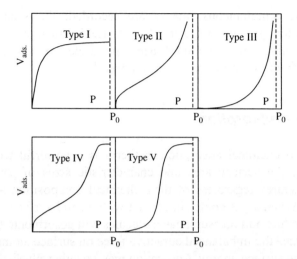

Fig. 8.9. Five basic types of gas sorption isotherms: (I), monolayer sorption in pores of molecular dimension; (II, IV and V), multilayer sorption in highly porous materials with pores up to ~100 nm; (III), multilayer sorption on a nonwetting material. [S. Brunauer, *The Adsoption of Gases and Vapors*, Princeton University Press, Princeton, NJ, 1945.]

radius, r, when the equality expressed in the equation is realized.[47] In practice, the measurement was performed when p/p_0 was adjusted to 0.99, corresponding to pore size up to 95 nm.[48] The amount of gas adsorbed as a function of pressure at constant temperature is termed an isotherm. Figure 8.9 shows five basic types of isotherms.[49] The surface area can be determined from the monolayer adsorption, when the area occupied by each adsorbed gas molecule is known, whereas the pore size distribution can be calculated based on Eq. (8.9).

Chemisorption has also been explored for the determination of surface area; however it takes place via specific chemical forces and is thus unique to the gas and solid in question.[50] In general, it is studied at temperatures much higher than the boiling point of the gas, so there would be no physical adsorption.

8.3. Chemical Characterization

Chemical characterization is to determine the surface and interior atoms and compounds as well as their spatial distributions. As mentioned in the introduction section, many chemical analysis methods have been developed for the surface analysis or thin films, but are readily applicable to the characterization of nanostructures and nanomaterials. Our discussion will be limited to the most popular methods; these techniques can be generally grouped into various optical and electron spectroscopy and ion spectrometry.

8.3.1. Optical spectroscopy

Optical spectroscopy has been widely used for the characterization of nanomaterials, and the techniques can be generally categorized into two groups: absorption and emission spectroscopy and vibrational spectroscopy. The former determines the electronic structures of atoms, ions, molecules or crystals through exciting electrons from the ground to excited states (absorption) and relaxing from the excited to ground states (emission). To illustrate the principles of the techniques, absorption and photoluminescence spectroscopy are discussed in this section. The vibrational techniques may be summarized as involving the interactions of photons with species in a sample that results in energy transfer to or from the sample via vibrational excitation or de-excitation. The vibrational frequencies provide the information of chemical bonds in the detecting samples. In this section, infrared spectroscopy and Raman spectroscopy will be used as examples to illustrate the principles of vibrational spectroscopy.

Absorption and transmission spectroscopy. The characteristic lines observed in the absorption and emission spectra of nearly isolated atoms and ions due to transitions between quantum levels are extremely sharp. As a result, their wavelengths or photon energies can be determined with great accuracy. The lines are characteristic of a particular atom or ion and can be used for identification purposes. Molecular spectra, while usually less sharp than atomic spectra, are also relatively sharp. Positions of spectral lines can be determined with sufficient accuracy to verify the electronic structure of molecules. In solids, the large degeneracy of the atomic levels is split by interactions into quasi-continuous bands (valence and conduction bands), and makes their optical spectra rather broad. The energy difference between the highest lying valence (the highest occupied molecular orbital, HOMO) and the lowest lying conduction (the lowest unoccupied molecular orbital, LUMO) bands is designated as the fundamental gap. Penetration depths of electromagnetic radiation are on the order of 50 nm through most of the optical spectrum (visible light). Such small penetration depths limit the applications of optical absorption spectroscopy for the characterization of bulk solids; however, this technique is readily applicable for the characterization of nanostructures and nanomaterials. Figure 8.10 shows optical absorption spectra of CdSe nanocrystals

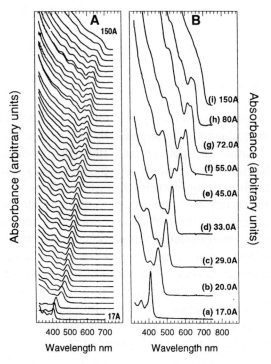

Fig. 8.10. Optical absorption spectra of CdSe nanocrystals with varying diameters. (A) This is seen spectroscopically as a blue shift in the absorption edge and a larger separation between electronic transitions for a homogeneous size series of CdSe nanocrystal dispersion, collected at room temperature. (B) Observation of discrete electronic transitions in optical absorption. [C.B. Murray, C.R. Kagan, and M.G. Bawendi, *Ann. Rev. Mater. Sci.* **30**, 545 (2000).]

with varying diameters and reveals the increased band gap as indicated by a blue shift in the absorption edge and discrete electronic transitions as the nanocrystals get smaller.[19]

Photoluminescence (PL). Luminescence refers to the emission of light by a material through any process other than blackbody radiation.[51] The emission of light can result from a variety of stimulations. For example, when the emission is resulted from electronic stimulation, it is referred to as cathodoluminescence (CL). Another example is X-ray fluorescence, when high-energy photons, i.e. X-ray, are used to excite the sample. In PL one measures physical and chemical properties of materials by using photons to induce excited electronic states in the material system and analyzing the optical emission as these states relax. Typically, light is directed onto the sample for excitation, and the emitted luminescence is collected by a lens and passed through an optical spectrometer onto a photon detector. The spectral distribution and time dependence of the emission are related to electronic transition probabilities within the sample, and can be used to provide qualitative and, sometimes, quantitative information about chemical composition, structure, impurities, kinetic process and energy transfer. Sensitivity is one of the strengths of the PL technique, allowing very small quantities (nanograms) or low concentrations (parts-per-trillion) of material to be analyzed. Precise quantitative concentration determinations are difficult unless conditions can be carefully controlled, and many applications of PL are primarily qualitative.

In PL, a material gains energy by absorbing photon at some wavelength by promoting an electron from a low to a higher energy level. This may be described as making a transition from the ground state to an excited state of an atom or molecule, or from the valence band to the conduction band of a semiconductor crystal or polymer (electron-hole creation). The system then undergoes a non-radiative internal relaxation involving interaction with crystalline or molecular vibrational and rotational modes, and the excited electron moves to a more stable excited level, such as the bottom of the conduction band or the lowest vibrational molecular state. After a characteristic lifetime in the excited state, electron will return to the ground state. In the luminescent materials some or all of the energy released during this final transition is in the form of light, in which case the relaxation is called radiative. The wavelength of the emitted light is longer than that of the incident light. It should be noted that depending on the characteristic life-time of emission, fast PL with life-time of submicrosecond is also called "fluorescence", whereas slow ones, 10^{-4} to 10 s, are referred to as "phosphorescence."

Optical absorption and photoluminescence spectra are commonly used in the characterization of the size of nanocrystals of semiconductors.[52,53] For example, Fig. 8.11 shows the optical absorption and PL spectra as a function of nanocrystal size and clearly demonstrates that the band gap of CdSe nanocrystals increases with a decreasing size.[52]

Infrared Spectroscopy. Molecules and crystals can be thought of as systems of balls (atoms or ions) connected by springs (chemical bonds). These systems can be set into vibration, and vibrate with frequencies determined by the mass of the balls (atomic weight) and by the stiffness of the springs (bond strengths). The mechanical molecular and crystal vibrations are at very high frequencies ranging from 10^{12} to 10^{14} Hz (3–300 μm wavelength), which is in the infrared (IR) regions of the electromagnetic spectrum. The oscillations induced by certain vibrational frequencies provide a means for matter to couple with an impinging beam of infrared electromagnetic radiation and to

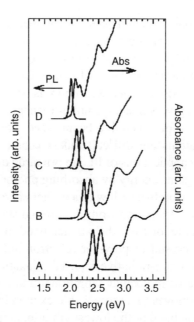

Fig. 8.11. 10-K optical absorption and photoluminescence spectra of optically thin and clear, close-packed nanocrystals of CdSe of (A) 30.3, (B) 39.4, (C) 48.0, and (D) 62.1 Å in diameter. [C.R. Kagan, C.B. Murray, and M.G. Bawendi, *Phys. Rev.* **B54**, 8633 (1996).]

exchange energy with it when the frequencies are in resonance. These absorption frequencies represent excitations of vibrations of the chemical bonds and, thus, are specific to the type of bond and the group of atoms involved in the vibration. In the infrared experiment, the intensity of a beam of infrared radiation is measured before and after it interacts with the sample as a function of light frequency. A plot of relative intensity versus frequency is the "infrared spectrum." A familiar term "FTIR" refers to Fourier Transform Infrared Spectroscopy, when the intensity-time output of the interferometer is subjected to a Fourier transform to convert it to the familiar infrared spectrum (intensity-frequency). The identities, surrounding environments or atomic arrangement, and concentrations of the chemical bonds that are present in the sample can be determined.

Raman spectroscopy[54] is another vibrational technique and differs from the infrared spectroscopy by an indirect coupling of high-frequency radiation, such as visible light, with vibrations of chemical bonds.[55] Raman spectrum is very sensitive to the lengths, strengths and arrangements of chemical bonds in a material, but less sensitive to the chemical composition. When the incident photon interacts with the chemical bond, the chemical bond is excited to a higher energy state. Most of the energy would be re-radiated at the same frequency as that of the incident exciting light, which is known as the Rayleigh scattering. A small portion of the energy is transferred and results in exciting the vibrational modes, and this Raman process is called Stokes scattering. The subsequent re-radiation has a frequency lower (a smaller wavenumber) than that of the incident exciting light. The vibrational energy is deducted by measuring the difference between the frequency of the Raman line and the Rayleigh line. Existing exciting vibrations, e.g. through thermal activation, can also couple with and add their energies to the incident beam, which is called anti-Stokes scattering. The resulting Raman lines appear at higher frequencies or larger wavenumbers. The Stokes and anti-Stokes scattering spectra are mirror images on opposite sides of the Rayleigh line. However, Stokes scattering spectra are mostly used, since they are less temperature sensitive. The Raman effect is extremely weak and, thus, intense monochromatic continuous gas lasers are used as the exciting light. It should be noted that Raman spectroscopy is more a structural characterization technique than a chemical analysis.

8.3.2. Electron spectroscopy

In this section we will briefly discuss the basic and methodologies of Energy Dispersive X-ray Spectroscopy (EDS), Auger Electron Spectroscopy (AES), and X-ray Photoelectron Spectroscopy (XPS). The electron spectroscopy relies on the unique energy levels of the emission of photons (X-ray) or electrons ejected from the atoms in question. As schematically shown in Fig. 8.12,[55] when an incident electron or photon, such as X-ray or γ-ray, strikes an unexcited atom, an electron from an inner shell is ejected and leaves a hole or electron vacancy in the inner shell (Fig. 8.12b). An electron from an outer shell fills the hole by lowering its energy, and simultaneously the excess energy is released through either emission of an X-ray (Fig. 8.12c), which is used in EDS, or ejection of a third electron that is known as an Auger electron (Fig. 8.12d), from a further outer shell, which is used in AES. If incident photons are used for excitation, the resulting characteristic X-rays are known as fluorescent X-rays. Since each atom in the Periodic Table has a unique electronic structure with a unique set of energy levels, both X-ray and Auger spectral lines are characteristic of the element in question. By measuring the energies of the X-rays and Auger electrons emitted by a material, its chemical compositions can be determined.

A similar discussion is applicable to XPS. In XPS, relatively low-energy X-rays are used to eject the electrons from an atom via the photoelectric

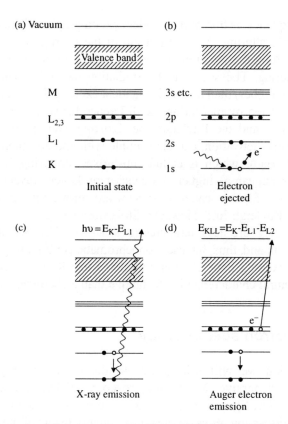

Fig. 8.12. Schematic of electron energy transitions: (a) initial state, (b) incident photon or electron ejects K shell electron, (c) X-ray emission when 2s electron fills electron hole, and (d) Auger electron emission with a KLL transition. [M. Orhring, *The Materials Science of Thin Films*, Academic Press, San Diego, CA, 1992.]

effect. The energy of the ejected electron, E_E, is determined by both the energy of the incident photon, $h\nu$ and the bound electron state, E_B:

$$E_E = h\nu - E_B \tag{8.10}$$

Since values of the binding energy are element-specific, atomic identification is possible through measurement of photoelectron energies.

8.3.3. Ionic spectrometry

Rutherford Backscattering Spectrometry (RBS) is a popular thin film characterization technique and relies on the use of very high-energy beams (MeV) of low mass ions.[56] Such ions can penetrate hundreds of nanometers deep into samples and lose their energies through electronic excitation and ionization of target atoms. Sometimes, such fast-moving light ions (usually ^{4}He$^+$) penetrate the atomic electron cloud shield and undergo close impact collisions with the nuclei of the much heavier target atoms. The resulting scattering from the Coulomb repulsion between ion and nucleus is known as Rutherford Backscattering. This collision is elastic and insensitive to the electronic configuration or chemical bonding of target atoms. The energy of the backscattered ion after such a collision, E_1, is solely dependent on the mass, M_0, and energy, E_0, of the incident ion, the mass of the target atom, M, and the scattering angle, θ, as given by:

$$E_1 = \left\{ \frac{\sqrt{M^2 - M_0^2 \sin^2\theta} + M_0 \cos\theta}{M_0 + M} \right\} 2E_0 \qquad (8.11)$$

With known mass and energy of incident ions and angular position of the ion detector (typically 170°), information on the nature of the elements present, their concentrations and depth distribution can all be simultaneously determined by measuring the number and energy of backscattered incident ions.

Secondary Ion Mass Spectrometry (SIMS) is capable of detecting an extremely low concentration in a solid, far exceeding any known analytical techniques.[57] In SIMS, a source of ions bombards the surface and sputters neutral atoms, for the most part, but also positive and negative ions from the outermost surface layer. Once in the gas phase, the ions are mass-analyzed in order to identify the species present as well as determine their abundance. SIMS can be further distinct as "static" and "dynamic" SIMS. Static SIMS requires that data be collected before the surface is appreciably modified by ion bombardment, and is well suited to surface analysis. Dynamic SIMS is operated with high sputtering rates, and thus enables depth profiling.

Table 8.1 summarizes the chemical characterization methods of electron spectroscopy and ionic spectrometry discussed above, and the

Table 8.1. Summary of some chemical characterization techniques.

Method	Element Sensitivity	Detection Limit (at %)	Lateral Resolution	Effective Probe Depth
SEM/EDS	Na-U	~0.1	~1 μm	~1 μm
AES	Li-U	~0.1 −1	50 nm	~1.5 nm
XPS	Li-U	~0.1 −1	~100 μm	~1.5 nm
RBS	He-U	~1	1 mm	~20 nm
SIMS	H-U	~10^{-4}%	~1 μm	1.5 nm

following summarizes the capabilities and limitations of each method as well as some of the distinctions among these techniques[56]:

(1) AES, XPS and SIMS are true surface analytical techniques, since the detected electrons and ions are emitted from surface layers less than ~1.5 nm deep. Provision is made to probe deeper, or depth profile, by sputter-etching and continuously analyzing the newly exposed surfaces.
(2) AES, XPS and SIMS are broadly applicable to detecting almost all the elements in the Periodic Table, with few exceptions, whereas EDS can only detect elements with $Z > 11$ and RBS is restricted to only selected combinations of elements whose spectra do not overlap.
(3) Only RBS is quantitatively precise to within an atomic percent without the use of composition standards. EDS is the second better choice for quantitative analysis of chemical compositions. AES, XPS and SIMS all require composition standards for quantitative analysis and have composition error of several atomic percent.
(4) XPS, and to a much lesser extent AES, are capable of readily providing information on the nature of chemical bonding and valence states.

8.4. Physical Properties of Nanomaterials

Between the dimensions on an atomic scale and the normal dimensions, which characterize bulk material is a size range where condensed matter exhibits some remarkable specific properties that may be significantly different from the physical properties of bulk materials. Some such peculiar properties are known, but there may be a lot more to be discovered. Some known physical properties of nanomaterials are related to different origins: for example, (i) large fraction of surface atoms, (ii) large surface energy, (iii) spatial confinement, and (iv) reduced imperfections. The following are just a few examples:

(1) Nanomaterials may have a significantly lower melting point or phase transition temperature and appreciably reduced lattice constants, due to a huge fraction of surface atoms in the total amount of atoms.
(2) Mechanical properties of nanomaterials may reach the theoretical strength, which are one or two orders of magnitude higher than that of single crystals in the bulk form. The enhancement in mechanical strength is simply due to the reduced probability of defects.
(3) Optical properties of nanomaterials can be significantly different from bulk crystals. For example, the optical absorption peak of a

semiconductor nanoparticle shifts to a short wavelength, due to an increased band gap. The color of metallic nanoparticles may change with their sizes due to surface plasmon resonance.
(4) Electrical conductivity decreases with a reduced dimension due to increased surface scattering. However, electrical conductivity of nanomaterials could also be enhanced appreciably, due to the better ordering in microstructure, e.g. in polymeric fibrils.
(5) Magnetic properties of nanostructured materials are distinctly different from that of bulk materials. Ferromagnetism of bulk materials disappears and transfers to superparamagnetism in the nanometer scale due to the huge surface energy.
(6) Self-purification is an intrinsic thermodynamic property of nanostructures and nanomaterials. Any heat treatment increases the diffusion of impurities, intrinsic structural defects and dislocations, and one can easily push them to the nearby surface. Increased perfection would have appreciable impact on the chemical and physical properties. For example, chemical stability would be enhanced.

Many such properties are size dependent. In other word, properties of nanostructured materials can be tuned considerably simply by adjusting the size, shape or extent of agglomeration. For example, the optical absorption peak, λ_{max} of metal particles can shift by hundreds of nanometers and particle charging energies altered by hundreds of millivolts via particle size and shape.

8.4.1. Melting points and lattice constants

Nanoparticles of metals, inert gases, semiconductors and molecular crystals are all found to have lower melting temperatures as compared with their bulk forms, when the particle size decreases below 100 nm. The lowering of the melting points is in general explained by the fact that the surface energy increases with a decreasing size. The decrease in the phase transition temperature can be attributed to the changes in the ratio of surface energy to volume energy as a function of particle size. One can apply the known methods of phenomenological thermodynamics to systems of nanoparticles with finite size by introducing the Gibbs model to account for the existence of a surface. Some assumptions are applied to develop a model or approximation to predict a size dependence of melting temperature of nanoparticles. The starting assumption is made of the simultaneous existence of a solid particle, of a liquid particle having the same mass, and of a vapor phase. The equilibrium conditions are then described based on these assumptions.[58,59] The relationship between the melting points of a bulk material, T_b, and a particle, T_m is given by[59,60]

$$T_b - T_m = \left[\frac{2T_b}{\Delta H \, \rho_s \, r_s}\right]\left[\gamma_s - \gamma_l \left(\frac{\rho_s}{\rho_l}\right)^{\frac{2}{3}}\right] \qquad (8.12)$$

Where r_s is the radius of the particle, ΔH is the molar latent heat of fusion, and γ and ρ are surface energy and density, respectively. It should be noted that the above theoretical description is based on classical thermodynamic considerations, in which the system dimensions are infinite, which is obviously inconsistent with the subject of nanoparticles in the range of a few nanometers. It should also be noted that the model is developed based on the assumption that nanoparticles all have the equilibrium shape and are perfect crystals. The equilibrium shape of a perfect crystal is given by the Wulff relationship[61,62] as discussed in detail in Chapter 2. However, small crystal particles are likely to consist of a multiple twinned structure, which may produce particles with energy less than a Wulff crystal. Further experimental results support that the multiple-twinned crystal particles possess well-defined and invariant shapes.[63,64] As will become clear in the following discussion, the approximation or the model discussed above has been found to be in a good agreement with the experimental results,[60] regardless of these assumptions.

It is not always easy to determine or define the melting temperature of nanoparticles. For example, the vapor pressure of a small particle is significantly higher than that of bulk counterpart, and the surface properties of nanoparticles are very different from the bulk materials. Evaporation from the surface would result in an effective reduction of particle size, and thus affect the melting temperature. Increased surface reactivity may promote oxidation of the surface layer and, thus, change the chemical composition on the particle surface through reactions with surrounding chemical species, leading to a change of melting temperature. However, it is possible to make an experimental determination of the size dependence of melting temperature of nanoparticles. Three different criteria have been explored for this determination: (i) the disappearance of the state of order in the solid, (ii) the sharp variation of some physical properties, such as evaporation rate, and (iii) the sudden change in the particle shape.[59] The melting point of bulk gold is of 1337 K and decreases rapidly for nanoparticles with sizes below 5 nm as shown in Fig. 8.13.[59] This figure shows both experimental data (the dots) and the results of a least-squares fits to Eq. (8.1) (the solid line). Such size dependence has also been found in other materials such as copper,[65] tin,[66] indium,[67] lead and bismuth[68] in the forms of particles and films.

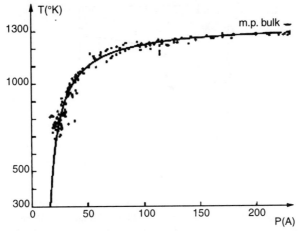

Fig. 8.13. The melting point of bulk gold is of 1337 K and decreases rapidly for nanoparticles with diameters below 5 nm. Both experimental data (the dots) and the results of a least-squares fits to Eq. (8.1) (the solid line) are included. [Ph. Buffat and J.-P. Borel, *Phys. Rev.* **A13**, 2287 (1976).]

Size dependence is found not to be limited to the melting points of metallic nanoparticles. Similar relationship has been reported in other materials including semiconductors and oxides. Furthermore, other phase transitions have similar size dependence. For example, the ferroelectric-paraelectric transition temperatures, or the Curie temperatures, of lead titanate and barium titanate decrease sharply below a certain size. For barium titanate ($BaTiO_3$), the Curie temperature of bulk material is 130°C and drops drastically at sizes below 200 nm, reaching 75°C at ~120 nm.[69] The bulk Curie temperature of lead titanate ($PbTiO_3$) is retained till the particle size drops below 50 nm and such size dependence of phase transition temperature has been summarized in Fig. 8.14.[70]

As expected, melting temperatures of various nanowires have also found to be lower than that of bulk forms. For example, gold nanorods were melted and transformed to spherical particles when heated by laser pulses.[71] Ge nanowires with diameters of 10–100 nm, prepared by VLS process and coated with carbon sheath demonstrated a significantly lowered melting temperature of ~650°C as compared to the melting point of bulk Ge of 930°C.[72,73] As driven by Rayleigh instability,[74] nanowires may spontaneously undergo a spheroidization process to break up into shorter segments and form spherical particles at a relatively low temperature to reduce the high surface energy of nanowires or nanorods, when their diameters are sufficiently thin or the bonding between constituent atoms are weak.

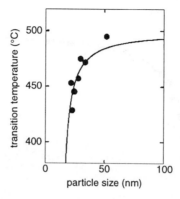

Fig. 8.14. The transition temperature as a function of lead titanate particle size. Experimental data are denoted by full circles and the solid curve is obtained by an empirical expression Tc = 500 − 588.5/(D − 12.6), where D is the particle diameter in nm. [K. Ishikawa, K. Yoshikawa, and N. Okada, *Phys. Rev.* **B37**, 5852 (1988).]

There is little information on the size dependence of the melting temperatures of thin films; however, thermal stability of thin films has been studied. Thin films of gold or platinum, which are commonly used as bottom electrodes, become discontinuous by forming holes and then isolated islands, when heated at elevated temperatures. However, there is no work reported on the thickness dependence of melting temperature of thin films.

Goldstein et al.[25] studied the dependence of melting points and lattice constant of spherical semiconductor CdS nanoparticles using TEM and XRD. The CdS nanoparticles were prepared by colloidal synthesis and the size of particles ranges from 24 to 76 Å in diameter with a standard deviation of ±7%. CdS nanoparticles were either bare surfaced or mercaptoacetic acid capped. The samples were heated using electron beams and the melting points were defined as the temperature at which the electron diffraction peaks associated with CdS crystalline structure disappear. Figure 8.15 shows the lattice constants and melting temperature of CdS nanoparticles as functions of particle size.[25] Figure 8.15a shows that the lattice constants of nanoparticles decrease linearly with an increasing reciprocal particle radius. It is further noticed that the CdS nanoparticles with surface modification demonstrate less reduction in lattice constant than that of bare nanoparticles. The increase of surface energy would explain the appreciable decrease in the melting temperatures of nanoparticles as shown in Fig. 8.15b. It should be noted that the change of the lattice constants is difficult to observe, and become measurable only at very small nanoparticles. In Chapter 3, the lattice constants of nanoparticles were generally found to have crystal structures and lattice constants the same as that of bulk materials.

Characterization and Properties of Nanomaterials

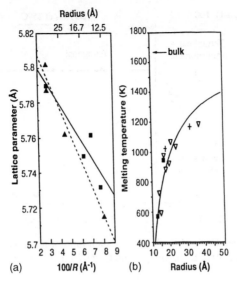

Fig. 8.15. (a) Lattice parameter of CdS nanocrystals as a function of the reciprocal particle radius, R. ▲: points from bare nanocrystals, the dashed line for bare nanocrystals yields a surface tension of 2.50 N.m., points from mercaptoacetic acid-capped nanocrystals, the solid line fit yields a surface tension of 1.74 N/m. (b) Size dependence of the melting point of CdS nanocrystals, and +: thiophenol or mercaptoacetic acid-capped nanocrystals determined by the disappearance of electron diffraction. ▽: determined by observing the change in dark field of a single CdS particle. [A.N. Goldstein, C.M. Echer, and A.P. Alivisatos, *Science* **256**, 1425 (1992).]

The change of crystal structure may occur when the dimension of materials is sufficiently small. For example, Arlt et al.[75] found that the crystal structure of $BaTiO_3$ changes with particle size at room temperature. Figure 8.16 shows the dependence of the lattice constant ratio on the average particle size.[77] At grain sizes larger than 1.5 μm, a constant ratio of the lattice parameters, $c/a - 1 = 1.02\%$, whereas at grain sizes smaller than 1.5 μm, the tetragonal distortion of the $BaTiO_3$ unit cell decreases at room temperature to $c/a - 1 < 1\%$. It is very interesting to observe that a phase change from tetragonal to pseudocubic structure gradually takes place as shown in Fig. 8.17.[77]

8.4.2. Mechanical properties

The mechanical properties of materials increase with a decreasing size. Many studies have been focused on the mechanical properties of one-dimensional structure; particularly a lot of work has been done on whiskers. It was found that a whisker can have a mechanical strength

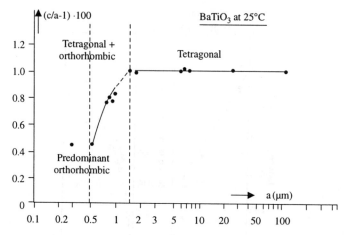

Fig. 8.16. The dependence of the lattice constant ratio of tetragonal $BaTiO_3$ on the average particle size. [G. Arlt, D. Hennings, and G. de With, *J. Appl. Phys.* **58**, 1619 (1985).]

approaching to the theoretical as first demonstrated by Herring and Galt in 1952.[76] It has been long known that the calculated strength of perfect crystals exceeds that of real ones by two or three orders of magnitude. It has also been found that the increase of mechanical strength becomes appreciable only when the diameter of a whisker is less than 10 microns. So the enhancement in mechanical strength starts in micron meter scale, which is noticeably different from other property size dependence.

Two possible mechanisms have been proposed to explain the enhanced strength of nanowires or nanorods (in reality with diameters less than 10 microns). One is to ascribe the increase of strength to the high internal perfection of the nanowires or whiskers. The smaller the cross-section of a whisker or nanowires, the less is the probability of finding in it any imperfections such as dislocations, micro-twins, impurity precipitates, etc.[77] Thermodynamically, imperfections in crystals are highly energetic and should be eliminated from the perfect crystal structures. Small size makes such elimination of imperfections possible. In addition, some imperfections in bulk materials, such as dislocations are often created to accommodate stresses generated in the synthesis and processing of bulk materials due to temperature gradient and other inhomogeneities. Such stresses are unlikely to exist in small structures, particularly in nanomaterials. Another mechanism is the perfection of the side faces of whiskers or nanowires. In general, smaller structures have less surface defects. It is particularly true when the materials are made through a bottom-up approach. For example, Nohara found that vapor grown whiskers with

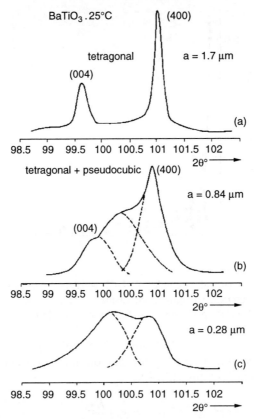

Fig. 8.17. A gradual phase transition from tetragonal structure of large sized BaTiO$_3$ particles of 1.7 μm in diameter to pseudocubic or orthorhombic structure of small size BaTiO$_3$ particles of 0.28 μm in diameter. [G. Arlt, D. Hennings, and G. de With, *J. Appl. Phys.* **58**, 1619 (1985).]

diameters of 10 microns or less had no detectable steps on their surfaces by electron microscopy, whereas irregular growth steps were revealed on whiskers with diameters above 10 microns.[78] Clearly, two mechanisms are closely related. When a whisker is grown at a low supersaturation, there is less growth fluctuation in the growth rate and both the internal and surface structures of the whiskers are more perfect. Figure 8.18 shows a typical dependence of strength on the diameter of a sodium chloride whisker,[79] and similar dependences are found in metals, semiconductors, and insulators.[80,81] In the last few years, AFM and TEM have been applied for measuring the mechanical property of nanowires or nanorods,[82–84] both AFM and TEM promise some direct evidence for the mechanical behavior of nanostructures and nanomaterials.

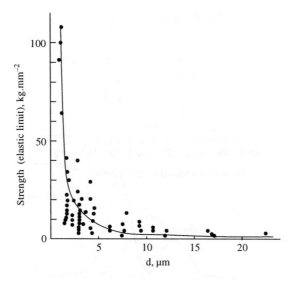

Fig. 8.18. The mechanical strength of NaCl whiskers increases significantly and approaches the theoretical strength as the diameters decrease below ~1 μm due to an increased bulk and surface perfections. [Z. Gyulai, *Z. Phys.* **138**, 317 (1954).]

Yield strength, σ_{TS}, and hardness, H, of polycrystalline materials are known to be dependent on the grain size on the micrometer scale, following the Hall–Petch relationship[85,86]:

$$\sigma_{TS} = \sigma_0 + \frac{K_{TS}}{\sqrt{d}} \quad (8.13)$$

or

$$H = H_0 + \frac{K_H}{\sqrt{d}} \quad (8.14)$$

where σ_0 and H_0 are constants related to the lattice friction stress, d the average grain size, and K_{TS} and K_H material-dependent constants.[87] The inverse square-root dependence on the average grain size follows a scaling of the length of the pile-up with the grain size.[88] The Hall–Petch model treats grain boundaries as barriers to dislocation motion, and thus dislocations pile up against the boundary. Upon reaching a critical stress, the dislocations will cross over to the next grain and induce yielding. As discussed above about the mechanical properties of fine whiskers, nanomaterials such as whiskers possess high perfection and no dislocations have been found in nanostructured materials as of 1992.[89] Therefore, the Hall–Petch model would be invalidated in the nanometer regime.[90]

Experimentally, it has been found that nanostructured metals have higher or lower strength and hardness compared to coarse-grained materials, depending on the methods used to vary the grain size.[89] For example, copper with an average grain size of 6 nm has 5 times higher microhardness over annealed sample having a grain size of 50 μm, and the hardness of palladium with 5–10 nm grains is also 5 times higher than that of 100 μm grained sample.[90] Pure nanocrystalline copper has a yield strength in excess of 400 MPa, approximately 6 times higher than that of coarse-grained copper.[91,92] However, opposite size dependence in copper and palladium was also reported, i.e. a decrease in hardness with decreasing grain size.[93] Various models have been proposed for predicting and explaining the size dependence of strength and hardness in nanomaterials. Two models are developed to predict opposite size dependence of hardness. Hahn et al.[94] proposed that grain boundary sliding is the rate-limiting step of deformation, which explains reasonably well the experimental data that strength and hardness decrease with a decreasing grain size. Another model used a rule of mixture approach, in which two phases—the bulk intragranular phase and the grain boundary phase are considered.[95,96] The model predicts an increasing hardness with a decreasing grain size, before reaching a maximum at a critical grain size of approximately 5 nm, below which the material begins to soften. This model fits very well with the experimental data that hardness increases with a decreasing grain size.[97,98] However, so far there is no experimental result to verify the presence of a critical grain size of 5 nm. Although mechanical properties of various nanostructured element metals have been studied, including silver,[97] copper,[92,95] palladium,[92,95] gold,[98] iron,[99] and nickel,[100] the actual role of grain size or grain boundaries on mechanical properties is not clear, and many factors can have significant influence on the measurement of mechanical properties of nanostructured materials, such as residual strains, flaw sizes and internal stresses. Compared to nanostructured bulk metals, there is even less research and, thus, understanding of the size effect on the mechanical properties of oxides, though some research has been reported on SnO_2,[101] TiO_2[102,103] and ZnO.[105] Other mechanical properties of nanostructured materials, such as Young's modulus, creep and superplasticity, have also been studied; however, there is no solid understanding on the size dependence that have been established.

Nanostructured materials may have different elastoplasticity from that of large-grained bulk materials. For example, near-perfect elastoelasticity was observed in pure nanocrystalline copper, prepared by means of powder metallurgy, as shown in Fig. 8.19.[104] Neither work hardening nor neck formation was observed in tensile tests, which are common characteristics of ductile metals and alloys. However, no explanation is available to this finding. Twinning is observed in nanosized aluminum grains, which have never been found in particles in micrometers or above.[105]

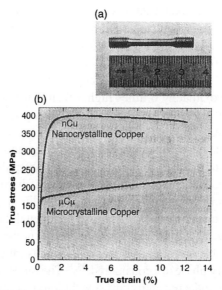

Fig. 8.19. (a) Tensile test specimens are machine cut from the nanocrystalline metal, prepared by powder metallurgy. (b) Comparison of stress and strain for nanocrystalline and microcrystalline copper. The tensile tests are carried out at room temperature and at the low strain rate of 5×10^{-6} s^{-1}. [Y. Champion, C. Langlois, S. Guérin-Mailly, P. Langlois, J. Bonnentien, and M.J. Hÿtch, *Science* **300**, 310 (2003).]

8.4.3. *Optical properties*

The reduction of materials' dimension has pronounced effects on the optical properties. The size dependence can be generally classified into two groups. One is due to the increased energy level spacing as the system becomes more confined, and the other is related to surface plasmon resonance.

8.4.3.1. *Surface plasmon resonance*

Surface plasmon resonance is the coherent excitation of all the "free" electrons within the conduction band, leading to an in-phase oscillation.[106,107] When the size of a metal nanocrystal is smaller than the wavelength of incident radiation, a surface plasmon resonance is generated[108] and Fig. 8.20 shows schematically how a surface plasmon oscillation of a metallic particle is created in a simple manner.[109] The electric field of an incoming light induces a polarization of the free electrons relative to the

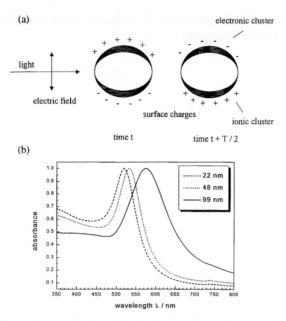

Fig. 8.20. Surface plasmon absorption of spherical nanoparticles and its size dependence. (a) A schematic illustrating the excitation of the dipole surface plasmon oscillation. The electric field of an incoming light wave induces a polarization of the (free) conduction electrons with respect to the much heavier ionic core of a spherical metal nanoparticle. A net charge difference is only felt at the nanoparticle surfaces, which in turn acts as a restoring force. In this way a dipolar oscillation of the electrons is created with period T. (b) Optical absorption spectra of 22, 48 and 99 nm spherical gold nanoparticles. The broad absorption band corresponds to the surface plasmon resonance. [S. Link and M.A. El-Sayed, *Int. Rev. Phys. Chem.* **19**, 409 (2000).]

cationic lattice. The net charge difference occurs at the nanoparticle boundaries (the surface), which in turn acts as a restoring force. In this manner a dipolar oscillation of electrons is created with a certain frequency. The surface plasmon resonance is a dipolar excitation of the entire particle between the negatively charged free electrons and its positively charged lattice. The energy of the surface plasmon resonance depends on both the free electron density and the dielectric medium surrounding the nanoparticle. The width of the resonance varies with the characteristic time before electron scattering. For larger nanoparticle, the resonance sharpens as the scattering length increases. Noble metals have the resonance frequency in the visible light range.

Mie was the first to explain the red color of gold nanoparticle colloidal in 1908 by solving Maxwell's equation for an electromagnetic light wave interacting with small metallic spheres.[110] The solution of this electrodynamic calculation leads to a series of multi-pole oscillation cross-section of the nanoparticles[110]:

$$\sigma_{ext} = \left[\frac{2\pi}{|\kappa|^2}\right]\Sigma(2L+1)\mathrm{Re}(a_L + b_L) \qquad (8.15)$$

$$\sigma_{sca} = \left[\frac{2\pi}{|\kappa|^2}\right]\Sigma(2L+1)(|a_L|^2 + |b_L|^2) \qquad (8.16)$$

with $\sigma_{abs} = \sigma_{ext} - \sigma_{sca}$ and

$$a_L = \frac{m\psi_L(mx)\psi'_L(x) - \psi'_L(mx)\psi_L(x)}{m\psi_L(mx)\eta'_L(x) - \psi'_L(mx)\eta_L(x)} \qquad (8.17)$$

$$b_L = \frac{\psi_L(mx)\psi'_L(x) - m\psi'_L(mx)\psi_L(x)}{\psi_L(mx)\eta'_L(x) - m\psi'_L(mx)\eta_L(x)} \qquad (8.18)$$

where $m = n/n_m$, where n is the complex refractive index of the particle and n_m is the real refractive index of the surrounding medium. k is the wave-vector and $x = k\,r$ with r being the radius of a metallic nanoparticle. ψ_L and η_L are the Ricatti–Bessel cylindrical functions. L is the summation index of the partial waves.

Equations (8.17) and (8.18) clearly indicate that the plasmon resonance depends explicitly on the particle size, r. The larger the particles, the more important the higher-order modes as the light can no longer polarize the nanoparticles homogeneously. These higher-order modes peak at lower energies. Therefore, the plasmon band red shifts with increasing particle size. At the same time, the plasmon bandwidth increases with increasing particle size. The increase of both absorption wavelength and peak width with increasing particle size has been clearly demonstrated experimentally, e.g. as shown in Fig. 8.21.[111] Such direct size dependence on the particle size is regarded as extrinsic size effects.

The situation concerning the size dependence of the optical absorption spectrum is more complicated for smaller nanoparticles for which only the dipole term is important. For nanoparticles much smaller than the wavelength of incident light ($2r \ll \lambda$, or roughly $2r < \lambda_{max}/10$), only the dipole oscillation contributes to the extinction cross-section.[108,109] The Mie theory can be simplified to the following relationship (dipole approximation):

$$\sigma_{ext}(\omega) = \frac{9\omega\varepsilon_m^{\frac{3}{2}} V \varepsilon_2(\omega)}{c\,\{[\varepsilon_1(\omega) + 2\varepsilon_m]^2 + \varepsilon_2(\omega)^2\}} \qquad (8.19)$$

where V is the particle volume, ω is the angular frequency of the exciting light, c is the speed of light, and ε_m and $\varepsilon(\omega) = \varepsilon_1(\omega) + i\varepsilon_2(\omega)$ are the bulk

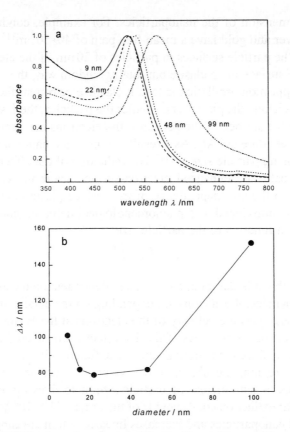

Fig. 8.21. (a) UV-Vis absorption spectra of 9, 22, 48 and 99 nm gold nanoparticles in water. All spectra are normalized at their absorption maxima, which are 517, 521, 533 and 575 nm, respectively. (b) The plasmon bandwidth Δλ as a function of particle diameter. [S. Link and M.A. El-Sayed, *J. Phys. Chem.* **B103**, 4212 (1999).]

dielectric constant of the surround material and the particle, respectively. While the first is assumed to be frequency independent, the latter is complex and is a function of energy. The resonance condition is fulfilled when $\varepsilon_1(\omega) = -2\varepsilon_m$, if ε_2 is small or weakly dependent on ω. Equation (8.19) shows that the extinction coefficient does not depend on the particle sizes; however, a size dependence is observed experimentally.[112–114] This discrepancy arises obviously from the assumption in the Mie theory, that the electronic structure and dielectric constant of nanoparticles are the same as those of its bulk form, which becomes no longer valid when the particle size becomes very small. Therefore, the Mie theory needs to be modified by introducing the quantum size effect in smaller particles.

In small particles, electron surface scattering becomes significant, when the mean free path of the conduction electrons is smaller than the physical dimension of the nanoparticles. For example, conduction electrons in silver and gold have a mean free path of 40–50 nm[115] and will be limited by the particle surfaces in particles of 20 nm. If the electrons scatter with the surface in an elastic but totally random way, the coherece of the overall plasmon oscillation is lost. Inelastic electron-surface collisions would also change the phase. The smaller the particles, the faster the electrons reach the surface of the particles, the electrons can scatter and lose the coherence more quickly. As a result, the plasmon bandwidth increases with decreasing particle size.[116,117] The reduction of the effective electron mean free path and enhanced electron-surface scattering can also correctly explain the size dependence of the surface plasmon absorption as follows. γ is introduced as a phenomenological damping constant and is found to be a function of the particle size[118,119]:

$$\gamma = \frac{\gamma_0 + A v_F}{r} \quad (8.20)$$

where γ_0 is the bulk damping constant and dependent on the electron scattering frequencies, A is a constant, depending on the details of the scattering processes, v_F is the velocity of the electrons at the Fermi energy, and r is the radius of the particles. This size effect is considered as an intrinsic size effect, since the materials' dielectric function itself is size dependent. In this region, the absorption wavelength increases, but the peak width decreases with increasing particle size (also shown in Fig. 8.21).

The molar extinction coefficient is of the order of 1×10^9 M^{-1} cm^{-1} for 20 nm gold nanoparticles and increases linearly with increasing volume of the particles.[116] These extinction coefficients are three to four orders of magnitude higher than those for the very strong absorbing organic dye molecules. The coloration of nanoparticles renders practical applications and some of the applications have been explored and practically used. For example, the color of gold ruby glass results from an absorption band at about 0.53 μm.[118] This band comes from the spherical geometry of the particles and the particular optical properties of gold according to Mie theory[112] as discussed above. The spherical boundary condition of the particles shifts the resonance oscillation to lower frequencies or longer wavelength. The size of the gold particles influences the absorption. For particle larger than about 20 nm in diameter, the band shifts to longer wavelength as the oscillation becomes more complex. For smaller particles, the bandwidth progressively increases because the mean free path of the free electrons in the particles is about 40 nm, and is effectively reduced.[119] Silver particles in glass color it yellow, resulting from a similar absorption band at 0.41 μm.[120] Copper has a plasma absorption band at 0.565 μm for copper particles in glass.[121]

Similar to nanoparticles, metal nanowires have surface plasmon resonance properties.[122] However, metal nanorods exhibited two surface plasmon resonance modes, corresponding to the transverse and longitudinal excitations. While the wavelength of transverse mode is essentially fixed around 520 nm for Au and 410 nm for Ag, their longitudinal modes can be easily tuned to span across the spectral region from visible to near infrared by controlling their aspect ratios. It was also demonstrated that gold nanorods with an aspect ratio of 2–5.4 could fluoresce with a quantum yield more than one million times that of the bulk metal.[123]

8.4.3.2. Quantum size effects

Unique optical property of nanomaterials may also arise from another quantum size effect. When the size of a nanocrystal (i.e. a single crystal nanoparticle) is smaller than the de Broglie wavelength, electrons and holes are spatially confined and electric dipoles are formed, and discrete electronic energy level would be formed in all materials. Similar to a particle in a box, the energy separation between adjacent levels increases with decreasing dimensions. Figure 8.22 schematically illustrates such discrete electronic configurations in nanocrystals, nanowires and thin films; the electronic

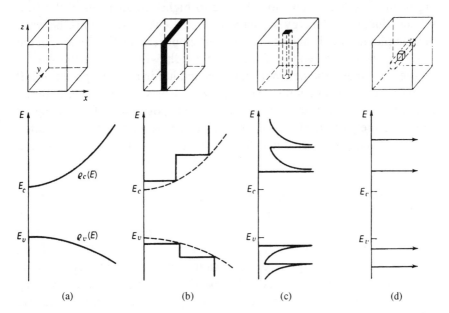

Fig. 8.22. Schematic illustrating discrete electronic configurations in nanocrystals, nanowires and thin films and enlarged band gap between valence band and conduction band.

configurations of nanomaterials are significantly different from that of their bulk counterpart. These changes arise through systematic transformations in the density of electronic energy levels as a function of the size, and these changes result in strong variations in the optical and electrical properties with size.[124,125] Nanocrystals lie in between the atomic and molecular limit of discrete density of electronic states and the extended crystalline limit of continuous band.[126] In any material, there will be a size below which there is substantial variation of fundamental electrical and optical properties with size, when energy level spacing exceeds the temperature. For a given temperature, this occurs at a very large size (in nanometers) in semiconductors as compared with metals and insulators. In the case of metals, where the Fermi level lies in the center of a band and the relevant energy level spacing is very small, the electronic and optical properties more closely resemble those of continuum, even in relatively small sizes (tens or hundreds of atoms).[127,128] In semiconductors, the Fermi level lies between two bands, so that the edges of the bands are dominating the low-energy optical and electrical behavior. Optical excitations across the gap depend strongly on the size, even for crystallites as large as 10,000 atoms. For insulators, the band gap between two bands is already too big in the bulk form.

The quantum size effect is most pronounced for semiconductor nanoparticles, where the band gap increases with a decreasing size, resulting in the interband transition shifting to higher frequencies.[129–132] In a semiconductor, the energy separation, i.e. the energy difference between the completely filled valence band and the empty conduction band is of the order of a few electrovolts and increases rapidly with a decreasing size.[131] Figure 8.23 shows the optical absorption and luminescence spectra of InP nanocrystals as a function of particle size.[13] It is very clear that both the absorption edge and the luminescence peak position shift to a higher energy as the particle size reduces. Such a size dependence of absorption peak has been widely used in determining the size of nanocrystals. Figure 8.24 shows the band gap of silicon nanowires as a function of the nanowire diameter, including both experimental results[133] and calculated data.[134,135]

The same quantum size effect is also known for metal nanoparticles[136,137]; however, in order to observe the localization of the energy levels, the size must be well below 2 nm, as the level spacing has to exceed the thermal energy (~26 meV). In a metal, the conduction band is half filled and the density of energy levels is so high that a noticeable separation in energy levels within the conduction band (intraband transition) is only observed when the nanoparticle is made up of ~100 atoms. If the size of metal nanoparticle is made small enough, the continuous density of electronic states is broken up into discrete energy levels. The spacing, δ,

Characterization and Properties of Nanomaterials

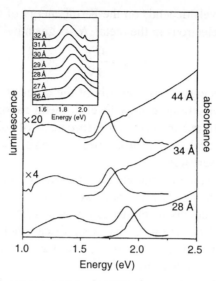

Fig. 8.23. Optical absorption and PL spectra of InP nanocrystals as a function of particle size. The PL spectra are composed of a high-energy band edge emission band and a low-energy trapped emission band. The insert shows additional scaled PL spectra of a sequence of samples with decreasing sizes exhibiting a smooth blue shift of the band edge emission feature with decreasing nanocrystal size. The samples have been treated with decylamine and were exposed to air. [A.A. Guzelian, J.E.B. Katari, A.V. Kadavanich, U. Banin, K. Hamad, E. Juban, A.P. Alivisatos, R.H. Wolters, C.C. Arnold, and J.R. Heath, *J. Phys. Chem.* **100**, 7212 (1996).]

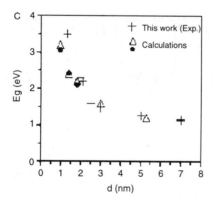

Fig. 8.24. The band gap of silicon nanowires as a function of the nanowire diameter, including both experimental results [D.D.D. Ma, C.S. Lee, F.C.K. Au, S.Y. Tong, and S.T. Lee, *Science* **299**, 1874 (2003)] and calculated data. [A.J. Read, R.J. Needs, K.J. Nash, L.T. Canham, P.D.J. Calcott, and A. Qteish, *Phys. Rev. Lett.* **69**, 1232 (1992) and B. Delley and E.F. Steigmeier, *Appl. Phys. Lett.* **67**, 2370 (1995).]

between energy levels depends on the Fermi energy of the metal, E_F, and on the number of electrons in the metal, N, as given by[138]:

$$\delta = \frac{4E_F}{3N} \qquad (8.21)$$

where the Fermi energy E_F is typically of the order of 5 eV in most metals. The discrete electronic energy level in metal nanoparticles has been observed in far-infrared absorption measurements of gold nanoparticle.[139] At finite size, the evolution of properties of metals from the atomic level to bulk solid is observable.

When the diameter of nanowires or nanorods reduces below the de Broglie wavelength, size confinement would also play an important role in determining the energy level just as for nanocrystals. For example, the absorption edge of Si nanowires has a significant blue shift with sharp, discrete features and silicon nanowires also has shown relatively strong "band-edge" photoluminescence.[140–142]

In addition to the size confinement, light emitted from nanowires is highly polarized along their longitudinal directions.[143–145] Figure 8.25 shows such a distinct anisotropy in the PL intensities recorded in the direction parallel and perpendicular to the long axis of an individual, isolated indium phosphide (InP) nanowires.[145] The magnitude of polarization anisotropy could be quantitatively explained in terms of the large dielectric contrast between the nanowire and the surrounding environment, as

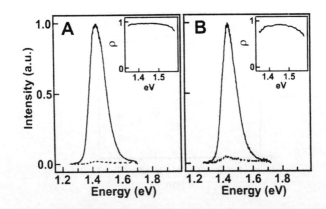

Fig. 8.25. (A) Excitation and (B) emission spectra recorded from an individual InP nanowires of 15 nm in diameter. The polarization of the exciting laser was aligned parallel (solid line) and perpendicular (dashed line) to the long axis of this nanowire, respectively. The inset plots the polarization ratio as a function of energy. [J.F. Wang, M.S. Gudiksen, X.F. Duan, Y. Cui, and C.M. Lieber, Science 293, 1455 (2001).]

Characterization and Properties of Nanomaterials

opposed to quantum mechanical effects such as mixing of valence bands. As noted in Chapter 3, there are other terms commonly used in describing nanoparticles. Nanocrystals are specifically denoted to single crystal nanoparticles. Quantum dots are used to describe small particles that exhibit quantum size effects. Similarly, quantum wires are referred to as quantum wires when exhibiting quantum effects.

8.4.4. Electrical conductivity

The effects of size on electrical conductivity of nanostructures and nanomaterials are complex, since they are based on distinct mechanisms. These mechanisms can be generally grouped into four categories: surface scattering including grain boundary scattering, quantized conduction including ballistic conduction, Coulomb charging and tunneling, and widening and discrete of band gap, and change of microstructures. In addition, increased perfection, such as reduced impurity, structural defects and dislocations, would affect the electrical conductivity of nanostructures and nanomaterials.

8.4.4.1. Surface scattering

Electrical conduction in metals or Ohmic conduction can be described by the various electron scattering, and the total resistivity, ρ_T, of a metal is a combination of the contribution of individual and independent scattering, known as Matthiessen's rule:

$$\rho_T = \rho_{Th} + \rho_D \tag{8.22}$$

ρ_{Th} is the thermal resistivity and ρ_D the defect resistivity. Electron collisions with vibrating atoms (phonons) displaced from their equilibrium lattice positions are the source of the thermal or phonon contribution, which increases linearly with temperature. Impurity atoms, defects such as vacancies, and grain boundaries locally disrupt the periodic electric potential of the lattice and effectively cause electron scattering, which is temperature independent. Obviously, the defect resistivity can be further divided into impurity resistivity, lattice defect resistivity, and grain boundary resistivity. Considering individual electrical resistivity directly proportional to the respective mean free path (λ) between collisions, the Matthiessen's rule can be written as:

$$\frac{1}{\lambda_T} = \frac{1}{\lambda_{Th}} + \frac{1}{\lambda_D} \tag{8.23}$$

Theory suggests that λ_T ranges from several tens to hundreds of nanometers. Reduction in material's dimensions would have two different effects on electrical resistivity. One is an increase in crystal perfection or reduction of defects, which would result in a reduction in defect scattering and, thus, a reduction in resistivity. However, the defect scattering makes a minor contribution to the total electrical resistivity of metals at room temperature, and thus the reduction of defects has a very small influence on the electrical resistivity, mostly unnoticed experimentally. The other is to create an additional contribution to the total resistivity due to surface scattering, which plays a very important role in determining the total electrical resistivity of nanosized materials. If the mean free electron path, λ_S, due to the surface scattering is the smallest, then it will dominate the total electrical resistivity:

$$\frac{1}{\lambda_T} = \frac{1}{\lambda_{Th}} + \frac{1}{\lambda_D} + \frac{1}{\lambda_S} \qquad (8.24)$$

In nanowires and thin films, the surface scattering of electrons results in reduction of electrical conductivity. When the critical dimension of thin films and nanowires is smaller than the electron mean-free path, the motion of electrons will be interrupted through collision with the surface. The electrons undergo either elastic or inelastic scattering. In elastic, also known as specular, scattering, the electron reflects in the same way as a photon reflects from a mirror. In this case, the electron does not lose its energy and its momentum or velocity along the direction parallel to the surface is preserved. As a result, the electrical conductivity remains the same as in the bulk and there is no size effect on the conductivity. When scattering is totally inelastic, or nonspecular or diffuse, the electron mean-free path is terminated by impinging on the surface. After the collision, the electron trajectory is independent of the impingement direction and the subsequent scattering angle is random. Consequently, the scattered electron loses its velocity along the direction parallel to the surface or the conduction direction, and the electrical conductivity decreases. There will be a size effect on electrical conduction.

Figure 8.26 depicts the Thompson model[146] for inelastic scattering of electrons from film surface with film thickness, d, less than the bulk free mean electron path, λ_0. The mean value of λ_f is given by:

$$\lambda = \frac{1}{2}d\int_0^d dz \int_0^\pi \lambda \sin\theta \, d\theta \qquad (8.25)$$

After integration, we have:

$$\lambda_f = \frac{d}{2}\left(\ln\left(\frac{\lambda_0}{d}\right) + \frac{3}{2}\right) \qquad (8.26)$$

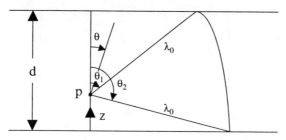

Fig. 8.26. Schematic illustrating the Thompson model for inelastic scattering of electrons from film surface with film thickness, d, less than the bulk free mean electron path, λ_0. [J.J. Thompson, *Proc. Cambridge Phil. Soc.* **11**, 120 (1901).]

Finally the film resistivity, ρ_f, relative to bulk values are given by:

$$\frac{\rho_0}{\rho_f} = \frac{d}{2\lambda_0}\left(\ln\left(\frac{\lambda_0}{d}\right) + \frac{3}{2}\right) \quad (8.27)$$

From Eqs. (8.15) and (8.16), it is clear that as d shrinks, λ_f decreases and ρ_f increases, and there is an obvious size dependence or size effect. It should be noted that the above model is based on an assumption that all surface scattering is inelastic and in terms of classical physics. A more accurate quantum theory, known as Fuchs–Sondheimer (F–S) theory, was also developed.[147,148] When the model is further improved by considering an admixture of both elastic and inelastic contributions, with P being the fraction of elastic surface scattering, an approximate formula for thin films, where $\lambda_0 \gg d$, is obtained:

$$\frac{\rho_0}{\rho_f} = \frac{3d}{4\lambda_0}(1+2P)\left(\ln\left(\frac{\lambda_0}{d}\right) + 0.423\right) \quad (8.28)$$

Clearly, the character of the Thompson equation [Eq. (8.27)] is preserved. The fraction of elastic scattering on a surface is very difficult to determine experimentally; however, it is known that surface impurity and roughness favor inelastic scattering. It should be noted that although the surface scattering discussed above is focused on metals, the general conclusions are equally applicable to semiconductors. An increased surface scattering would result in reduced electron mobility and, thus, an increased electrical resistivity. Increased electrical resistivity of metallic nanowires with reduced diameters due to surface scattering has been widely reported.[149]

Figure 8.27 shows the thickness dependence of electrical resistivities of thin films as a function of temperature.[150] In this experiment, epitaxial films were prepared by first depositing Co on atomically clean silicon substrates under ultrahigh vacuum conditions and followed by vacuum annealing to promote the formation of cobalt silicide. The stoichiometry was well controlled; the film substrate interface was found to be nearly atomically perfect; and the outer surface was extremely smooth. A little dependence

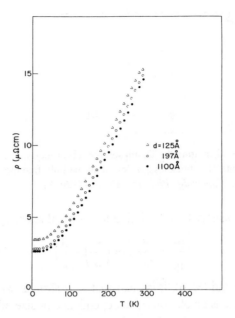

Fig. 8.27. The thickness dependence of electrical resistivities of thin films as a function of temperature. [J.C. Hensel, R.T. Tung, J.M. Poate, and F.C. Unterwald, *Phys. Rev. Lett.* **54**, 1840 (1985).]

on film thickness down to 6 nm was found. Further, it was found that the average λ_o was 97 nm by independent low temperature magnetoresistance measurements, and an average degree of specularity of about 90% from both the free surface and $CoSi_2$–Si interface was suggested. In a polycrystalline material, as the crystallite size becomes smaller than the electron mean-free path, a contribution to electrical resistivity from grain boundary scattering arises. Proton conductivity of polycrystalline hydrated antimony oxide films was found to decrease with small grain sizes as shown in Fig. 8.28, which was attributed to grain boundary scattering.[151]

It should also be noted that the surface inelastic scattering of electrons and phonons would result in a reduced thermal conductivity of nanostructures and nanomaterials, similar to the surface inelastic scattering on electrical conductivity, though very little research has been reported so far. Theoretical studies suggest that thermal conductivity of silicon nanowires with a diameter less than 20 nm would be significantly smaller than the bulk value.[152–154]

8.4.4.2. Change of electronic structure

As shown in Fig. 8.22, a reduction in characteristic dimension below a critical size, i.e. the electron de Broglie wavelength, would result in a change

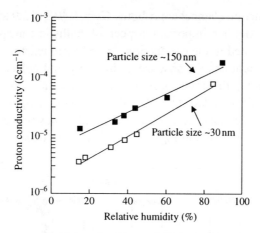

Fig. 8.28. The proton conductivity of polycrystalline hydrated antimony oxide discs at 19.5 °C as a function of relative humidity. Throughout the entire measurement region of humidity, the disc consisting of larger grains has a larger proton conductivity than of small grained sample, which was attributed to grain boundary scattering. [K. Ozawa, Y. Sakka, and M. Amano, *J. Sol-Gel Sci. Technol.* **19**, 595 (2000).]

of electronic structure, leading to widening and discrete band gap. Effects of such a change of band gap on the optical properties has been extensively studied and discussed in the previous section. Such a change generally would also result in a reduced electrical conductivity. Some metal nanowires may undergo a transition to become semiconducting as their diameters are reduced below certain values, and semiconductor nanowires may become insulators. Such a change can be partially attributed to the quantum size effects, i.e. increased electronic energy levels when the dimensions of materials are below a certain size as discussed in the previous section. For example, single crystalline Bi nanowires undergo a metal-to-semiconductor transition at a diameter of ~52 nm[155] and the electrical resistance of Bi nanowires of ~40 nm was reported to decrease with decreasing temperature.[156] GaN nanowires of 17.6 nm in diameter was found to be still semiconducting,[157,158] however, Si nanowires of ~15 nm became insulating.[159]

8.4.4.3. *Quantum transport*

Quantum transport in small devices and materials has been studied extensively.[160,161] Only a brief summary is presented below including discussions on ballistic conduction, Coulomb charging and tunneling conduction.

Ballistic conduction occurs when the length of conductor is smaller than the electron mean-free path.[162–165] In this case, each transverse waveguide mode or conducting channel contributes $G_0 = 2e^2/h = 12.9 \text{ k}\Omega^{-1}$ to the total conductance. Another important aspect of ballistic transport is that no energy is dissipated in the conduction,[162] and there exist no elastic scattering. The latter requires the absence of impurity and defects. When elastic scattering occurs, the transmission coefficients, and thus the electrical conductance will be reduced,[162,166] which is then no longer precisely quantized.[167] Ballistic conduction of carbon nanotubes was first demonstrated by Frank and his co-workers in Fig. 8.29.[168] The conductance of arc-produced

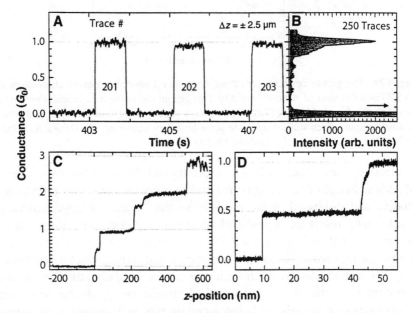

Fig. 8.29. (A) Conductance of a nanotube contact that is moved at constant speed into and out of the mercury contact as a function of time. The period of motion is 2 s and the displacement $\Delta z = \pm 2.5$ μm. The conductance "jumps" to $\sim 1 G_0$ and then remains constant for ~ 2 μm of its dipping depth. The direction of motion is then reversed and the contact is broken after 2 μm. The cycle is repeated to show its reproducibility; cycles 201 through 203 are displayed as an example. (B) Histogram of the conductance data of all 250 traces in the sequence. The plateaus at $1 G_0$ and at 0 produce peaks in the histogram. The relative areas under the peaks correspond to the relative plateau lengths. Because the total displacement is known, the plateau lengths can be accurately determined; in this case, the $1 G_0$ plateau corresponds to a displacement of 1880 nm. Plateau lengths thus determined are insensitive to random oscillations of the liquid level and hence are more accurate than measurements from individual traces. (C) A trace of a nanotube contact with two major plateaus, each with a minor pre-step. This trace is interpreted as resulting from a nanotube that is bundled with a second one (as in Fig. (A), inset). The second tube comes into contact with the metal ~ 200 nm after the first. Shorter plateaus (from ~ 10 to 50 nm long) with noninteger conductance are often seen and are interpreted to result from the nanotube tips. A clear example of this effect is shown in (D). [S. Frank, P. Poncharal, Z.L. Wang, and W.A. de Heer, *Science* **280**, 1744 (1998).]

multi-wall carbon nanotubes is one unit of the conductance quantum G_0, and no heat dissipation is observed. Extremely high stable current densities, $J > 10^7$ A/cm^2 have been attained.

Coulomb blockade or Coulomb charging occurs when the contact resistance is larger than the resistance of nanostructures in question and when the total capacitance of the object is so small that adding a single electron requires significant charging energy.[169] Metal or semiconductor nanocrystals of a few nanometers in diameter exhibit quantum effects that give rise to discrete charging of the metal particles. Such a discrete electronic configuration permits one to pick up the electric charge one electron at a time, at specific voltage values. This Coulomb blockade behavior, also known as "Coulombic staircase", has originated the proposal that nanoparticles with diameters below 2–3 nm may become basic components of single electron transistors (SETs).[170] To add a single charge to a semiconductor or metal nanoparticle requires energy, since electrons can no longer be dissolved into an effectively infinite bulk material. For a nanoparticle surrounded by a dielectric with a dielectric constant of ε_r, the capacitance of the nanoparticle is dependent on its size as:

$$C(r) = 4\pi r \varepsilon_0 \varepsilon_r \qquad (8.29)$$

where r is the radius of the nanoparticle and ε_0 is the permittivity of vacuum. The energy required to add a single charge to the particle is given by the charging energy[171,172]:

$$E_c = \frac{e^2}{2C(r)} \qquad (8.30)$$

Tunneling of single charges onto metal or semiconductor nanoparticles can be seen at temperatures of $k_B T < E_c$, in the I–V characteristics from devices containing single nanoparticles[173,174] or from STM measurements of nanoparticles on conductive surfaces.[175] Such Coulomb staircase is also observed in individual single-wall carbon nanotubes.[176] It should be noted that Eqs. (8.29) and (8.30) clearly indicate that the charging energy is independent of materials. Figure 8.30 shows a characteristic I–V curve for such a device with a single gold nanoparticle, where the charging energy gives rise to a barrier to a current flow known as Coulomb blockade.[177] When a gate electrode is added to the structure, the chemical potential of the nanoparticle as well as the voltage of current flow can be modulated. Such a three-terminal device known as a single-electron transistor has received great attention as an exploratory device structure.[177,178]

Tunneling conduction is another charge transport mechanism important in the nanometer range and has been briefly discussed in Chapter 7.

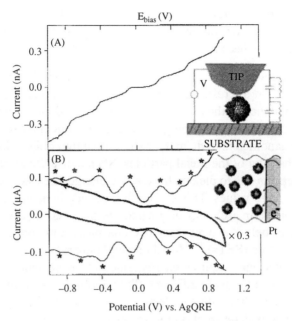

Fig. 8.30. (A) Au STM tip addressing a single cluster adsorbed on an Au-on-mica substrate (inset) and Coulomb staircase I–V curve at 83 K; potential is tip-substrate bias; equivalent circuit of the double tunnel junction gives capacitances $C_{upper} = 0.59$ aF and $C_{lower} = 0.48$ aF. (B) voltammetry (CV -, 100 mV/s; DPV _, ★ are current peaks, 20 mV/s, 25 mV pulse, top and bottom are negative and positive scans, respectively) of a 0.1 mM 28 kDa cluster solution in 2:1 toluene: acetonitrile/0.05 M Hx$_4$NClO$_4$ at a 7.9×10^{-3} cm^2 Pt electrode, 298 K, Ag wire pseudoreference electrode. [R.S. Ingram, M.J. Hostetler, R.W. Murray, T.G. Schaaff, J.T. Khoury, R.L. Whetten, T.P. Bigioni, D.K. Guthrie, and P.N. First, *J. Am. Chem. Soc.* **119**, 9279 (1997).]

Tunneling involves charge transport through an insulating medium separating two conductors that are extremely closely spaced. It is because the electron wave functions from two conductors overlap inside the insulating material, when its thickness is extremely thin. Figure 8.31 gives the tunneling conductivity as a function of C_{14} to C_{23} fatty acid monolayers and is demonstrated that the electrical conductivity decreases exponentially with increasing thickness of insulating layer.[179] Under such conditions, electrons are able to tunnel through the dielectric material when an electric field is applied. It should be noted that Coulomb charging and tunneling conduction, strictly speaking, are not material properties. They are system properties. More specifically, they are system properties dependent on the characteristic dimension.

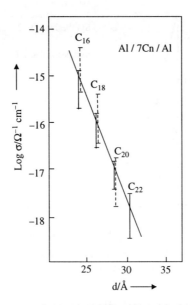

Fig. 8.31. The tunneling conductivity of a self-assembled monolayer as a function of molecular length or film thickness, when changing from C_{14} to C_{23} fatty acid monolayers, demonstrating that the electrical conductivity decreases exponentially with increasing thickness of insulating layer. [H. Kuhn, *J. Photochem.* **10**, 111 (1979).]

8.4.4.4. Effect of microstructure

Electrical conductivity may change due to the formation of ordered microstructure, when the size is reduced to a nanometer scale. For example, polymer fibers demonstrated an enhanced electrical conductivity.[180] The enhancement was explained by the ordered arrangement of the polymer chains. Within nanometer fibris, polymers are aligned parallel to the axis of the fibris, which results in increased contribution of intramolecular conduction and reduced contribution of intermolecular conduction. Since intermolecular conduction is far smaller than intramolecular conduction, ordered arrangement of polymers with polymer chains aligned parallel to the conduction direction would result in an increased electrical conduction. Figure 8.32 shows the electrical conductivity of polyheterocyclic fibris as a function of diameter.[182] A drastic increase in electrical conductivity with a decreasing diameter was found at diameters less than 500 nm. Smaller the diameter, the better alignment of polymer is expected. A lower synthesis temperature also favors a better alignment and thus a higher electrical conductivity.

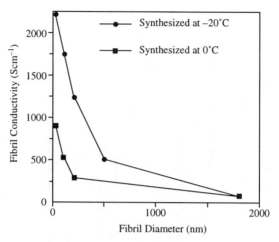

Fig. 8.32. The electrical conductivity of polyheterocyclic fibris as a function of diameter. [Z. Cai, J. Lei, W. Liang, V. Menon, and C.R. Martin, *Chem. Mater.* **3**, 960 (1991).]

8.4.5. Ferroelectrics and dielectrics

Ferroelectric materials are polar compound crystals with reversible spontaneous polarization,[181] and are now candidate dielectrics for integration in the microelectronic systems.[182,183] Ferroelectrics are also pyroelectrics and piezoelectrics and, thus, can be used in infrared imaging systems[184] and microelectromechanical systems.[185] The causes of size effects on ferroelectrics are numerous, and it is difficult to separate true size effects from other factors such as defect chemistry and mechanical strain, which makes the discussion on size effects difficult.

Ferroelectricity differs from other cooperative phenomena in that a surface requires termination of electrical polarization, which forms a depolarizing field. The strength of such a depolarization field is obviously dependent on the size and can influence the ferroelectric–paraelectric transition.[186] Such a size effect has been explained as follows by Mehta et al.[187] Assuming the ferroelectric is a perfect insulator and is homogeneously poled, and the polarization charge is localized at the surface, the depolarization electric field, E_{FE}, in the ferroelectric is constant and given by:

$$E_{FE} = -\frac{P(1-\theta)}{\varepsilon_o \varepsilon_r} \quad (8.31)$$

where ε_o is the permittivity of vacuum, ε_r is the ferroelectric's dielectric constant, P is the saturation polarization of the ferroelectric, and θ is dependent on the ferroelectric size:

$$\theta = \frac{L}{2\varepsilon_r C + L} \quad (8.32)$$

where L is the characteristic size of the ferroelectric, i.e. the distance separating oppositely charged surfaces of the ferroelectric as a result of spontaneous polarization, and C is a constant dependent only on the material that is in contact with the charged ferroelectric surfaces. It is clear that the depolarization field is strongly size dependent, i.e. a decreasing size would result in an increasing depolarization field. Size effects on the ferroelectric phase transition have been investigated both theoretically[188–192] and experimentally[193–196]. Batra et al. have pointed out that the ferroelectric phase in ferroelectric films, for example, 10^{-5} cm or 100 nm thick, becomes unstable by the presence of strong depolarization fields if the surface charges due to the polarization are not fully compensated and that below "transition length", the polarizations are not stable.[190,191]

In a polycrystalline ferroelectric, the ferroelectric properties may disappear when the particles are smaller than a certain size. Such a relation could be understood considering the phase transition temperature reduces with the particle size. A reduction in particle size results in a high temperature crystal structure stable at low temperatures. Consequently, the Curie temperature, or the ferroelectric–paraelectric transition temperature decreases with a reduced particle size. When the Curie temperature drops below room temperature, ferroelectrics lose its ferroelectricity at room temperature. Size effects on ferroelectric and dielectric properties of bulk and thin film ferroelectrics have been summarized in excellent review articles.[197,198]

Ishikawa and coworkers[199] studied the size effect on the ferroelectric phase transition in $PbTiO_3$ nanoparticles that were synthesized by wet chemical routes from alkoxide precursors, and with the particle size controlled by firing temperature. They found when the particle size is less than 50 nm, the transition temperature determined by Raman scattering, decreases from its bulk value of 500°C as the size decreases. They further derived the size dependence of the Curie temperature by an empirical expression for $PbTiO_3$ nanoparticles:

$$T_c (°C) = \frac{500 - 588.5}{D - 12.6} \tag{8.33}$$

where D is the average size of $PbTiO_3$ nanoparticles (nm) and their experimental results are plotted in Fig. 8.14 with the solid line from Eq. (8.33).[201]

In polycrystalline ferroelectrics, there are other factors that may influence the ferroelectric properties. For example, residual strains may actually stabilize the ferroelectric state to smaller sizes.[200,201] Dielectric constant or relative permittivity of ferroelectrics would increase with a decreasing grain size, and such a decrease becomes much more profound when the size is smaller than 1 micrometer as predicted by simulation,[202]

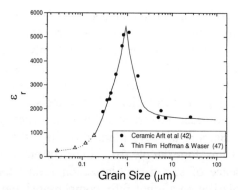

Fig. 8.33. Dielectric constant or relative permittivity of ferroelectrics would increase with a decreasing grain size to 1 μm in diameter and then decrease with further decrease in grain size or film thickness. [T.M. Shaw, S. Trolier-McKinstry, and P.C. McIntyre, *Ann. Rev. Mater. Sci.* **30**, 263 (2000).]

however, the experimental results show that the dielectric constant would increase with a decreasing particle size and reach a maximum at ~1 μm in diameter and then decrease as the particle size or film thickness reduces further as shown in Fig. 8.33.[199]

The mechanical boundary conditions can also affect the stability of the ferroelectric phase and impact the equilibrium domain structures, since many ferroelectrics are also ferroelastic. This makes the size dependence of ferroelectricity very complicated, since the elastic boundary conditions of isolated particles, grains within a ceramic, and thin films are different. It should be noted that no direct observation of superparaelectric behavior in ferroelectrics has been documented yet.

8.4.6. Superparamagnetism

Ferromagnetic particles become unstable when the particle size reduces below a certain size, since the surface energy provides a sufficient energy for domains to spontaneously switch polarization directions. As a result, ferromagnetics become paramagnetics. However, nanometer sized ferromagnetic turned to paramagnetic behaves differently from the conventional paramagnetic and is referred to as superparamagnetics.

Nanometer sized ferromagnetic particles of up to $N = 10^5$ atoms ferromagnetically coupled by exchange forces, form a single domain,[203] with a large single magnetic moment μ with up to 10^5 Bohr magnetons, μ_B. Bean and Livingston demonstrated that these clusters or particles at elevated temperatures can be analogously described as paramagnetic atoms or molecules, however with much larger magnetic moments.[204] The magnetization behavior of single domain particles in thermodynamic equilibrium at all fields is identical with that of atomic paramagnetism, except that an extremely large moment is involved, and thus large susceptibilities are involved. An operational definition of superparamagnetism would include at least two requirements. First, the magnetization curve must show no hysteresis, since that is not a thermal equilibrium property. Second, the magnetization curve for an isotropic sample must be temperature dependent to the extent that curves taken at different temperatures must approximately superimpose when plotted against H/T after correction for the temperature dependence of the spontaneous magnetization.

Superparamagnetism was first predicted to exist in small ferromagnetic particles below a critical size by Frankel and Dorfman.[205] This critical size was estimated to be 15 nm in radius for a spherical sample of the common ferromagnetic materials.[206] The first example of superparamagnetic property was reported in the literature as early as 1954 on nickel particles dispersed in silica matrix.[207] Figure 8.34 shows the typical magnetization

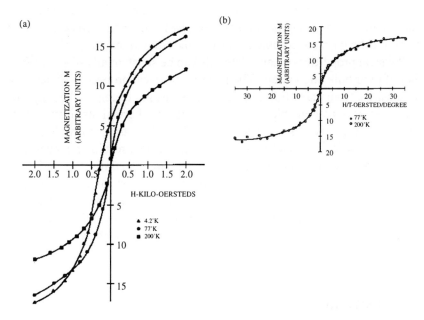

Fig. 8.34. Typical magnetization curves of 2.2 nm iron particles suspended in mercury at various temperatures and the approximate H/T superposition observed for their 77 K and 200 K data. [C.P. Bean and I.S. Jacobs, *J. Appl. Phys.* **27**, 1448 (1956).]

curves of 2.2 nm iron particles suspended in mercury at various temperatures and the approximate H/T superposition observed for their 77 K and 200 K data.[208] For low temperatures, the coupling of the spin to the magnetic anisotropy axis of the system becomes important.[209] The spin tends to align to a certain crystal axis. For example, bulk HCP cobalt has a uniaxial crystal anisotropy.

8.5. Summary

Many characterization and analytical techniques have been applied for the study of nanomaterials and nanostructures; only a few widely used methods are reviewed in this chapter. Both bulk and surface characterization techniques find applications in the study of nanomaterials. However, bulk methods are used to characterize the collective information of nanomaterials such as XRD and gas sorption isotherms. They do not provide information of individual nanoparticles or mesopores. Since most nanomaterials have uniform chemical composition and structures, bulk characterization methods are extensively used. Surface characterization methods such as SPM and TEM offer the possibilities to study individual nanostructures. For example, the surface and inner crystallinity and chemical compositions of nanocrystals can be studied using high resolution TEM. Bulk and surface characterization techniques are complementary in the study of nanomaterials.

Physical properties of nanomaterials can be substantially different from that of their bulk encounters. The peculiar physical properties of nanomaterials arise from many different fundamentals. For example, the huge surface energy is responsible for the reduction of thermal stability and the superparamagnetism. Increased surface scattering is responsible for the reduced electrical conductivities. Size confinement results in a change of both electronic and optical properties of nanomaterials. The reduction of size favors an increase in perfection and, thus, enhanced mechanical properties of individual nanosized materials; however, the size effects on mechanical properties of bulk nanostructured materials is far more complicated, since there are other mechanisms involved, such as grain boundary phase and stresses.

References

1. B.D. Cullity and S.R. Stock, *Elements of X-Ray Diffraction*, 3rd edition, Prentice Hall, Upper Saddle River, NJ, 2001.

2. L.H. Schwartz and J.B. Cohen, *Diffraction From Materials*, Springer-Verlag, Berlin, 1987.
3. L. Reimer, *Scanning Electron Microscopy*, Springer-Verlag, Berlin, 1985.
4. A.S. Nowick (ed.), *Electron Microscopy of Materials: An Introduction*, Academic Press, New York, 1980.
5. J.W. Edington, *Practical Electron Microscopy in Materials*, van Nostrand Reinhold, New York, 1976.
6. Z.L. Wang, *Reflected Electron Microscopy and Spectroscopy for Surface Analysis*, Cambridge University Press, Cambridge, 1996.
7. D. Bonnell (ed.), *Scanning Probe Microscopy and Spectroscopy*, Wiley-VCH, New York, 2001.
8. C.R. Brundle, C.A. Evans Jr., and S. Wilson (eds.), *Encyclopedia of Materials Characterization*, Butterworth-Heinemann, Stoneham, MA, 1992.
9. J.C. Vickerman, *Surface Analysis: The Principle Techniques*, John Wiley & Sons, New York, 1997.
10. A. Segmuller and M. Murakami, in *Analytical Techniques for Thin Films*, eds. K.N. Tu and R. Rosenberg, Academic Press, San Diego, CA, p. 143, 1988.
11. L.S. Birks and H. Friedman, *J. Appl. Phys.* **17**, 687 (1946).
12. A. Segmuller and M. Murakami, in *Thin Films From Free Atoms and Particles*, ed. K.J. Klabunde, Academic Press, Orlando, FL, p. 325, 1985.
13. A.A. Guzelian, J.E.B. Katari, A.V. Kadavanich, U. Banin, K. Hamad, E. Juban, A.P. Alivisatos, R.H. Wolters, C.C. Arnold, and J.R. Heath, *J. Phys. Chem.* **100**, 7212 (1996).
14. O. Glatter and O. Kratky, *Small Angle X-ray Scattering*, Academic Press, New York, 1982.
15. A. Guinier and G. Fournet, *Small Angle Scattering of X-Rays*, John Wiley & Sons, New York, 1955.
16. N.S. Andreev, E.A. Porai-Koshits, and O.V. Mazurin, in *Phase Separation in Glass*, eds., O.V. Mazurin and E.A. Porai-Koshits, North-Holland, Amsterdam, p. 67, 1984.
17. C.R. Kagan, C.B. Murray, and M.G. Bawendi, *Phys. Rev.* **B54**, 8633 (1996).
18. B.A. Korgel and D. Fitzmaurice, *Phys. Rev.* **B59**, 14191 (1999).
19. C.B. Murray, C.R. Kagan, and M.G. Bawendi, *Ann. Rev. Mater. Sci.* **30**, 545 (2000).
20. G. Porod, *Kolloid.-Z.* **124**, 83 (1951).
21. G. Porod, *Kolloid.-Z.* **125**, 51 (1952).
22. P. Debye and A.M. Bueche, *J. Appl. Phys.* **20**, 518 (1949).
23. S.J. Limmer, S. Vince Cruz, and G.Z. Cao, *Adv. Mater.*, submitted (2003).
24. M. von Heimendahl, in *Electron Microscopy of Materials: An Introduction*, ed. A.S. Nowick, Academic Press, New York, p. 1, 1980.
25. A.N. Goldstein, C.M. Echer, and A.P. Alivisatos, *Science* **256**, 1425 (1992).
26. Z.L. Wang, *Adv. Mater.* **12**, 1295 (2000).
27. P. Poncharal, Z.L. Wang, D. Ugarte, and W.A. de Heer, *Science* **283**, 1516 (1999).
28. Z.L. Wang, P. Poncharal, and W.A. de Heer, *J. Phys. Chem. Solids* **61**, 1025 (2000).
29. G. Binnig, H. Rohrer, C. Gerber, and E. Weibel, *Phys. Rev. Lett.* **49**, 57 (1982).
30. G. Binnig, C.F. Quate, and Ch. Gerber, *Phys. Rev. Lett.* **56**, 930 (1986).
31. R.S. Howland and M.D. Kirk, in *Encyclopedia of Materials Characterization*, eds. C.R. Brundle, C.A. Evans Jr., and S. Wilson, Butterworth-Heinemann, Stoneham, MA, p. 85, 1992.
32. G. Binnig, H. Rohrer, C. Gerber, and E. Weibel, *Phys. Rev. Lett.* **50**, 120 (1983).

33. H.P. Lang, M. Hegner, E. Meyer, and Ch. Gerber, *Nanotechnology* **13**, R29 (2002).
34. U. Hartmann, *Ann. Rev. Mater. Sci.* **29**, 53 (1999).
35. A. Majumdar, *Ann. Rev. Mater. Sci.* **29**, 505 (1999).
36. C.C. Williams, *Ann. Rev. Mater. Sci.* **29**, 471 (1999).
37. M. Fujihira, *Ann. Rev. Mater. Sci.* **29**, 353 (1999).
38. E. Betzig and J.K. Trautman, *Science* **257**, 189 (1992).
39. R. Kopelman and W.H. Tan, *Appl. Spec. Rev.* **29**, 39 (1994).
40. H. Heiselmann and D.W. Pohl, *Appl. Phys.* **A59**, 89 (1994).
41. J.W.P. Hsu, *MRS Bull.* **22**, 27 (1997).
42. P.F. Barbara, D.M. Adams, and D.B. O'Connor, *Ann. Rev. Mater. Sci.* **29**, 433 (1999).
43. E. Betzig, J.K. Trautman, T.D. Harris, J.S. Weiner, and L. Kostelak, *Science* **251**, 1469 (1991).
44. F. Zenhausern, Y. Martin, and H.K. Wickramasinghe, *Science* **269**, 1083 (1995).
45. D.M. Young and A.D. Crowell, *Physical Adsorption of Gases*, Butterworths, London, 1962.
46. C. Orr Jr. and J.M. Dallavalle, *Fine Particle Measurement: Size, Surface, and Pore Volume*. MacMillan, New York, 1959.
47. G.M. Pajonk, *Appl. Cata.* **72**, 217 (1991).
48. C.D. Volpe, S. Dire, and Z. Pagani, *J. Non-Cryst. Solids* **209**, 51 (1997).
49. S. Brunauer, *The Adsorption of Gases and Vapors*, Princeton University Press, Princeton, NJ, 1945.
50. J.R. Anderson, *Structure of Metallic Catalysts*, Academic Press, London, 1975.
51. C. Colvard, in *Encyclopedia of Materials Characterization*, eds., C.R. Brundle, C.A. Evans Jr., and S. Wilson, Butterworth-Heinemann, Stoneham, MA, p. 373, 1992.
52. C.R. Kagan, C.B. Murray, and M.G. Bawendi, *Phys. Rev.* **B54**, 8633 (1996).
53. Al. L. Efros and M. Rosen, *Ann. Rev. Mater. Sci.* **30**, 475 (2000).
54. W.B. White, in *Encyclopedia of Materials Characterization*, eds., C.R. Brundle, C.A. Evans Jr., and S. Wilson, Butterworth-Heinemann, Stoneham, MA, p. 428, 1992.
55. M. Orhring, *The Materials Science of Thin Films*, Academic Press, San Diego, CA, 1992.
56. J.R. Bird and J.S. Williams (eds.), *Ion Beams for Materials Analysis*, Academic Press, San Diego, CA, 1989.
57. A.W. Benninghoven, F.G. Rudenauer, and H.W. Werner, *Secondary Ion Mass Spectrometry-Basic Concepts, Instrumental Aspects, Applications and Trends*, Wiley, New York, 1987.
58. P. Pawlow, *Z. Phys. Chem.* **65**, 1 (1909) and **65**, 545 (1909).
59. K.J. Hanszen, *Z. Phys.* **157**, 523 (1960).
60. Ph. Buffat and J.-P. Borel, *Phys. Rev.* **A13**, 2287 (1976).
61. C. Herring, *Structure and Properties of Solid Surfaces*, University of Chicago, Chicago, IL, p. 24, 1952.
62. W.W. Mullins, *Metal Surfaces: Structure Energetics and Kinetics*, The American Society for Metals, Metals Park, OH, p. 28, 1962.
63. S. Ino and S. Ogawa, *J. Phys. Soc. Jpn.* **22**, 1365 (1967).
64. S. Ogawa and S. Ino, *J. Vac. Sci. Technol.* **6**, 527 (1969).
65. N.T. Gladkich, R. Niedermayer, and K. Spiegel, *Phys. Stat. Sol.* **15**, 181 (1966).
66. M. Blackman and A.E. Curzon, *Structure and Properties of Thin Films*, Wiley, New York, 1959.
67. B.T. Boiko, A.T. Pugachev, and Y.M. Bratsykhin, *Sov. Phys. Sol. State* **10**, 2832 (1969).

68. M. Takagi, *J. Phys. Soc. Jpn.* **9**, 359 (1954).
69. R.E. Newnham, K.R. Udayakumar, and S. Trolier-McKinstry, in *Chemical Processing of Advanced Materials*, eds., L.L. Hench and J.K. West, John Wiley and Sons, New York, p. 379, 1992.
70. K. Ishikawa, K. Yoshikawa, and N. Okada, *Phys. Rev.* **B37**, 5852 (1988).
71. S. Link, C. Burda, M.B. Mohamed, B. Nikoobakht, and M.A. El-Sayed, *Phys. Rev.* **B61**, 6086 (2000).
72. Y. Wu and P. Yang, *Appl. Phys. Lett.* **77**, 43 (2000).
73. Y. Wu and P. Yang, *Adv. Mater.* **13**, 520 (2001).
74. D. Quere, J.-M.D. Meglio, and F. Brochard-Wyart, *Science* **249**, 1256 (1990).
75. G. Arlt, D. Hennings, and G. de With, *J. Appl. Phys.* **58**, 1619 (1985).
76. C. Herring and J.K. Galt, *Phys. Rev.* **85**, 1060 (1952).
77. V.G. Lyuttsau, Yu.M. Fishman, and I.L. Svetlov, *Sov. Phys. – Crystallogr.* **10**, 707 (1966).
78. A. Nohara, *Jpn. J. Appl. Phys.* **21**, 1287 (1982).
79. Z. Gyulai, *Z. Phys.* **138**, 317 (1954).
80. S.S. Brenner, in *Growth and Perfection of Crystals*, eds., R.H. Doremus, B.W. Roberts, and D. Turnbull, John Wiley & Sons, New York, p. 157, 1958.
81. P.D. Bayer and R.E. Cooper, *J. Mater. Sci.* **2**, 233 (1967).
82. E.W. Wong, P.E. Sheehan, and C.M. Lieber, *Science* **277**, 1971 (1997).
83. P.E. Marszalek, W.J. Greenleaf, H. Li, A.F. Oberhauser, and J.M. Fernandez, *PNAS* **97**, 6282 (2000).
84. P. Poncharal, Z.L. Wang, D. Ugarte, and W.A. de Heer, *Science* **283**, 1513 (1999).
85. E.O. Hall, *Proc. Phys. Soc. London* **64B**, 747 (1951).
86. N.J. Petch, *J. Iron Steel Inst.* **174**, 25 (1953).
87. C. Suryanarayana, D. Mukhopadhyay, S.N. Patankar, and F.H. Froes, *J. Mater. Res.* **7**, 2114 (1992).
88. J.R. Weertman, M. Niedzielka, and C. Youngdhl, *Mechanical Properties and Deformation Behavior of Materials Having Ultra-Fine Microstructures*, Kluwer, Boston, MA, p. 241, 1993.
89. G.E. Fougere, J.R. Weertman, and R.W. Siegel, *NanoStructured Mater.* **3**, 379 (1993).
90. R.W. Siegel, *Mater. Sci. Engr.* **A168**, 189 (1993).
91. Y. Wang, M. Chen, F. Zhou, and E. Ma, *Nature* **419**, 912 (2003).
92. R.Z. Valiev, I.V. Alexandrov, Y.T. Zhu, and T.C. Lowe, *J. Mater. Res.* **17**, 5 (2002).
93. A.H. Chokshi, A. Rosen, J. Karch, and H. Gleiter, *Scripta Metallurgica* **23**, 1679 (1989).
94. H. Hahn, P. Mondal, and K.A. Padmanabhan, *NanoStructured Mater.* **9**, 603 (1997).
95. J.E. Carsley, J. Ning, W.W. Milligan, S.A. Hackney, and E.C. Aifantis, *NanoStructured Mater.* **5**, 441 (1995).
96. D.A. Konstantinidis and E.C. Aifantis, *NanoStructured Mater.* **10**, 1111 (1998).
97. X.Y. Qin, X.J. Wu, and L.D. Zhang, *NanoStructured Mater.* **5**, 101 (1995).
98. A. Kumpmann, B. Günther, and H.D. Kunze, *Mechanical Properties and Deformation Behavior of Materials Having Ultra-Fine Microstructures*, Kluwer, Boston, MA, p. 309, 1993.
99. J.C.S. Jang and C.C. Koch, *Scripta Metallurgica et Materialia* **24**, 1599 (1990).
100. G.D. Hughes, S.D. Smith, C.S. Pande, H.R. Johnson, and R.W. Armstrong, *Scripta Metallurgica* **20**, 93 (1986).
101. K.A. Padmanabhan, *Mater. Sci. Engr.* **A304**, 200 (2001).

102. H. Höfler and R.S, Averback, *Scripta Metallurgica et Materialia* **24**, 2401 (1990).
103. M.J. Mayo, R.W. Siegel, Y.X. Liao, and W.D. Nix, *J. Mater. Res.* **7**, 973 (1992).
104. Y. Champion, C. Langlois, S. Guérin-Mailly, P. Langlois, J. Bonnentien, and M.J. Hÿtch, *Science* **300**, 310 (2003).
105. M. Chen, E. Ma, K.J. Hemker, H. Sheng, Y. Wang, and X. Chen, *Science* **300**, 1275 (2003).
106. M. Kerker, *The Scattering of Light and Other Electromagnetic Radiation*, Academic Press, New York, 1969.
107. C.F. Bohren and D.R. Huffman, *Adsorption and Scattering of Light by Small Particles*, Wiley, New York, 1983.
108. U. Kreibeg and M. Vollmer, *Optical Properties of Metal Clusters*, Vol. 25, Springer-Verlag, Berlin, 1995.
109. S. Link and M.A. El-Sayed, *Int. Rev. Phys. Chem.* **19**, 409 (2000).
110. G. Mie, *Am. Phys.* **25**, 377 (1908).
111. S. Link and M.A. El-Sayed, *J. Phys. Chem.* **B103**, 4212 (1999).
112. U. Kreibig and U. Genzel, *Surf. Sci.* **156**, 678 (1985).
113. P. Mulvaney, *Langmuir* **12**, 788 (1996).
114. S. Link and M.A. El-Sayed, *J. Phys. Chem.* **B103**, 8410 (1999).
115. N.W. Ashcroft and N.D. Mermin, *Solid State Physics*, Saunders College, Philadelphia, PA, 1976.
116. U. Kreibig and C. von Fragstein, *Z. Phys.* **224**, 307 (1969).
117. U. Kreibig, *Z. Phys.* **234**, 307 (1970).
118. R.H. Doremus, *Glass Science*, 2nd edition, Wiley, New York, 1994.
119. R.H. Doremus, *J. Chem. Phys.* **40**, 2389 (1964).
120. R.H. Doremus, *J. Chem. Phys.* **41**, 414 (1965).
121. R.H. Doremus, S.C. Kao, and R. Garcia, *Appl. Opt.* **31**, 5773 (1992).
122. M.A. El-Sayed, *Acc. Chem. Res.* **34**, 257 (2001).
123. M.B. Mohamed, V. Volkov, S. Link, and M.A. El-Sayed, *Chem. Phys. Lett.* **317**, 517 (2000).
124. A.I. Ekimov and A.A. Onushchenko, *Sov. Phys. – Semicond.* **16**, 775 (1982).
125. R. Rossetti, S. Nakahara, and L.E. Brus, *J. Chem. Phys.* **79**, 1086 (1983).
126. A.P. Alivisatos, *J. Phys. Chem.* **100**, 13226 (1996).
127. M.L. Chen, M.Y. Chou, W.D. Knight, and W.A. de Heer, *J. Phys. Chem.* **91**, 3141 (1987).
128. C.R.C. Wang, S. Pollack, T.A. Dahlseid, G.M. Koretsky, and M. Kappes, *J. Chem. Phys.* **96**, 7931 (1992).
129. A.J. Nozik and R. Memming, *J. Phys. Chem.* **100**, 13061 (1996).
130. A.P. Alivisatos, *J. Phys. Chem.* **100**, 13226 (1996).
131. Y. Wang and N. Herron, *J. Phys. Chem.* **95**, 525 (1991).
132. L.E. Brus, *Appl. Phys.* **A53**, 465 (1991).
133. D.D.D. Ma, C.S. Lee, F.C.K. Au, S.Y. Tong, and S.T. Lee, *Science* **299**, 1874 (2003).
134. A.J. Read, R.J. Needs, K.J. Nash, L.T. Canham, P.D.J. Calcott, and A. Qteish, *Phys. Rev. Lett.* **69**, 1232 (1992).
135. B. Delley and E.F. Steigmeier, *Appl. Phys. Lett.* **67**, 2370 (1995).
136. J.A.A. Perenboom, P. Wyder, and P. Meier, *Phys. Rep.* **78**, 173 (1981).
137. W.P. Halperin, *Rev. Mod. Phys.* **58**, 533 (1986).
138. R. Kubo, A. Kawabata, and S. Kobayashi, *Ann. Rev. Mater. Sci.* **14**, 49 (1984).
139. M.M. Alvarez, J.T. Kjoury, T.G. Schaaff, M.N. Shafigullin, I. Vezmarm, and R.L. Whetten, *J. Phys. Chem.* **B101**, 3706 (1997).

140. X. Lu, T.T. Hanrath, K.P. Johnston, and B.A. Korgel, *Nano Lett.* **3**, 93 (2003).
141. T.T. Hanrath and B.A. Korgel, *J. Am. Chem. Soc.* **124**, 1424 (2001).
142. J.D. Holmes, K.P. Johston, R.C. Doty, and B.A. Korgel, *Science* **287**, 1471 (2000).
143. J.F. Wang, M.S. Gudiksen, X.F. Duan, Y. Cui, and C.M. Lieber, *Science* **293**, 1455 (2001).
144. M. Huang, S. Mao, H. Feick, H. Yan, Y. Wu, H. Kind, E. Weber, R. Russo, and P. Yang, *Science* **292**, 1897 (2001).
145. Y. Xia, P. Yang, Y. Sun, Y. Wu, B. Mayers, B. Gates, Y. Yin, F. Kim, and H. Yan, *Adv. Mater.* **15**, 353 (2003).
146. J.J. Thompson, *Proc. Cambridge Phil. Soc.* **11**, 120 (1901).
147. K. Fuchs, *Proc. Cambridge Phil. Soc.* **34**, 100 (1938).
148. E.H. Sondheimer, *Adv. Phys.* **1**, 1 (1951).
149. M.J. Skove and E.P. Stillwell, *Appl. Phys. Lett.* **7**, 241 (1965).
150. J.C. Hensel, R.T. Tung, J.M. Poate, and F.C. Unterwald, *Phys. Rev. Lett.* **54**, 1840 (1985).
151. K. Ozawa, Y. Sakka, and M. Amano, *J. Sol-Gel Sci. Technol.* **19**, 595 (2000).
152. K. Schwab, E.A. Henriksen, J.M. Worlock, and M.L. Roukes, *Nature* **404**, 974 (2000).
153. A. Buldum, S. Ciraci, and C.Y. Fong, *J. Phys. Condens. Matter* **12**, 3349 (2000).
154. S.G. Volz and G. Chen, *Appl. Phys. Lett.* **75**, 2056 (1999).
155. Z. Zhang, X. Sun, M.S. Dresselhaus, and J.Y. Ying, *Phys. Rev.* **B61**, 4850 (2000).
156. S.H. Choi, K.L. Wang, M.S. Leung, G.W. Stupian, N. Presser, B.A. Morgan, R.E. Robertson, M. Abraham, S.W. Chung, J.R. Heath, S.L. Cho, and J.B. Ketterson, *J. Vac. Sci. Technol.* **A18**, 1326 (2000).
157. Y. Cui and C.M. Lieber, *Science* **291**, 851 (2001).
158. Y. Wang, X. Duan, Y. Cui, and C.M. Lieber, *Nano Lett.* **2**, 101 (2002).
159. S.W. Chung, J.Y. Yu, and J.R. Heath, *Appl. Phys. Lett.* **76**, 2068 (2000).
160. S. Datta, *Electronic Transport in Mesoscopic Systems*, Cambridge University Press, Cambridge, 1995.
161. D.K. Ferry, H.L. Grubin, C.L. Jacoboni, and A.P. Jauho, (eds.), *Quantum Transport in Ultrasmall Devices*, Plenum Press, New York, 1994.
162. B.J. van Wees, H. van Houten, C.W.J. Beenakker, J.G. Williamson, L.P. Kouwenhoven, D. van der Marel, and C.T. Foxon, *Phys. Rev. Lett.* **60**, 848 (1988).
163. D.P.E. Smith, *Science* **269**, 371 (1995).
164. D.S. Fisher and P.A. Lee, *Phys. Rev.* **B23**, 6851 (1981).
165. H. van Houten and C. Beenakker, *Phys. Today*, p. 22, (July 22, 1996).
166. R. Landauer, *Philos. Mag.* **21**, 863 (1970).
167. W.A. de Heer, S. Frank, and D. Ugarte, *Z. Phys.* **B104**, 468 (1997).
168. S. Frank, P. Poncharal, Z.L. Wang, and W.A. de Heer, *Science* **280**, 1744 (1998).
169. H. Grabert and M.H. Devoret (eds.), *Single Charge Tunneling*, Plenum, New York, 1992.
170. D.L. Feldheim and C.D. Keating, *Chem. Soc. Rev.* **27**, 1 (1998).
171. M.A. Kastner, *Phys. Today* **46**, 24 (1993).
172. H. Grabert, in *Single Charge Tunneling*, eds. M.H. Devoret, and H. Grabert, Plenum, New York, p. 1, 1992.
173. D.L. Klein, P.L. McEuen, J.E.B. Katari, R. Roth, and A.P. Alivisatos, *Appl. Phys. Lett.* **68**, 2574 (1996).
174. C.T. Black, D.C. Ralph, and M. Tinkham, *Phys. Rev. Lett.* **76**, 688 (1996).
175. R.S. Ingram, M.J. Hostetler, R.W. Murray, T.G. Schaaff, J.T. Khoury, R.L. Whetten, T.P. Bigioni, D.K. Guthrie, and P.N. First, *J. Am. Chem. Soc.* **119**, 9279 (1997).

176. S.J. Tans, M.H. Devoret, H.J. Dai, A. Thess, R.E. Smalley, L.J. Geerligs, and C. Dekker, *Nature* **386**, 474 (1997).
177. T.A. Fulton and D.J. Dolan, *Phys. Rev. Lett.* **59**, 109 (1987).
178. T. Sato, H. Ahmed, D. Brown, and B.F.G. Johnson, *J. Appl. Phys.* **82**, 696 (1997).
179. H. Kuhn, *J. Photochem.* **10**, 111 (1979).
180. Z. Cai, J. Lei, W. Liang, V. Menon, and C.R. Martin, *Chem. Mater.* **3**, 960 (1991).
181. F. Jona and G. Shirane, *Ferroelectric Crystals*, Dover Pub. Inc., New York, 1993.
182. J.F. Scott and C.A. de Araujo, *Science* **246**, 1400 (1989).
183. O. Auciello, J.F. Scott, and R. Ramesh, *Phys. Today* **51**, 22 (1998).
184. L.E. Cross and S. Trolier-McKinstry, *Encycl. Appl. Phys.* **21**, 429 (1997).
185. D.L. Polla and L.F. Francis, *Mater. Res. Soc. Bull.* **21**, 59 (1996).
186. K. Binder, *Ferroelectrics* **35**, 99 (1981).
187. R.R. Mehta, B.D. Silverman, and J.T. Jacobs, *J. Appl. Phys.* **44**, 3379 (1973).
188. I.P. Batra, P. Würfel, and B.D. Silverman, *Phys. Rev. Lett.* **30**, 384 (1973).
189. I.P. Batra, P. Würfel, and B.D. Silverman, *Phys. Rev.* **B8**, 3257 (1973).
190. R. Kretschmer and K. Binder, *Phys. Rev.* **B20**, 1065 (1979).
191. A.J. Bell and A.J. Moulson, *Ferroelectrics* **54**, 147 (1984).
192. K. Binder, *Ferroelectrics* **73**, 43 (1987).
193. V.V. Kuleshov, M.G. Radchenko, V.P. Dudkevich, and Eu. G. Fesenko, *Cryst. Res. Technol.* **18**, K56 (1983).
194. P. Würfel and I.P. Batra, *Ferroelectrics* **12**, 55 (1976).
195. T. Kanata, T. Yoshikawa, and K. Kubota, *Solid State Commun.* **62**, 765 (1987).
196. A. Roelofs, T. Schneller, K. Szot, and R. Waser, *Nanotechnology* **14**, 250 (2003).
197. T.M. Shaw, S. Trolier-McKinstry, and P.C. McIntyre, *Ann. Rev. Mater. Sci.* **30**, 263 (2000).
198. R.E. Newnham, K.R. Udayakumar, and S. Trolier-McKinstry, in *Chemical Processing of Advanced Materials*, eds. L.L. Hench and J.K. West, John Wiley and Sons, New York, p. 379, 1992.
199. K. Ishikawa, K. Yoshikawa, and N. Okada, *Phys. Rev.* **B37**, 5852 (1988).
200. W. Kanzig, *Phys. Rev.* **98**, 549 (1955).
201. R. Bachmann and K. Barner, *Solid State Commun.* **68**, 865 (1988).
202. G. Arlt and N.A. Pertsev, *J. Appl. Phys.* **70**, 2283 (1991).
203. J.P. Bucher, D.C. Douglass, and L.A. Bloomfield, *Phys. Rev. Lett.* **66**, 3052 (1991).
204. C.P. Bean and J.D. Livingston, *J. Appl. Phys.* **30**, 120S (1959).
205. J. Frankel and J. Dorfman, *Nature* **126**, 274 (1930).
206. C. Kittel, *Phys. Rev.* **70**, 965 (1946).
207. W. Heukelom, J.J. Broeder, and L.L. van Reijen, *J. Chim. Phys.* **51**, 474 (1954).
208. C.P. Bean and I.S. Jacobs, *J. Appl. Phys.* **27**, 1448 (1956).
209. P.W. Selwood, *Chemisorption and Magnetization*, Academic Press, New York, 1975.

Chapter 9

Applications of Nanomaterials

9.1. Introduction

Nanotechnology offers an extremely broad range of potential applications from electronics, optical communications and biological systems to new materials. Many possible applications have been explored and many devices and systems have been studied. More potential applications and new devices are being proposed in literature. It is obviously impossible to summarize all the devices and applications that have been studied and it is impossible to predict new applications and devices. This chapter will simply provide some examples to illustrate the possibilities of nanostructures and nanomaterials in device fabrication and applications. It is interesting to note that the applications of nanotechnology in different fields have distinctly different demands, and thus face very different challenges, which require different approaches. For example, for applications in medicine, or in nanomedicine, the major challenge is "miniaturization": new instruments to analyze tissues literally down to the molecular level, sensors smaller than a cell allowing to look at ongoing functions, and small machines that literally circulate within a human body pursuing pathogens and neutralizing chemical toxins.[1]

Applications of nanostructures and nanomaterials are based on (i) the peculiar physical properties of nanosized materials, e.g. gold nanoparticles used as inorganic dye to introduce colors into glass and as low temperature catalyst, (ii) the huge surface area, such as mesoporous titania for photoelectrochemical cells, and nanoparticles for various sensors, and (iii) the small size that offers extra possibilities for manipulation and room for accommodating multiple functionalities. For many applications, new materials and new properties are introduced. For example, various organic molecules are incorporated into electronic devices, such as sensors.[2] This chapter intends to provide some examples that have been explored to illustrate the vast range of applications of nanostructures and nanomaterials.

9.2. Molecular Electronics and Nanoelectronics

Tremendous efforts and progress have been made in the molecular electronics and nanoelectronics.[3–12] In molecular electronics, single molecules are expected to be able to control electron transport, which offers the promise of exploring the vast variety of molecular functions for electronic devices, and molecules can now be crafted into a working circuit as shown schematically in Fig. 9.1.[3] When the molecules are biologically active, bioelectronic devices could be developed.[2,13] In molecular electronics, control over the electronic energy levels at the surface of conventional semiconductors and metals is achieved by assembling on the solid

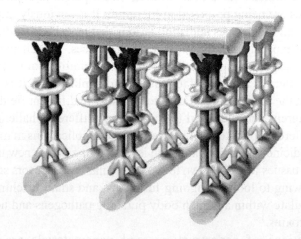

Fig. 9.1. Schematic showing that molecules can now be crafted into working circuit, though constructing real molecular chips remains a big challenge. [R. F. Service, *Science* **293**, 782 (2001).]

surfaces, poorly organized, partial monolayers of molecules instead of the more commonly used ideal ones. Once those surfaces become interfaces, these layers exert electrostatic rather than electrodynamic control over the resulting devices, based on both electrical monopole and dipole effects of the molecules. Thus electronic transport devices, incorporating organic molecules, can be constructed without current flow through the molecules. The simplest molecular electronics are sensors that translate unique molecular properties into electrical signals. Sensors using a field effect transistor (FET) configuration with its gate displaced into a liquid electrolyte, and an active layer of molecules for molecular recognition were reported in early 1970's.[14] A selective membrane is inserted on the insulator surface of the FET, and this permits the diffusion of specific analyte ions and construction of a surface dipole layer at the insulator surface. Such a surface dipole changes the electric potential at the insulator surface and, thus, permits the current going through the device. Such devices are also known as ion-selective FET (ISFET) or chemical FET (CHEMFET).[2,15,16] Thin films attached to metal nanoparticles have been shown to change their electrical conductivity rapidly and reproducibly in the presence of organic vapors, and this has been exploited for the development of novel gas sensors.[17,18] The monolayer on metal nanoparticles can reversibly adsorb and desorb the organic vapor, resulting in swelling and shrinking of the thickness of the monolayer, thus changing the distance between the metal cores. Since the electron hopping conductivity through the monolayers is sensitively dependent on the distance, the adsorption of organic vapor increases the distance and leads to a sharp decrease in electrical conductivity.

Many nanoscale electronic devices have been demonstrated: tunneling junctions,[19–21] devices with negative differential resistance,[22] electrically configurable switches,[23,24] carbon nanotube transistors,[25,26] and single molecular transistors.[27,28] Devices have also been connected together to form circuits capable of performing single functions such as basic memory[23,24,29] and logic functions.[30–33] Ultrahigh density nanowires lattices and circuits with metal and semiconductor nanowires have also been demonstrated.[34] Computer architecture based on nanoelectronics (also known as nanocomputers) has also been studied,[35,36] though very limited. Various processing techniques have been applied in the fabrication of nanoelectronics such as focused ion beam (FIB),[37–39] electron beam lithography,[34,40] and imprint lithography.[33] Major obstacles preventing the development of such devices include addressing nanometer-sized objects such as nanoparticles and molecules, molecular vibrations, robustness and the poor electrical conductivity.

Au nanoparticles have been widely used in nanoelectronics and molecular electronics using its surface chemistry and uniform size. For example, Au nanoparticles function as carrier vehicles to accommodate multiple functionalities through attaching various functional organic molecules or biocomponents.[41] Au nanoparticles can also function as mediators to connect different functionalities together in the construction of nanoscale electronics for the applications of sensors and detectors. Various electronic devices based on Au nanoparticles and Au_{55} clusters have been explored.[42–44] In particular, single electron transistor action has been demonstrated for systems that contain ideally only one nanoparticle in the gap between two electrodes separated by only a few nanometers. This central metal particle represents a Coulomb blockade and exhibits single electron charging effects due to its extremely small capacitance. It can also act as a gate if it is independently addressable by a third terminal. An electrochemically addressable nano-switch, consisting of a single gold particle covered with a small number of dithiol molecules containing a redox-active viologen moiety has been demonstrated, and the electron transfer between the gold substrate and the gold nanoparticle depend strongly on the redox-state of the viologen.[45]

Single-walled carbon nanotubes have also been intensively studied for nanoelectronic devices, due to the semiconducting behavior of different allotropes.[46] Examples of single-walled carbon nanotube nanoelectronic devices include single-electron transistors,[47–49] FET,[30,50,51] sensors,[52,53] circuits,[54] and a molecular electronics toolbox.[55] Carbon nanotubes have been explored for many other applications, such as actuators,[56,57] sensors,[58,59] and thermometers made of multiple-walled carbon nanotubes filled with gallium.[60]

9.3. Nanobots

A very promising and fast growing field for the applications of nanotechnology is in the practice of medicine, which, in general, is often referred to as nanomedicine. One of the attractive applications in nanomedicine has been the creation of nanoscale devices for improved therapy and diagnostics. Such nanoscale devices are known as nanorobots or more simply as nanobots.[61] These nanobots have the potential to serve as vehicles for delivery of therapeutic agents, detectors or guardians against early disease and perhaps repair of metabolic or genetic defects. Similar to the conventional or macroscopic robots, nanobots would be programmed to perform specific functions and be remotely controlled, but possess a much smaller size, so that they can travel and perform desired functions inside the human body. Such devices were first described by Drexler in his book, Engines of Creation in 1986.[62]

Haberzettl[61] described what nanobots would do in practice in medicine, which is briefly summarized below. Nanobots applied to medicine would be able to seek out a target within the body such as a cancer cell or an invading virus, and perform some function to fix the target. The fix delivered by the nanobots may be that of releasing a drug in a localized area, thus minimizing the potential side effects of generalized drug therapy, or it may bind to a target and prevent it from further activity thus, for example, preventing a virus from infecting a cell. Further in the future, gene replacement, tissue regeneration or nanosurgery are all possibilities as the technology becomes more mature and sophisticated.

Although such capable and sophisticated nanobots are not yet realized, many functions in much simplified nanobots are being investigated and tested in the lab. It is also being argued that the nanobots would not take the conventional approach of macroscopic robots. Examples include:

(1) Architecture or structure to carry the payload, known as carrier: Three groups of nanoscale materials have been studied extensively as a structure or vehicle to carry various payloads. The first group is carbon nanotubes or buckyballs, the second group various dendrimers, and the third various nanoparticles and nanocrystals.
(2) Targeting mechanisms to guide the nanobots to the desired site of action: The most likely mechanisms to be employed are based on antigen or antibody interactions or binding of target molecules to membrane-bound receptors. The navigation system for nanobots would be most likely to use the same method that the human body uses, going with the flow and "dropping anchor" when the nanobots reached its target.
(3) Communication and information processing: Single molecular electronics may offer simple switch function of on and off, optical labeling would be a more readily achievable reality.
(4) Retrieve of nanobots from human body: Retrieve of nanobots from human body would be another challenge in the development of nanobots. Most nanodevices could be eliminated from the body through natural mechanisms of metabolism and excretion. Nanodevices made of biodegradable or naturally occurring substances, such as calcium phosphate, would be another favorable approach. "Homing" nanobots would be ideal, which can be collected and removed after performing the desire function. The possible negative impacts of nanobots include the pollution and clog of systems in human body, and the nanobots may become "out of control" when some functions are lost or nanobots malfunction.

9.4. Biological Applications of Nanoparticles

Biological applications of colloidal nanocrystals have been summarized in an excellent review article, and the following text is mainly based on this article.[63] One important branch of nanotechnology is nanobiotechnology. Nanobiotechnology includes (i) the use of nanostructures as highly sophisticated scopes, machines or materials in biology and/or medicine, and (ii) the use of biological molecules to assemble nanoscale structures.[63] The following will briefly describe one of the important biological applications of colloidal nanocrystals: molecular recognition. But there are many more biological applications of nanotechnology.[64–66]

Molecular recognition is one of the most fascinating capabilities of many biological molecules.[67,68] Some biological molecules can recognize and bind to other molecules with extremely high selectivity and specificity. For molecular recognition applications, antibodies and oligonucleotides are widely used as receptors. Antibodies are protein molecules created by the immune systems of higher organisms that can recognize a virus as a hostile intruder or antigen, and bind to it in such a way that the virus can be destroyed by other parts of the immune system.[67] Oligonucleotides, known as single stranded deoxyribonucleic acid (DNA), are linear chains of nucleotides, each of which is composed of a sugar backbone and a base. There are four different bases: adenine (A), cytosine (C), guanine (G), and thymine (T).[67] The molecular recognition ability of oligonucleotides arises from two characteristics. One is that each oligonucleotide is characterized by the sequence of its bases, and another is that base A only binds to T and C only to G. That makes the binding of oligonucleotides highly selective and specific.

Antibodies and oligonucleotides are typically attached to the surface of nanocrystals via (i) thiol-gold bonds to gold nanoparticles,[69,70] (ii) covalent linkage to silanized nanocrystals with bifunctional crosslinker molecules,[71–73] and (iii) a biotin-avidin linkage, where avidin is adsorbed on the particle surface.[74,75] When a nanocrystal is attached or conjugated to a receptor molecules, it is "tagged". Nanocrystals conjugated with a receptor can now be "directed" to bind to positions where ligand molecules are present, which "fit" the molecular recognition of the receptor[76] as schematically shown in Fig. 9.2. This facilitates a set of applications including molecular labeling.[63,77–79] For example, when gold nanoparticles aggregate, a change of color from ruby-red to blue is observed, and this phenomenon has been exploited for the development of very sensitive colorimetric methods of DNA analysis.[80] Such devices are capable of detecting trace amounts of a particular oligonucleotide sequence and distinguishing between perfectly

Applications of Nanomaterials

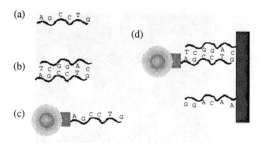

Fig. 9.2. DNA as a molecular template to arrange nanoscale objects. (a) One oligonucleotide composed of six bases (A,G,C,C,T,G). (b) One oligonucleotide (AGCCTG) bound to a complementary oligonucleotide. (c) Conjugate formed between a silanized CdSe/ZnS nanocrystal and an oligonucleotide with six bases. (d) The nanocrystal-oligonucleotide conjugate binds to an oligonucleotide with complementary sequence that is immobilized on a surface, but does not bind to oligonucleotides with different sequences. [W.J. Parak, D. Gerion, T. Pellegrino, D. Zanchet, C. Micheel, S.C. Williams, R. Boudreau, M.A. Le Gros, C.A. Larabell, and A.P. Alivisatos, *Nanotechnology* 14, R15 (2003).]

complementary DNA sequences and those that exhibit different degrees of base pair mismatches.

9.5. Catalysis by Gold Nanoparticles

Bulk gold is chemically inert and thus considered to be not active or useful as a catalyst.[81,82] However, gold nanoparticles can have excellent catalytic properties as first demonstrated by Haruta.[83] For example, gold nanoparticles with clean surface have demonstrated to be extremely active in the oxidation of carbon monoxide if deposited on partly reactive oxides, such as Fe_2O_3, NiO and MnO_2. γ-alumina,[84] and titania[85,86] are also found to be reactive. Figure 9.3 shows a STM image of Au nanoparticles $TiO_2(110)$-(1×1) substrate as prepared before a $CO:O_2$ reaction.[85] The Au coverage is 0.25 ML, and the sample was annealed at 850 K for 2 min. The size of the images is 50 nm by 50 nm.[85] Au nanoparticles also exhibit extraordinary high activity for partial oxidation of hydrocarbons, hydrogenation of unsaturated hydrocarbons and reduction of nitrogen oxides.[83]

The excellent catalytic property of gold nanoparticles is a combination of size effect and the unusual properties of individual gold atom. The unusual properties of gold atom are attributable to the so-called relativistic effect that stabilizes the $6s^2$ electron pairs.[81,87] The relativistic effect is briefly described below. As the atomic number increases, so does the mass of nucleus. The speed of the innermost $1s^2$ electrons has to increase to maintain their position, and for gold, they attain a speed of 60% light

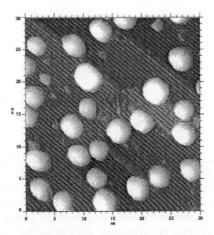

Fig. 9.3. A STM image of Au on TiO_2 (110)-(1 × 1) substrate as prepared before a $CO:O_2$ reaction. The Au coverage is 0.25 ML, and the sample was annealed at 850 K for 2 min. The size of the images is 30 nm by 30 nm. [Courtesy of Prof. D. Wayne Goodman at Texas A&M University, detailed information seen M. Valden, X. Lai, and D.W. Goodman, *Science* **281**, 1647 (1998).]

speed. A relativistic effect on their mass results in the 1*s* orbital contraction. Then all the outer *s* orbitals have to contract in sympathy, but *p* and *d* electrons are much less affected. In consequence, the 6 s^2 electron pair is contracted and stabilized, and the actual size of Au is ~15% smaller than it would be in the absence of the relativistic effect. Further, much of the chemistry of gold, including the catalytic properties, is therefore determined by the high energy and reactivity of the 5 *d* electrons. This relativistic effect explains why gold differs so much from its neighbors. Essential requirements for high oxidation activity of gold particles include: small particle size (not larger than 4 nm),[88] use of "reactive" support, and a preparative method that achieves the desired size of particle in intimate contact with the support. As the size of gold nanoparticles is sufficiently small, (i) the fraction of surface atoms increases, (ii) the band structure is week, so surface atoms on such small particles behave more like individual atoms, and a greater fraction of atoms are in contact with the support, and the length of the periphery per unit mass of metals rises.

Thiol-stabilized gold nanoparticles have also been exploited for catalysis applications. Examples include asymmetric dihydroxylation reactions,[89] carboxylic ester cleavage,[90] electrocatalytic reductions by anthraquinone functionalized gold particles[91] and particle-bound ring opening metathesis polymerization.[92] It should be noted that the above-mentioned catalytic applications are based on the carefully designed

9.6. Band Gap Engineered Quantum Devices

Band gap engineering is a general term referring to the synthetic tailoring of band gaps[93,94] with the intent to create unusual electronic transport and optical effects, and novel devices. Obviously, most of the devices based on semiconductor nanostructures are band gap engineered quantum devices. However, the examples discussed in this section are focused mainly on the device design and fabrication of quantum well and quantum dot lasers by vapor deposition and lithography techniques.

9.6.1. *Quantum well devices*

Lasers fabricated using single or multiple quantum wells based on III–V semiconductors as the active region have been extensively studied over the last two decades. Quantum well lasers offer improved performance with lower threshold current and lower spectra width as compared to that of regular double heterostructure lasers. Quantum wells allow the possibility of independently varying barriers and cladding layer compositions and widths, and thus separate determination of optical confinement and

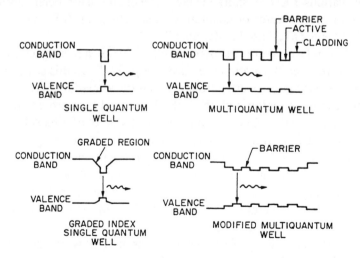

Fig. 9.4. Schematic energy band diagrams of different types of quantum well structures used to optimize the laser performance. [P.K. Bhattacharya and N.K. Dutta, *Ann. Rev. Mater. Sci.* **23**, 79 (1993).]

electrical injection. Quantum well lasers were first fabricated using the GaAs/AlGaAs material systems,[95,96] and Fig. 9.4 shows schematic energy band diagrams of different types of quantum well structures used to optimize the laser performance.[97] One of the main differences between the single quantum well and the multiple quantum well lasers is that the confinement factor of the optical mode is significantly smaller for the former. This results in higher threshold carrier and current densities for single quantum well lasers; however the confinement factor of single quantum well lasers can be significantly increased using a graded-index cladding structure.[98] InGaAsP/InP is another material system used in the fabrication of quantum well lasers.[99,100] InGaAsN/GaAs quantum wells are yet another example.[101] Strain has been explored and introduced into quantum well lasers, since strain can alter the band structure parameters significantly to produce many desirable features such as better high temperature performance resulting from reduced Auger recombination, small chirp, and high bandwidth.[97] Other quantum well optical devices have also been extensively studied and include quantum well electroabsorption and electro-optic modulators, quantum well infrared photodetectors, avalanche photodiodes and optical switching and logic devices.

Blue/green light-emitting diodes (LED) have been developed based on nanostructures of wide-band gap II–VI semiconductor materials.[102] Such devices take direct advantages of quantum well heterostructure configurations and direct energy band gap to achieve high internal radiative efficiency. Various LED at short visible wavelengths have been fabricated based on nanostructures or quantum well structures of ZnSe-based materials[103,104] and ZnTe-based materials.[105]

Blue/green lasers were first demonstrated[106,107] in a p–n injection diode that employed a configuration sketched in Fig. 9.5.[102] In this structure, the Zn(S,Se) ternary layers were introduced to serve as cladding layers for the optical waveguide region with the ZnSe layers and thus provide the electronic barriers for the (Zn,Cd)Se quantum wells. A lot of effects have been devoted to the improvement of materials and structure-design from the above structure.[108,109] The typical blue/green lasers operate continuously at room temperature and emit a significant amount of power with wavelengths ranging from 463 to 514 nm depending on the actual structure. The various laser structures are composed of (Zn,Mg)(Se,S) and Zn(Se,S) cladding layers with (Zn,Cd)Se quantum wells and possess a graded ohmic contact consisting of Au metal on a pseudo-alloy of Zn(Se,Te).

Heterojunction bipolar transistor (HBT) is an example of nanostructured devices based on GeSi/Si nanostructures.[110,111] For this structure, the GeSi layer is thick enough so that no quantum confinement occurs. In

Applications of Nanomaterials

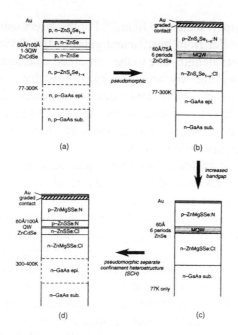

Fig. 9.5. Schematic diagrams of key blue/green laser diode configurations and their evolution from the initial laser design to the later laser design. [L.A. Kolodziejski, R.L. Gunshor, and A.V. Nurmikko, *Ann. Rev. Mater. Sci.* **25**, 711 (1995).]

the operation of a bipolar transistor, by applying a small current to the base, a large amount of current can flow from the emitter to the collector if the gain is high. Comparing to the conventional bipolar junction transistor, the HBT offers an advantage of reduction of hole injection into the emitter, due to the valence band discontinuity. The barrier to the hole injection is exponentially sensitive to the valence band offset, ΔE_v.

9.6.2. Quantum dot devices

The key parameter that controls the wavelength is the size of the dot. Large sized dots emit at longer wavelengths than small sized ones. Quantum dot heterostructures are commonly synthesized by molecular beam epitaxy at the initial stages of strained heteroepitaxial growth via the layer-island or Stranski–Krastanov growth mode.[112,113]

Quantum dots have been established their use in lasers and detectors. Quantum dot lasers with ultralow-threshhold current densities and low sensitivity to temperature variations have been demonstrated.[114,115] Intersublevel detectors made of quantum dot nanostructures were found not sensitive to normal-incidence light.[116] For the lasers using the quantum dot media often suffer from insufficient gain for the device to operate at the ground state wavelength, due to the combined consequence of the low density of states and the low area density of dots that is normally used. Several techniques have been developed to overcome this barrier. For example, several layers of quantum dots are used to increase the modal gain. Other methods include coating the laser facets to increase their reflectivity and lengthen the laser cavity.

The efficiency of luminescence from quantum dot structures depends on a number of factors including the capture of the carriers within the dots, the minimization of nonradiative recombination channels within the dots and in the surrounding matrix, and the elimination of defects at the hetero-interfaces. Embedding quantum dots inside an appropriate quantum well structure (also referred to as active region) demonstrated dramatically enhanced emission efficiency and low threshold current, due to the improved structural and optical properties of the embedding layers, and the enhanced ability of capturing and confining carriers to the vicinity of the dots.[117,118] Further structural improvement can be achieved by sandwiching quantum dots in a compositionally graded quantum well.[119] When the quantum dots of InAs are inserted at the center of compositionally graded $In_xGa_{1-x}As$ layers, the relative emission efficiency has been increased by nearly an order of magnitude over the emission of dots inside a constant composition (In,Ga)As structure.

9.7. Nanomechanics

In the previous two chapters, we have discussed the applications of SPM in the field of imaging surface topography and measurement of local properties of sample surface (Chapter 8) and nano manipulation and nanolithography in fabrication and processing of nanodevices. In this chapter, we will briefly introduce another important application of SPM, i.e. nanodevices derived from SPM. Although many devices are being investigated and more are to be developed in the conceivable future, we will take two examples to illustrate the possibilities and general approaches, specifically, nanosensors and nanotwizers.

Lang et al.[120] made an excellent summary of the applications of AFM cantilever based sensors in their tutorial article. When the surface of a cantilever or a tip is functionalized in such a way that a chemically active and a chemically inactive surface is obtained, chemical or physical processes on the active cantilever surface can be observed using the temporal evolvement of the cantilever's response. Cantilevers can be used as a nanomechanical sensor device for detecting chemical interactions between binding partners on the cantilever surface and in its environment. Such interactions might be produced by electrostatic or intermolecular forces. At the interface between an active cantilever surface and the surrounding medium, the formation of induced stress, the production of heat or a change in mass can be detected. In general, detection modes can be grouped into three strands: static mode, dynamic mode and heat mode as illustrated in Fig. 9.6.[120]

In the static mode, the static bending of the cantilever beam due to external influences and chemical/physical reactions on one of the cantilever's surfaces is investigated. The asymmetric coating with a reactive layer on one surface of the cantilever favors preferential adsorption of molecules on this surface. In most cases, the intermolecular forces in the adsorbed molecule layer produce a compressive stress, i.e. the cantilever bends. If the reactive coating is polymer and adsorbing molecules can diffuse, the reactive coating will swell and the cantilever beam will bend. Similarly, if the cantilever beam emerges into a chemical or biochemical solution, the asymmetric interaction between the cantilever beam and the surrounding environment results in bending of the cantilever beam. Many new concepts and devices have been explored.[121–124]

In dynamic mode, the cantilever is driven at its resonance frequency. If the mass of the oscillating cantilever changes owing to additional mass

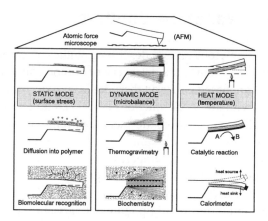

Fig. 9.6. AFM cantilever based sensors with detection modes being grouped into three strands: static mode, dynamic mode and heat mode. [H.P. Lang, M. Hegner, E. Meyer, and Ch. Gerber, *Nanotechnology* **13**, R29 (2002).]

deposited on the cantilever, or if mass is removed from the cantilever, its resonance frequency changes. Using electronics designed to track the resonance frequency of the oscillating cantilever, the mass changes on the cantilever are derived from shifts of resonance frequency. The cantilever can be regarded as a tiny microbalance, capable of measuring mass changes of less than 1 pg.[125] In dynamic mode, active coatings should apply on both surfaces of the cantilever to increase the active surface where the mass change takes places. Dynamic mode works better in gas than in liquid, which complicates the exact determination of the resonance frequency of the cantilever. More examples are available in Ref. 126.

In heat mode, the cantilever is coated asymmetrically, one surface with a layer having a different thermal expansion coefficient than that of the cantilever itself. When such a cantilever is subjected to a temperature change, it will bend. Deflections corresponding to temperature changes in the micro-Kelvin range can be easily measured. If the coating is catalytically active, e.g. a platinum layer facilitates the reaction of hydrogen and oxygen to form water. In such a case, heat is generated on the active surface and will result in bending of the cantilever. Such a method can also be used in the study of phase transition and measurement of thermal properties of a very small amount of materials.[127,128]

Although the above discussion has been limited on the single cantilever nanosensors, the same principle is readily applicable to multiple cantilever nanosensors. For example, a SPM cantilever array consisting of more than 1000 cantilevers have been fabricated.[129]

9.8. Carbon Nanotube Emitters

There have been numerous reports describing studies on carbon nanotubes as field emitters,[130–136] since the discovery of carbon nanotubes. Standard electron emitters are based either on thermionic emission of electrons from heated filaments with low work functions or field emission from sharp tips. The latter generates monochromatic electron beams; however, ultrahigh vacuum and high voltages are required. Further, the emission current is typically limited to several microamperes. Carbon fibers, typically 7 μm in diameter, have been used as electron emitters; however, they suffer from poor reproducibility and rapid deterioration of the tip.[137] Carbon nanotubes have high aspect ratios and small tip radius of curvature. In addition, their excellent chemical stability and mechanical strength are advantageous for application in field emitters. Rinzler et al.[132] demonstrated laser-irradiation-induced electron field emission from an individual multiwall nanotube. Although the emission current of a single tube is constrained because of its very small dimensions, an array of nanotubes oriented perpendicular to an electrode would make an efficient field emitter.

De Heer and co-workers first demonstrated a high-intensity electron gun based on field emission from an array of oriented carbon nanotubes.[130] Field emission current densities of ~ 0.1 mA/cm^2 were observed when a voltage of 200 V was applied, and a current density of >100 mA/cm^2 was realized at 700 V. The gun was reported to be air stable and inexpensive to fabricate, and functions stably and reliably for long time. However, later research found a gradual degradation with time of the emission performances on both single-wall carbon nanotube and multi-wall carbon nanotube emitters.[135] The degradation was explained by the destruction of nanotubes by ion bombardment with ions either from gas phase ionization or anode emission. It was also found that the degradation of single-wall carbon nanotube emitter is significantly faster (a factor ≥ 10), since they are more sensitive to electron or ion bombardment.

A flat panel display based on nanotube field emission was also demonstrated.[134] A 32 × 32 matrix-addressable diode nanotube display prototype was fabricated and a steady emission was produced in 10^{-6} torr vacuum. Pixels were well defined and switchable under a half-voltage "off-pixel" scheme. A fully sealed field emission display of 4.5 inch in size has been fabricated using single-wall carbon nanotube—organic binders.[138] The nanotubes were vertically aligned using paste squeeze and surface rubbing techniques, and fabricated displays were fully scalable at temperatures as low as 415°C. The turn-on field of 1 V/μm and brightness of 1800 cd/m^2 at 3.7 V/μm was observed on the entire 4.5 inch area from the green phosphor-indium-tin-oxide glass. Figure 9.7 shows a CRT lighting element equipped with aligned CNT emitters and the electron tube is 20 mm in diameter and 75 mm long.[139] A test of this cathode-ray tube lighting element suggested a lifetime of exceeding 10,000 h.[139]

Field emission properties of carbon nanotubes have been studied extensively. It was found that both aligned[130,134,140] and randomly[133,135,141,142] oriented nanotubes have impressive emission capabilities. Chen et al.[143] compared field emission data from aligned high-density carbon nanotubes with orientations parallel, 45°, and perpendicular to the substrate. The different orientations were obtained by changing the angle between the substrate and the bias electrical field direction. It was found that carbon nanotubes all demonstrated efficient field emission regardless of their orientations. The nanotube arrays oriented parallel to the substrate have a lower onset applied field, and a higher emission current density under the same electric field than those

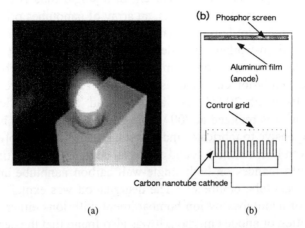

(a)   (b)

Fig. 9.7. A CRT lighting element equipped with aligned CNT emitters on SUS304 (a) operating device and (b) structure. The electron tube is 20 mm in diameter and 75 mm long. [H. Murakami, M. Hirakawa, C. Tanaka, and H. Yamakawa, *Appl. Phys. Lett.* **76**, 1776 (2000).]

oriented perpendicular to the substrate. The result indicates that electrons can emit from the body of nanotubes and carbon nanotubes can be used as linear emitter. The ability to emit electrons from the body of nanotubes was attributed to the small radius of the tubes and the presence of defects on the surface of carbon nanotubes. Saito and co-workers[144,145] have conducted field emission microscopy of single-wall nanotubes and open multiwall nanotubes. In addition to field emitters, carbon nanotubes have been explored for many other applications including sensors, scanning probe tips, hydrogen storage and Li batteries as summarized in an excellent review paper by Terrones.[146]

9.9. Photoelectrochemical Cells

The development of photoelectrochemical cells, also commonly known as photovoltaic cells or solar cells, emphasizes the need for a higher conversion efficiency of solar energy to electrical power. Photoelectrochemical devices consisting of silicon-based p–n junction materials[147,148] and other heterojunction materials,[148–150] most notably indium-gallium-phosphide/ gallium-arsenide and cadmium-telluride/cadmium-sulfide, have been extensively studied for efficient light conversion, and have obtained the highest efficiency close to 20%,[147,148] as compared to cells based on other materials. However, the high cost of production, expensive equipment, and necessary clean-room facilities associated with the development of these devices have directed exploration of solar energy conversion to cheaper materials and devices.

Sol-gel-derived titania films with a crystal structure of anatase and a mesoporous structure have been demonstrated as an excellent material for photoelectrochemical cells and have gained a lot of attention since its introduction by O'Regan and Grätzel.[151] Such devices are commonly referred to as dye-sensitized solar cells consisting of porous nanocrystalline titania (TiO_2) film in conjunction with an efficient light-absorbing dye, and have shown an impressive energy conversion efficiency of >10% at lower production costs.[152–155] Figure 9.8 shows the operation schematic of such a dye-sensitized mesoporous titania photovoltaic cell and a SEM micrograph of a mesoporous anatase titania film.[155] In such devices, TiO_2 functions as a suitable electron-capturing and electron-transporting material with a conduction band at 4.2 eV and an energy band gap of 3.2 eV, corresponding to an absorption wavelength of 387 nm.[156] In this process, the dye adsorbed to TiO_2 is exposed to a light source, absorbs photons upon exposure, and injects electrons into the conduction band of the TiO_2 electrode. Regeneration of the dye is initiated by subsequent hole-transfer to the electrolyte and electron capture after the completion of the I^-/I_3^- redox couple at the solid electrode–liquid electrolyte interface.

Nanostructures are advantageous for photoelectrochemical cell devices for high efficient conversion of light to electrical power due to its large

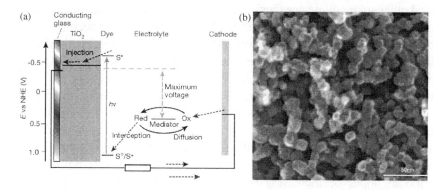

Fig. 9.8. (a) The operation schematic of such a dye-sensitized mesoporous titania photovoltaic cell and (b) a SEM micrograph of a mesoporous anatase titania film. [M. Grätzel, *Nature* **414**, 338 (2001).]

surface area at which photoelectrochemical processes take place. Many techniques have been investigated to synthesize TiO_2 electrodes to improve the structure for more efficient electron transport and good stability. Chemical vapor deposition of Ti_3O_5 has been utilized to deposit layered crystalline anatase TiO_2 thin films that are optically responsive and stable.[156] Gas-phase hydrothermal crystallization of $TiCl_4$ in aqueous mixed paste has been done to obtain crack-free porous nanocrystalline TiO_2 thick film through low-temperature processing.[157] Compression techniques of TiO_2 powder have also been used to form porous and stable films.[158] The most common and widely used technique for the preparation of crack-free TiO_2 thick film for use as suitable electron-transporting electrodes involves the preparation of TiO_2 paste by way of sol-gel processing of commercially-available TiO_2 colloidal precursors containing an amount of organic additives and followed with hydrothermal treatment. This conventional method requires the deposition of the prepared paste by either doctor-blading, or spin coating, or screen-printing on a transparent conducting substrate.[159–161] Moderate temperature sintering is utilized to remove the organic species and to connect the colloidal particles. Typical thickness of mesoporous TiO_2 film[153–155] using this method ranges from 2 μm to 20 μm, depending on the colloidal particle size and the processing conditions, and the maximum porosity obtained by this technique has reported to be ~50% with an average pore size around 15 nm and internal surface area of $>100\,m^2/g$.

Although various techniques have been utilized and explored to synthesize a more efficient structure of TiO_2 film to enhance the electrical and photovoltaic properties of solar cell devices, the capability of these devices to surpass the 10% light conversion efficiency has been hindered. Efforts to find other solar cell devices with various broad-band semiconducting oxide materials, including ZnO[162–164] and SnO_2,[164,165] films, have been made for possible improvement of the current state of TiO_2-based dye-sensitized solar cell devices. Composite structures consisting of a combination of TiO_2 and SnO_2, ZnO or Nb_2O_5 materials,[164,166,167] or a combination of other oxides,[168–170] have also been examined in an attempt to enhance the overall light conversion efficiency. In addition, hybrid structures comprised of a blend of semiconducting oxide film and polymeric layers for solid-state solar cell devices have been explored in an effort to eliminate the liquid electrolyte completely for increased electron transfer and electron regeneration in hopes of increasing the overall efficiency.[171–173] So far, these devices have achieved an overall light conversion efficiency of up to 5% for ZnO devices,[162] up to 1% for SnO_2 devices,[165] up to 6% for composite devices,[165] and up to 2% for hybrid devices,[171] all of which are still less efficient than solar cell devices based on dye-sensitized TiO_2 mesoporous film.

9.10. Photonic Crystals and Plasmon Waveguides

9.10.1. *Photonic crystals*

Photonic crystals have a broad range of applications.[174,175] Examples readily for commercialization include waveguides and high-resolution spectral filters. Photonic crystals allow for guiding geometries such as 90° corners.[176] Potential applications are photonic crystal lasers, light emitting diodes and photonic crystal thin films to serve as anticounterfeit protection on credit cards. Ultimately, it is hoped that photonic crystal diodes and transistors will eventually enable the construction of an all-optical computer.

A photonic-band-gap (PBG) crystal, or simply referred to as photonic crystal, is a spatially periodic lattice consisting of alternating regions of dielectric materials with different refractive indices.[177] The concept of PBG crystals was first proposed by Yablonovitch[178] and John[179] in 1987, and the first experimental realization of 3D photonic crystal was reported in 1991.[180] Figure 9.9 shows a schematic of one-, two-, and three-dimensional photonic crystals. Because of its long-range order, a photonic crystal is capable of controlling the propagation of photons in much the same way as a semiconductor does for electrons: that is, there exists a forbidden gap in the photonic band structure that can exclude the existence of optical modes within a specific range of frequencies. A photonic band gap provides a powerful means to manipulate and control photons, and can find many applications in photonic structures or systems. For example, photonic crystals can be used to block the propagation of photons irrespective of their polarization direction, localize photons to a specific area at restricted frequencies, manipulate the dynamics of a spontaneous or stimulated emission process, and serve as a lossless waveguide to confine or

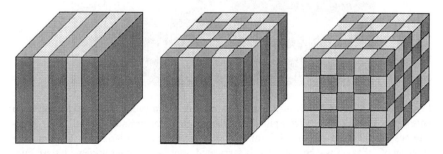

Fig. 9.9. Schematic representing one-, two-, and three-dimensional photonic crystals consisting of alternating regions of dielectric materials.

direct the propagation of light along a specific channel. It should also be noted that photonic crystals work at all wavelengths and, thus, find applications in the near-infrared telecommunication window or visible region if the size of the periodic structures (lattice constants) is appropriately chosen. A number of methods have been explored for the fabrication of photonic crystals.[181] Examples include layer-by-layer stacking techniques[182,183] electrochemical etching,[184] chemical vapor deposition,[185] holographic lithography,[186] and self-assembly of monodispersed spherical colloids.[187,188] Figure 9.10 shows SEM micrograph of a periodic array of silicon pillars fabricated using deep anisotropic etching. The silicon pillars are 205 nm in diameter and 5 μm tall. This structure possesses a band gap of ~1.5 μm for transverse magnetic polarization. By removing an array of pillars, a waveguide bend is fabricated. Input and output waveguides are integrated with the two-dimensional photonic crystal.[189]

A complete or full band gap is defined as the one that can extend over the entire Brillouin zone in the photonic band structure.[190] An incomplete band gap is often referred to as a pseudo gap, because it appears only in the transmission spectrum along a certain direction of propagation. A complete band gap can be considered as a set of pseudo gaps that overlap for a certain range of frequencies over all three dimensions of space.

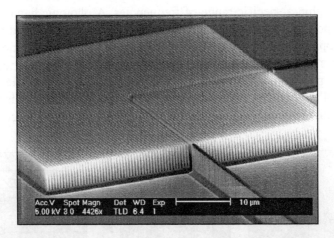

Fig. 9.10. SEM micrograph of a periodic array of silicon pillars fabricated using deep anisotropic etching. The silicon pillars are 205 nm in diameter and 5 μm tall. This structure possesses a band gap of ~1.5 μm for transverse magnetic polarization. By removing an array of pillars, a waveguide bend is fabricated. Input and output waveguides are integrated with the photonic crystal. [T. Zijlstra, E. van der Drift, M. J. A. de Dood, E. Snoeks, and A. Polman, *J. Vac. Sci. Technol.* **B17**, 2734 (1999).]

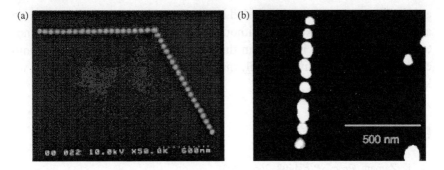

Fig. 9.11. (a) SEM image of a 60° corner in a plasmon waveguide, fabricated using electron beam lithography. The gold dots are ~50 nm in diameter and spaced by ~75 nm (center-to-center). (b) Straight plasmon waveguide made using 30 nm diameter colloidal gold nanoparticles. The particles were assembled on a straight line using an AFM in contact mode, and subsequently imaged in non-contact mode. [S.A. Maier, M.L. Brongersma, P.G. Kik, S. Meltzer, A.A.G. Requicha, and H.A. Atwater, *Adv. Mater.* **13**, 1501 (2001).]

9.10.2. Plasmon waveguides

Plasmon waveguides are optical devices based on surface plasmon resonance of noble metal nanoparticles. The surface plasmon resonance is due to the strong interaction between the electric field of light and free electrons in the metal particle, which has been discussed in the previous chapter. Arrays of closely spaced metal nanoparticles set up coupled plasmon modes that give rise to coherent propagation of electromagnetic energy along the array via near-field coupling between adjacent particles.[191–193] The dipole field resulting from a plasmon oscillation in a single metal nanoparticle can induce a plasmon oscillation in a closely spaced neighboring particle due to near field electrodynamic interactions.[193,194] It has been shown that electromagnetic wave can be guided on a scale below the diffraction limit and around 90° corners or bending radius ≪ wavelength of light as shown in Fig. 9.11.[191] Electron beam lithography and AFM nanomanipulation have been applied to fabricate plasmon waveguides with gold nanoparticles of 30 and 50 nm in diameter, and the center-to-center space was three times of the particle radius.[191]

9.11. Summary

This chapter provided some examples to illustrate some applications of nanostructures and nanomaterials. It is apparent that many more applications have not been included in the discussion here, and many more are being or will be explored. Although it is not sure which pathway nanotechnology will take, it is certain that nanotechnology is penetrating into every aspects of our life and will make the world different from what we know now.

References

1. K. Zamani, *Proc. SPIE* **4608**, 266 (2002).
2. A. Vilan and D. Cahen, *Trends in Biotechnology* **20**, 22 (2002).
3. R.F. Service, *Science* **293**, 782 (2001).
4. G.Y. Tseng and J. C. Ellenbogen, *Science* **294**, 1293 (2001).
5. J.H. Schön, H. Meng, and Z. Bao, *Nature* **413**, 713 (2001).
6. J.H. Schön, H. Meng, and Z. Bao, *Science* **294**, 2138 (2001).
7. A. Aviram and M. A. Ratner, *Chem. Phys. Lett.* **29**, 277 (1974).
8. A. Bachtold, P. Hadley, T. Nakanishi, and C. Dekker, *Science* **294**, 1317 (2001).
9. J. Chen, M.A. Reed, A.M. Rawlett, and J.M. Tour, *Science* **286**, 1550 (1999).
10. S.W. Chung, J. Yu, and J.R. Heath, *Appl. Phys. Lett.* **76**, 2068 (2000).
11. Y. Huang, X.F. Duan, Q. Wei, and C.M. Lieber, *Science* **291**, 630 (2001).
12. G. Mahler, V. May, and M. Schreiber, (eds.), *Molecular Electronics: Properties, Dynamics, and Applications*, Marcel Dekker, New York, 1996.
13. D.L. Klein, R. Roth, A.K.L. Lim, A.P. Alivisatos, and P.L. McEuen, *Nature* **389**, 699 (1997).
14. P. Bergveld, *IEEE Trans. Biomed. Eng.* **BME19**, S342 (1972).
15. P. Bergveld and A. Sibbald, *Analytical and Biomedical Applications of Ion-selective Field Effect Transistors*, Elsevier, Amsterdam, 1988.
16. J. Janata, *Analyst* **119**, 2275 (1994).
17. H. Wohltjen and A.W. Snow, *Anal. Chem.* **70**, 2856 (1998).
18. S.D. Evans, S.R. Johnson, Y.L. Cheng, and T. Shen, *J. Mater. Chem.* **10**, 183 (2000).
19. M.A. Reed, C. Zhou, C.J. Muller, T.P. Burgin, and J.M. Tour, *Science* **278**, 252 (1997).
20. X.D. Cui, A. Primak, X. Zarate, J. Tomfohr, O.F. Sankey, A.L. Moore, T.A. Moore, D. Gust, G. Harris, and S.M. Lindsay, *Science* **294**, 571 (2001).
21. R. Compañó, *Nanotechnology* **12**, 85 (2001).
22. J. Chen, M.A. Reed, A.M. Rawlett, and J.M. Tour, *Science* **286**, 1550 (1999).
23. C.P. Collier, E.W. Wong, M. Belohradský, F.M. Raymo, J.F. Stoddart, P.J. Kuekes, R.S. Williams, and J.R. Heath, *Science* **285**, 391 (1999).
24. C.P. Collier, G. Mattersteig, E.W. Wong, Y. Luo, K. Beverly, J. Sampaio, F.M. Raymo, J.F. Stoddart, and J.R. Heath, *Science* **289**, 1172 (2000).
25. S.J. Tans, A.R.M. Verschueren, and C. Dekker, *Nature* **393**, 49 (1998).
26. S.J. Wind, J. Appenzeller, R. Martel, V. Derycke, and P. Avouris, *Appl. Phys. Lett.* **80**, 3817 (2002).
27. W. Liang, M.P. Shores, M. Bockrath, J.R. Long, and H. Park, *Nature* **417**, 725 (2002).
28. J. Park, A.N. Pasupathy, J.I. Goldsmith, C. Chang, Y. Yaish, J.R. Petta, M. Rinkoski, J.P. Sethna, H.D. Abruna, P.L. McEuen, and D.C. Ralph, *Nature* **417**, 722 (2002).
29. Y. Luo, C.P. Collier, J.O. Jeppesen, K.A. Nielsen, E. Delonno, G. Ho, J. Perkins, H.R. Tseng, T. Yamamoto, J.F. Stoddart, and J.R. Heath, *Chemphyschem.* **3**, 519 (2002).

30. A. Bachtold, P. Hadley, T. Nakanishi, and C. Dekker, *Science* **294**, 1317 (2001).
31. Y. Huang, X. Duan, Y. Cui, L.J. Lauhon, K.H. Kim, and C.M. Lieber, *Science* **294**, 1313 (2001).
32. X.F. Duan, Y. Huang, and C.M. Lieber, *Nano Lett.* **2**, 487 (2002).
33. Y. Chen, G.Y. Jung, D.A.A. Ohlberg, X. Li, D.R. Stewart, J.O. Jeppesen, K.A. Nielsen, J.F. Stoddart, and R.S. Williams, *Nanotechnology* **14**, 462 (2003).
34. N.A. Melosh, A. Boukai, F. Diana, B. Gerardot, A. Badolato, P.M. Petroff, and J.M. Heath, *Science* **300**, 112 (2003).
35. J.R. Heath, P.J. Kuekes, G.S. Snider, and R.S. Williams, *Science* **280**, 1716 (1998).
36. F. Peper, J. Lee, S. Adachi, and S. Mashiko, *Nanotechnology* **14**, 469 (2003).
37. S.J. Kim, Yu I. Latyshev, and T. Yamashita, *Appl. Phys. Lett.* **74**, 1156 (1999).
38. R.W. Mosley, W.E. Booij, E.J. Tarte, and M.G. Blamire, *Appl. Phys. Lett.* **75**, 262 (1999).
39. C. Bell, G. Burnell, D.J. Kang, R.H. Hadfield, M.J. Kappers, and M.G. Blamire, *Nanotechnology* **14**, 630 (2003).
40. C. Vieu, F. Carcenac, A. Pepin, Y. Chen, M. Mejias, A. Lebib, L. Manin-Ferlazzo, L. Couraud, and H. Launois, *Appl. Surf. Sci.* **164**, 111 (2000).
41. D.L. Feldheim and C.D. Keating, *Chem. Soc. Rev.* **27**, 1 (1998).
42. T. Sato, H. Ahmed, D. Brown, and B.F.G. Johnson, *J. Appl. Phys.* **82**, 1007 (1997).
43. S.H.M. Persson, L. Olofsson, and L. Hedberg, *Appl. Phys. Lett.* **74**, 2546 (1999).
44. M. Brust and C.J. Kiely, *Colloids Surf.* **A202**, 175 (2002).
45. D.L. Gittins, D. Bethekk, D.J. Schiffrin, and R.J. Nichols, *Nature* **408**, 67 (2000).
46. M.S. Dresselhaus, G. Gresselhaus, and P.C. Eklund, *Science of Fullerences and Carbon Nanotubes*, Academic Press, San Diego, CA, 1996.
47. S.J. Tans, M.H. Devoret, H. Dai, A. Thess, R.E. Smalley, L.J. Geerligs, and C. Dekker, *Nature* **386**, 474 (1997).
48. J. Kong, C. Zhou, E. Yenilmez, and H. Dai, *Appl. Phys. Lett.* **77**, 3977 (2000).
49. H.W.C. Postma, T. Peepen, Z. Yao, M. Grifoni, and C. Dekker, *Science* **293**, 76 (2001).
50. R. Martel, T. Schmidt, H.R. Shea, T. Hertel, and P. Avouris, *Appl. Phys. Lett.* **73**, 2447 (1998).
51. X. Liu, C. Lee, C. Zhou, and J. Han, *Appl. Phys. Lett.* **79**, 3329 (2001).
52. J. Kong, N.R. Franklin, C. Zhou, M.G. Chapline, S. Peng, K. Cho, and H. Dai, *Science* **287**, 622 (2000).
53. R.J. Chen, Y. Zhang, D. Wang, and H. Dai, *J. Am. Chem. Soc.* **123**, 3838 (2001).
54. P.G. Collins, M.S. Arnold, and P. Avouris, *Science* **292**, 706 (2001).
55. A.M. Rawlett, T.J. Hopson, I. Amlani, R. Zhang, J. Tresek, L.A. Nagahara, R.K. Tsui, and H. Goronkin, *Nanotechnology* **14**, 377 (2003).
56. R.H. Baughman, C. Cui, A.A. Zakhidov, Z. Iqbal, J.N. Barisci, G.M. Spinks, G.G. Wallace, A. Mazzoldi, D. De Rossi, A.G. Rinzler, O. Jaschinski, S. Roth, and M. Kertesz, *Science* **284**, 1340 (1999).
57. A.M. Fennimore, T.D. Yuzvinsky, W.Q. Han, M.S. Fuhrer, J. Cumings, and A. Zettl, *Nature* **424**, 408 (2003).
58. J. Kong, N.R. Franklin, C. Zhou, M.G. Chapline, S. Peng, K. Cho, and H. Dai, *Science* **287**, 622 (2000).
59. S. Ghosh, A.K. Sood, and N. Kumar, *Science* **299**, 1042 (2003).
60. Y. Gao and Y. Bando, *Nature* **415**, 599 (2002).
61. C.A. Haberzettl, *Nanotechnology* **13**, R9 (2002).
62. K.E. Drexler, *Engines of Creation: The Coming Era of Nanotechnology*, Anchor Press/Doubleday, New York, 1986.

63. W.J. Parak, D. Gerion, T. Pellegrino, D. Zanchet, C. Micheel, S.C. Williams, R. Bourdreau, M.A. Le Gros, C.A. Larabell, and A.P. Alivisatos, *Nanotechnology* **14**, R15 (2003).
64. T.A. Taton, *Nature Mater.* **2**, 73 (2003).
65. M. Han, X. Gao, J.Z. Su, and S. Nie, *Nat. Biotechnol.* **19**, 631 (2001).
66. T.P. De and A. Maitra, in *Handbook of Surface and Colloid Chemistry*, ed., K.S. Birdi, CRC Press, Boca Raton, FL, p. 603, 1997.
67. L. Stryer, *Biochemistry*, 4th edition, Freeman, New York, 1995.
68. J. Fritz, M.K. Baller, H.P. Lang, H. Rothuizen, P. Vettiger, G. Meyer, H.J. Guntherodt, C. Gerber, and J.K. Gimzewski, *Science* **288**, 316 (2000).
69. A.P. Alivisatos, K.P. Johnsson, X. Peng, T.E. Wilson, C.J. Loweth, M.P. Bruchez, Jr., and P.G. Schultz, *Nature* **382**, 609 (1996).
70. R. Elghanian, J.J. Storhoff, R.C. Mucic, R.L. Letsinger, and C.A. Mirkin, *Science* **277**, 1078 (1997).
71. W.J. Parak, D. Gerion, D. Zanchet, A.S. Woerz, T. Pellegrino, C. Micheel, S.C. Williams, M. Seitz, R.E. Bruehl, Z. Bryant, C. Bustamante, C.R. Bertozzi, and A.P. Alivisatos, *Chem. Mater.* **14**, 2113 (2002).
72. M.J. Bruchez, M. Moronne, P. Gin, S. Weiss, and A.P. Alivisatos, *Science* **281**, 2013 (1998).
73. W.C.W. Chan and S. Nie, *Science* **281**, 2016 (1998).
74. W.L. Shaiu, D.D. Larson, J. Vesenka, and E. Henderson, *Nucl. Acids Res.* **21**, 99 (1993).
75. E.L. Florin, V.T. Moy, and H.E. Gaub, *Science* **264**, 415 (1994).
76. W.J. Parak, D. Gerion, T. Pellegrino, D. Zanchet, C. Micheel, S.C. Williams, R. Boudreau, M.A. Le Gros, C.A. Larabell, and A.P. Alivisatos, *Nanotechnology* **14**, R15 (2003).
77. C.M. Niemeyer, *Angew. Chem. Int. Ed. Engl.* **40**, 4128 (2001).
78. C.A. Mirkin, *J. Nanoparticle Res.* **2**, 121 (2000).
79. A.A. Taton, *Trends in Biotechnology* **20** (7), 277 (2002).
80. J.J. Storhoff, R. Elghanian, R.C. Mucic, C.A. Mirkin, and R.L. Letsinger, *J. Am. Chem. Soc.* **120**, 1959 (1998).
81. G.C. Bond, *Catal. Today* **72**, 5 (2002).
82. R. Grisel, K.J. Weststrate, A. Gluhoi, and B.E. Nieuwenhuys, *Gold Bull.* **35**, 39 (2002).
83. M. Haruta, *Catal. Today* **36**, 153 (1997).
84. R.J.H. Grisel and B.E. Nieuwenhuys, *J. Catal.* **199**, 48 (2001).
85. M. Valden, X. Lai, and D.W. Goodman, *Science* **281**, 1647 (1998).
86. V. Bondzie, S.C. Parker, and C.T. Campbell, *Catal. Lett.* **63**, 143 (1999).
87. P. Pyykkö, *Chem. Rev.* **88**, 563 (1988).
88. G.C. Bond and D.T. Thompson, *Catal. Rev. Sci. Eng.* **41**, 319 (1999).
89. H. Li, Y.Y. Luk, and M. Mrksich, *Langmuir* **15**, 4957 (1999).
90. L. Pasquato, F. Rancan, P. Scrimin, F. Mancin, C. Frigeri, *Chem. Commun.*, 2253 (2000).
91. J.J. Pietron and R.W. Murray, *J. Phys. Chem.* **B103**, 4440 (1999).
92. M. Bartz, J. Kuther, R. Seshadri, and W. Tremel, *Angew. Chem. Int. Ed. Engl.* **37**, 2466 (1998).
93. F. Capasso, *Science* **235**, 172 (1987).
94. F. Capasso and S. Datta, *Phys. Today* **43**, 74 (1990).
95. R. Dingle, W. Wiegmann, and C.H. Henry, *Phys. Rev. Lett.* **33**, 827 (1974).

96. N. Holonyak Jr., R.M. Kolbas, W.D. Laidig, B.A. Vojak, and K. Hess, *J. Appl. Phys.* **51**, 1328 (1980).
97. P.K. Bhattacharya and N.K. Dutta, *Ann. Rev. Mater. Sci.* **23**, 79 (1993).
98. S.D. Hersee, B. DeCremoux, and J.P. Duchemin, *Appl. Phys. Lett.* **44**, 476 (1984).
99. N.K. Dutta, T. Wessel, N.A. Olsson, R.A. Logan, R. Yen, and P.J. Anthony, *Electron. Lett.* **21**, 571 (1985).
100. W.T. Tsang, L. Yang, M.C. Wu, Y.K. Chen, and A.M. Sergent, *Electron. Lett.* **26**, 2035 (1990).
101. M. Kondow, K. Uomi, A. Niwa, T. Kitatani, S. Watahiki, and Y. Yazawa, *Jpn. J. Appl. Phys.* **35**, 1273 (1996).
102. L.A. Kolodziejski, R.L. Gunshor, and A.V. Nurmikko, *Ann. Rev. Mater. Sci.* **25**, 711 (1995).
103. J. Ding, M. Hagerott, P. Kelkar, A.V. Nurmikko, D.C. Grillo, L. He, J. Han, and R.L. Gunshor, *Phys. Rev.* **B50**, 5787 (1994).
104. M. Hagerott, J. Ding, H. Jeon, A.V. Nurmikko, Y. Fan, L. He, J. Han, J. Saraie, R.L. Gunshor, C.G. Hua, and N. Otsuka, *Appl. Phys. Lett.* **62**, 2108 (1993).
105. E.T. Yu, M.C. Phillips, J.O. McCaldin, and T.C. McGill, *Appl. Phys. Lett.* **61**, 1962 (1992).
106. M.A. Haase, J. Qiu, J.M. DePuydt, and H. Cheng, *Appl. Phys. Lett.* **59**, 1272 (1991).
107. H. Jeon, J. Ding, W. Patterson, A.V. Nurmikko, W. Xie, D.C. Grillo, M. Kobayashi, and R.L. Gunshor, *Appl. Phys. Lett.* **59**, 3619 (1991).
108. H. Okuyama, T. Miyajima, Y. Morinaga, F. Hiei, M. Ozawa, and K. Akimoto, *Electron. Lett.* **28**, 1798 (1992).
109. J.M. Gaines, R.R. Drenten, K.W. Haberern, T. Marshall, P. Mensz, and J. Petruzzello, *Appl. Phys. Lett.* **62**, 2462 (1993).
110. C.A. King, *Heterojunction Bipolar Transistors with GeSi Alloys in Heterostructures and Quantum Devices*, Academic Press, San Diego, CA, 1994.
111. E.A. Fitzgerald, *Ann. Rev. Mater. Sci.* **25**, 417 (1995).
112. D. Bimberg, M. Grundmann, and N.N. Ledentsov, *Quantum Dot Heterostructures*, Wiley, New York, 1999.
113. G. Park, O.B. Shchekin, S. Csutak, D.L. Huffaker, and D.G. Peppe, *Appl. Phys. Lett.* **75**, 3267 (1999).
114. V.M. Ustinov, A.E. Zhukov, A.R. Kovsh, S.S. Mikhrin, N.A. Maleev, B.V. Volovik, Yu G. Musikhin, Yu M. Shernyakov, E. Yu Kondat'eva, M.V. Maximov, A.F. Tsatsul'nikov, N.N. Ledentsov, Zh I. Alferov, J.A. Lott, and D. Bimberg, *Nanotechnology* **11**, 406 (2000).
115. O.B. Shchekin and D.G. Deppe, *Appl. Phys. Lett.* **80**, 3277 (2002).
116. D. Pan, E. Towe, and S. Kennerly, *Appl. Phys. Lett.* **73**, 1937 (1998).
117. L.F. Lester, A. Stintz, H. Li, T.C. Newell, E.A. Pease, B.A. Fuchs, and K.J. Malloy, *IEEE Photon. Technol. Lett.* **11**, 931 (1999).
118. G.T. Liu, A. Stintz, H. Li, K.J. Malloy, and L.F. Lester, *Electron. Lett.* **35**, 1163 (1999).
119. L. Chen, V.G. Stoleru, and E. Towe, *IEEE J. Selected Topics in Quant. Electron.* **8**, 1045 (2002).
120. H.P. Lang, M. Hegner, E. Meyer, and Ch. Gerber, *Nanotechnology* **13**, R29 (2002).
121. R. Berger, E. Delamarche, H.P. Lang, Ch. Gerber, J.K. Gimzewski, E. Meyer, and H.J. Güntherodt, *Science* **276**, 2021 (1997).
122. H.P. Lang, R. Berger, C. Andreoli, J. Brugger, M. Despont, P. Vettiger, Ch. Gerber, J.K. Gimzewski, J.P. Ramseyer, E. Meyer, and H.J. Güntherodt, *Appl. Phys. Lett.* **52**, 383 (1998).

123. H.P. Lang, M.K. Baller, R. Berger, Ch. Gerber, J.K. Gimzewski, F.M. Battiston, P. Fornaro, J.P. Ramseyer, E. Meyer, and H.J. Güntherodt, *Anal. Chim. Acta* **393**, 59 (1999).
124. M.K. Baller, H.P. Lang, J. Fritz, Ch. Gerber, J.K. Gimzewski, U. Drechsler, H. Rothuizen, M. Despont, P. Vettiger, F.M. Battiston, J.P. Ramseyer, P. Fornaro, E. Meyer, and H.J. Güntherodt, *Ultramicroscopy* **82**, 1 (2000).
125. R. Berger, H.P. Lang, Ch. Gerber, J.K. Gimzewski, J.H. Fabian, L. Scandella, E. Meyer, and H.J. Güntherodt, *Chem. Phys. Lett.* **294**, 393 (1998).
126. F.M. Battiston, J.P. Ramseyer, H.P. Lang, M.K. Baller, Ch. Gerber, J.K. Gimzewski, E. Meyer, and H.J. Güntherodt, *Sensors Actuators* **B77**, 122 (2001).
127. R. Berger, Ch. Gerber, J.K. Gimzewski, E. Meyer, and H.J. Güntherodt, *Appl. Phys. Lett.* **69**, 40 (1996).
128. R. Berger, Ch. Gerber, H.P. Lang, and J.K. Gimzewski, *Microelectron. Engr.* **35**, 373 (1997).
129. M.I. Lutwyche, M. Despont, U. Drechsler, U. Durig, W. Harberle, H. Rothuizen, R. Stutz, R. Widmer, G.K. Binnig, and P. Vettiger, *Appl. Phys. Lett.* **77**, 329 (2000).
130. Y. Saito, S. Uemura, and K. Hamaguchi, *Jpn. J. Appl. Phys.* **37**, L346 (1998).
131. W.A. de Heer, A. Châtelain, and D. Ugarte, *Science* **270**, 1179 (1995).
132. A.G. Rinzler, J.H. Hafner, P. Nokolaev, L. Lou, S.G. Kim, D. Tomanek, P. Nordlander, D.T. Colbert, and R.E. Smalley, *Science* **269**, 1550 (1995).
133. P.G. Collins and A. Zettl, *Appl. Phys. Lett.* **69**, 1969 (1996).
134. Q.H. Wang, A.A. Setlur, J.M. Lauerhaas, J.Y. Dai, E.W. Seeling, and R.P.H. Chang, *Appl. Phys. Lett.* **72**, 2912 (1998).
135. J.M. Bonard, J.P. Salvetat, T. Stochli, W.A. de Heer, L. Forro, and A. Châtelain, *Appl. Phys. Lett.* **73**, 918 (1998).
136. J.A. Misewich, R. Martel, Ph. Avouris, J.C. Tsang, S. Heinze, and J. Tersoff, *Science* **300**, 783 (2003).
137. C. Lee, *J. Phys.* **D6**, 1105 (1973).
138. W.B. Choi, D.S. Chung, J.K. Kang, H.Y. Kim, Y.W. Jin, I.T. Han, Y.H. Lee, J.E. Jung, N.S. Lee, G.S. Park, and J.M. Kim, *Appl. Phys. Lett.* **75**, 3129 (1999).
139. H. Murakami, M. Hirakawa, C. Tanaka, and H. Yamakawa, *Appl. Phys. Lett.* **76**, 1776 (2000).
140. L.A. Chernozatonskii, Y.V. Gulyaev, Z.J. Kasakovskaja, and N.I. Sinityn, *Chem. Phys. Lett.* **233**, 63 (1995).
141. P.G. Collins and A. Zettl, *Phys. Rev.* **B55**, 9391 (1997).
142. O.M. Kuttel, O. Groening, C. Emmenegger, and L. Schlapbach, *Appl. Phys. Lett.* **73**, 2113 (1998).
143. Y. Chen, D.T. Shaw, and L. Guo, *Appl. Phys. Lett.* **76**, 2469 (2000).
144. Y. Saito, K. Hamaguchi, T. Nishino, K. Hata, K. Tohji, A. Kasuya, and Y. Nishina, *Jpn. J. Appl. Phys. Part 2* **36**, L1340 (1997).
145. Y. Saito, K. Hamaguchi, K. Hata, K. Uchida, Y. Tasaka, F. Ikazaki, M. Yumura, A. Kasuya, and Y. Nishina, *Nature* **389**, 554 (1997).
146. M. Terrones, *Ann. Rev. Mater. Res.* **33**, 419 (2003).
147. M.A. Green, *Prog. Photovolt. Res. Appl.* **9**, 123 (2001).
148. A. Shah, P. Torres, R. Tscharner, N. Wyrsch, and H. Keppner, *Science* **285**, 692 (1999).
149. S.A. Ringel, J.A. Carlin, C.L. Andre, M.K. Hudait, M. Gonzalez, D.M. Wilt, E.B. Clark, P. Jenkins, D. Scheiman, A. Allerman, E.A. Fitzgerald, and C.W. Leitz, *Prog. Photovolt. Res. Appl.* **10**, 417 (2002).

150. A. Romeo, D.L. Batzner, H. Zogg, C. Vignali, and A.N. Tiwari, *Sol. Energ. Mater. Sol. Cells* **67**, 311 (2001).
151. B. O'Regan and M. Grätzel, *Nature* **353**, 737 (1991).
152. U. Bach, D. Lupo, P. Comte, J.E. Moser, F. Weissörtel, J. Salbeck, H. Spreitzer, and M. Grätzel, *Nature* **395**, 583 (1998).
153. A. Hagfeldt and M. Grätzel, *Acc. Chem. Res.* **33**, 269 (2000).
154. M. Grätzel, *Prog. Photovolt. Res. Appl.* **8**, 171 (2000).
155. M. Grätzel, *Nature* **414**, 338 (2001).
156. M. Thelakkat, C. Schmitz, and H.W. Schmidt, *Adv. Mater.* **14**, 577 (2002).
157. D. Zhang, T. Yoshida, and H. Minoura, *Chem. Lett.* **874** (2002).
158. G. Boschloo, H. Lindström, E. Magnusson, A. Holmberg, and A. Hagfeldt, *J. Photochem. Photobiology A: Chem.* **148**, 11 (2002).
159. F. Pichot, J.R. Pitts, and B.A. Gregg, *Langmuir* **16**, 5626 (2000).
160. Y.V. Zubavichus, Y.L. Slovokhotov, M.K. Nazeeruddin, S.M. Zakeeruddin, M. Grätzel, and V. Shklover, *Chem. Mater.* **14**, 3556 (2002).
161. S. Nakade, M. Matsuda, S. Kambe, Y. Saito, T. Kitamura, T. Sakata, Y. Wada, H. Mori, and S. Yanagida, *J. Phys.Chem.* **B106**, 10004 (2002).
162. K. Keis, C. Bauer, G. Boschloo, A. Hagfeldt, K. Westermark, H. Rensmo, and H. Siegbahn, *J. Photochem. Photobiology A: Chem.* **148**, 57 (2002).
163. S. Karuppuchamy, K. Nonomura, T. Yoshida, T. Sugiura, and H. Minoura, *Solid State Ionics* **151**, 19 (2002).
164. K. Tennakone, P.K.M. Bandaranayake, P.V.V. Jayaweera, A. Konno, and G.R.R.A. Kumara, *Physica* **E14**, 190 (2002).
165. S. Chappel and A. Zaban, *Sol. Energ. Mater. Sol. Cells* **71**, 141 (2002).
166. S. Chappel, S.G. Chen, and A. Zaban, *Langmuir* **18**, 3336 (2002).
167. S.G. Chen, S. Chappel, Y. Diamant, and A. Zaban, *Chem. Mater.* **13**, 4629 (2001).
168. T.S. Kang, S.H. Moon, and K.J. Kim, *J. Electrochem. Soc.* **149**, E155 (2002).
169. E. Palomares, J.N. Clifford, S.A. Haque, T. Lutz, and J.R. Durrant, *J. Am. Chem. Soc.* **125**, 475 (2003).
170. P.K.M. Bandaranayake, P.V.V. Jayaweera, and K. Tennakone, *Sol. Energ. Mater. Sol. Cells* **76**, 57 (2003).
171. W.U. Huynh, X. Peng, and A.P. Alivisatos, *Adv. Mater.* **11**, 923 (1999).
172. W. Huynh, J.J. Dittmer, and A.P. Alivisatos, *Science* **295**, 2425 (2002).
173. D. Gebeyehu, C.J. Brabec, and N.S. Sariciftci, *Thin Solid Films* **403–404**, 271 (2002).
174. M. Ibanescu, Y. Fink, S. Fan, E.L. Thomas, and J.D. Joannopoulos, *Science* **289**, 415 (2000).
175. J. Ouellette, *The Industry Physicist*, p.14, December 2001/January 2002.
176. A. Mekis, J.C. Chen, I. Kurland, S. Fan, P.R. Villeneuve, and J.D. Joannopoulos, *Phys. Rev. Lett.* **77**, 3787 (1999).
177. J.D. Joannopoulos, R.D. Meade, and J.N. Winn, *Photonic Crystals*, Princeton University Press, Princeton, NJ, 1995.
178. E. Yablonovitch, *Phys. Rev. Lett.* **58**, 2059 (1987).
179. S. John, *Phys. Rev. Lett.* **58**, 2486 (1987).
180. E. Yablonovitch, T.J. Gmitter, and K.M. Leung, *Phys. Rev. Lett.* **67**, 2295 (1991).
181. A. Polman and P. Wiltzius, *MRS Bull.* **26**, 608 (2001).
182. S.Y. Lin, J.G. Fleming, D.L. Hetherington, B.K. Smith, R. Biswas, K.M. Ho, M.M. Sigalas, W. Zubrzycki, S.R. Kurtz, and J. Bur, *Nature* **394**, 251 (1998).
183. S. Noda, K. Tomoda, N. Yamamoto, and A. Chutinan, *Science* **289**, 604 (2000).

184. A. Birner, R.B. Wehrspohn, U. Gösele, and K. Busch, *Adv. Mater.* **13**, 377 (2001).
185. M.C. Wanke, O. Lehmann, K. Müller, Q.Z. Wen, and M. Stuke, *Science* **275**, 1284 (1997).
186. M. Campbell, D.N. Sharp, M.T. Harrison, R.G. Denning, and A.J. Turberfield, *Nature* **404**, 53 (2000).
187. J.E.G.J. Wijnhoven and W. Vos, *Science* **281**, 802 (1998).
188. Y. Xia, B. Gates, Y. Yin, and Y. Lu, *Adv. Mater.* **12**, 693 (2000).
189. T. Zijlstra, E. van der Drift, M. J. A. de Dood, E. Snoeks, and A. Polman, *J. Vac. Sci. Technol.* **B17**, 2734 (1999).
190. J.D. Joannopoulos, P.R. Villeneuve, and S. Fan, *Nature* **386**, 143 (1997).
191. S.A. Maier, M.L. Brongersma, P.G. Kik, S. Meltzer, A.A.G. Requicha, and H.A. Atwater, *Adv. Mater.* **13**, 1501 (2001).
192. M. Quinten, A. Leitner, J.R. Krenn, and F.R. Aussenegg, *Opt. Lett.* **23**, 1331 (1998).
193. M.L. Brongersma, J.W. Hartman, and H.A. Atwater, *Phys. Rev.* **B62**, R16356 (2000).
194. J.R. Krenn, A. Dereux, J.C. Weeber, E. Bourillot, Y. Lacroute, J.P. Goudonnet, G. Schider, W. Gotschy, A. Leitner, F.R. Aussenegg, and C. Girard, *Phys. Rev. Lett.* **82**, 2590 (1999).

Appendix 1

Periodic Table of the Elements

1	2	3	4	5	6	7	8	9	10	11	12	13	14	15	16	17	18
1 H 1.0079																	2 He 4.0026
3 Li 6.941	4 Be 9.0122											5 B 10.811	6 C 12.011	7 N 14.007	8 O 15.999	9 F 18.998	10 Ne 20.180
11 Na 22.990	12 Mg 24.305											13 Al 26.982	14 Si 28.086	15 P 30.974	16 S 32.065	17 Cl 35.453	18 Ar 39.948
19 K 39.098	20 Ca 40.078	21 Sc 44.956	22 Ti 47.867	23 V 50.942	24 Cr 51.996	25 Mn 54.938	26 Fe 55.845	27 Co 58.933	28 Ni 58.693	29 Cu 63.546	30 Zn 65.409	31 Ga 69.723	32 Ge 72.64	33 As 74.922	34 Se 78.96	35 Br 79.904	36 Kr 83.798
37 Rb 85.468	38 Sr 87.62	39 Y 88.906	40 Zr 91.224	41 Nb 92.906	42 Mo 95.94	43 Tc (98)	44 Ru 101.07	45 Rh 102.91	46 Pd 106.42	47 Ag 107.87	48 Cd 112.41	49 In 114.82	50 Sn 118.71	51 Sb 121.76	52 Te 127.60	53 I 126.90	54 Xe 131.29
55 Cs 132.91	56 Ba 137.33	57-71 *	72 Hf 178.49	73 Ta 180.95	74 W 183.84	75 Re 186.21	76 Os 190.23	77 Ir 192.22	78 Pt 195.08	79 Au 196.97	80 Hg 200.59	81 Tl 204.38	82 Pb 207.2	83 Bi 208.98	84 Po (209)	85 At (210)	86 Rn (222)
87 Fr (223)	88 Ra (226)	89-103 #	104 Rf (261)	105 Db (262)	106 Sg (266)	107 Bh (264)	108 Hs (277)	109 Mt (268)	110 Ds (281)	111 Uuu (272)	112 Uub (285)		114 Uuq (289)				

* Lanthanide series

57 La 138.91	58 Ce 140.12	59 Pr 140.91	60 Nd 144.24	61 Pm (145)	62 Sm 150.36	63 Eu 151.96	64 Gd 157.25	65 Tb 158.93	66 Dy 162.50	67 Ho 164.93	68 Er 167.26	69 Tm 168.93	70 Yb 173.04	71 Lu 174.97

\# Actinide series

89 Ac (227)	90 Th 232.04	91 Pa 231.04	92 U 238.03	93 Np (237)	94 Pu (244)	95 Am (243)	96 Cm (247)	97 Bk (247)	98 Cf (251)	99 Es (252)	100 Fm (257)	101 Md (258)	102 No (259)	103 Lr (262)

Appendix 2

The International System of Units

Quantity	Name	Symbol	SI base units
Length	Meter	m	m
Mass	Kilogram	kg	kg
Time	Second	s	s
Current	Ampere	A	A
Temperature	Kelvin	K	K
Luminous intensity	Candela	Cd	Cd
Force	Newton	N	$kg.m/s^2$
Energy	Joule	J	$kg.m^2/s^2$
Pressure	Pascal	Pa	$kg.m/s^2$
El. charge	Coulomb	C	$A.s$
Power	Watt	W	$kg.m^2/s^3$
Voltage	Volt	V	$kg/m^2/A.s^3$
El. resistance	Ohm	Ω	$kg.m^2/A^2.s^3$
El. conductance	Siemens	S	$A^2.s^3/kg.m^2$
Magn. flux	Weber	Wb	$kg.m^2/A.s^2$
Magn. induction	Tesla	T	$kg/A.s^2$
Inductance	Henry	H	$kg/A^2.s^2$
Capacitance	Farad	F	$A^2.s^4/kg.m^2$

Appendix 3

List of Fundamental Physical Constants

Quantity	Symbol	Value with units
Avogadro's number	N_A	6.023×10^{23} molecules/mole
Boltzmann's constant	κ	1.38×10^{-23} J/atom.K
Bohr magneton	μ_B	9.27×10^{-24} A.m^2
Constant of gravitation	G	6.67×10^{-11} m^3/kg.s^2
Electron charge	e	1.602×10^{-19} C
Electron mass	m_e	9.11×10^{-31} kg
Faraday's constant	F	96,500 C/mol
Gas constant	R_g	8.31 J/mol.K
Permeability of a vacuum	μ_o	1.257×10^{-6} H/m
Permittivity of a vacuum	ε_o	8.85×10^{-12} F/m
Planck's constant	h	6.63×10^{-34} J.s
Velocity of light in a vacuum	c_o	3×10^8 m/s

Appendix 4

The 14 Three-Dimensional Lattice Types

CUBIC
$a = b = c$
$\alpha = \beta = \gamma = 90°$

simple   body-centered   face-centered

TETRAGONAL
$a = b \neq c$
$\alpha = \beta = \gamma = 90°$

simple   body-centered

ORTHORHOMBIC
$a \neq b \neq c$
$\alpha = \beta = \gamma = 90°$

simple   body-centered   end-centered   face-centered

RHOMBOHEDRAL
$a = b = c$
$\alpha = \beta = \gamma \neq 90° < 120°$

HEXAGONAL
$a = b \neq c$
$\alpha = \beta = 90°$
$\gamma = 120°$

MONOCLINIC
$a \neq b \neq c$
$\alpha = \gamma = 90° \neq \beta$

simple   end-centered

TRICLINIC
$a \neq b \neq c$
$\alpha \neq \beta \neq \gamma$

Appendix 5

The Electromagnetic Spectrum

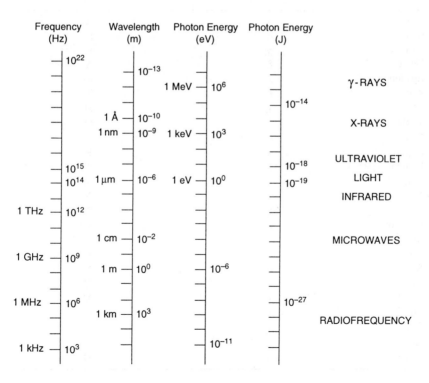

Appendix 6

The Greek Alphabet

Name	Lower case	Upper case
Alpha	α	A
Beta	β	B
Gamma	γ	Γ
Delta	δ	Δ
Epsilon	ε	E
Zeta	ζ	Z
Eta	η	H
Theta	θ	Θ
Iota	ι	I
Kappa	κ	K
Lambda	λ	Λ
Mu	μ	M
Nu	ν	N
Xi	ξ	Ξ
Omicron	o	O
Pi	π	Π
Rho	ρ	P
Sigma	σ	Σ
Tau	τ	T
Upsilon	υ	Y
Phi	φ	Φ
Chi	χ	X
Psi	φ	Ψ
Omega	ω	Ω

Index

III–V semiconductors   399
II–VI semiconductor   400
7×7 reconstruction on Si (111) surface   340

α-Fe$_2$O$_3$ nanoparticles   86, 88
absorption and emission spectroscopy   345
accommodation coefficient   114
active surfactant   207
adsorbing polymer   44
adsorption isotherm   343
adsorption limited process   114
aerogel   245
aerosol   98
aerosol-assisted CVD or AACVD   194
AES   185, 349
AFM   313, 340, 411
AFM based nanolithography   306
Ag nanoparticles   92, 66, 70, 72, 89, 90, 95, 100
agglomeration   11, 25, 31
AgTe nanoparticles   100
aligned carbon nanotubes   235
alkanethiols   210
alkylsilanes   208
all-optical computer   409
alternating layer deposition   203
amphiphilic molecules   213
amphoteric surfactants   239
anchored polymer   44

anionic surfactants   239
anisotropic growth   112
anodic oxidation   245
anodized alumina membrane   143
anti-Stokes scattering   349
antibodies   396
aprotic solvent   43
arc discharge   231, 233
arc evaporation   185, 233
atomic force microscopy (AFM)   7, 292, 294
atomic layer CVD (ALCVD)   199
atomic layer deposition (ALD)   199
atomic layer epitaxy (ALE)   199
atomic layer growth (ALG)   199
Au$_{55}$ clusters   394
Au nanoparticles   4, 63, 67, 89, 90, 95, 363, 366, 370, 377, 394, 397, 411
Auger electron   349
Auger electron spectroscopy (AES)   185, 349
avalanche photodiodes   400

β-FeO(OH) nanoparticles   88
ballistic conduction   375
band gap engineering   399
barium titanate   355, 357
base growth   237
basic memory   393
BaTiO$_3$ 355, 357
BaTiO$_3$ nanowires   126

BCF theory 117
Bessel function 316
Bi nanowires 375
biological cell electrodes 238
biotin-avidin linkage 396
block copolymers 242
block polymers 97
blue/green lasers 400
blue/green light-emitting diodes 400
bottom-up approaches 4, 7
Bragg's law 331
Brownian motion 33, 34, 36
buckminster fullerene 230
buckyball 230, 395

C_{60} molecule 230
cantilever based sensors 402
capillary condensation 343
capillary forces 158, 207, 246, 315
capping material 74
carbon aerogels 249
carbon fibers 404
carbon fullerenes 6
carbon nanotube composites 268
carbon nanotube transistors 393
carbon nanotubes 6, 143, 230, 232, 318, 376, 394, 395, 404
catalyst growth mechanism 237
catalytic carbon filament growth 237
cathode deposition 147
cathodoluminescence 346
cationic surfactants 239
CdS nanoparticles 80, 99, 100, 356
CdSe nanoparticles 75, 97, 100, 101, 334, 346
CdSe/CdS core/shell nanostructure 103
CdSe/ZnS core/shell nanostructure 102
CdTe nanoparticles 75, 100
$Cd_xZn_{1-x}S$ nanoparticles 91
centrifugation deposition 161
ceramic processing 25
charge determining ions 32
charge-exchange model 305
chemical solution deposition (CSD) 223
chemical vapor deposition (CVD) 161, 189, 408, 410
chemical vapor infiltration (CVI) 196, 268
chemisorption 207, 344
citrate reduction 258

CMC 240
Co film 373
co-ions or coions 32
coating 220, 257
coating thickness 220
colloidal dispersions 151
composition segregation 19
condensation reaction 82, 256
conformal near field photolithography 284
constant current mode 294
constant voltage mode 294
consumable templates 162
contact-mode photolithography 281
controlled release of ions 87
core-shell structure 257
Coulomb blockade 377, 394
Coulomb charging 375, 377
Coulombic force 34
Coulombic staircase 377
counter ions 33
covalently linked assembly 319
critical coating thickness 222
critical energy barrier 55, 94, 175
critical micellar concentration (CMC) 240
critical nucleus size 55, 94, 175
Cu nanoparticles 90
Curie temperature 381
Czochraski crystal growth 23, 129

dc sputtering 187
d-spacing 331
de Broglie relationship 338
de Broglie's wavelength 5, 290, 367, 370, 374
Debye–Hückel screening strength 35
deep Ultra-Violet lithography (DUV) 282
defect scattering 372
delivery of therapeutic agents 394
dendrimers 395
depolarization field 380
deprotonation 85
diamond films 197
diffusion barrier 42
diffusion layer 152
diffusion-limited growth 42, 58, 86
dip-coating 220
dip-pen nanolithography 313

dipolar oscillation 363
discrete charging 377
discrete electronic configuration 367, 377
dislocations 178
dislocation–diffusion theory 120
dispersion interactions 316
dissolution-condensation growth 112, 123
DLVO theory 38
DNA templating 164
domain structures 382
double layer structure 34, 152
dye-sensitized solar cells 407
dynamic force microscopy 342
dynamic mode 403
dynamic SIMS 351

effusion cells 185
elastic modulus microscopy 342
elastic scattering 336, 372
elastomeric stamp 308
elastoplasticity 361
electric-field assisted assembly 318
electrical self-bias field 235
electrically configurable switches 393
electrochemical cell 267
electrochemical deposition 144
electrochemical etching 410
electrochemical methods 233
electrochemical vapor deposition (EVD) 196
electrochemically induced sol-gel deposition 156
electrodeposition 144
electrointercalation 267
electroless deposition 149
electroless electrolysis 149
electrolysis 147
electrolytic cell 147
electromagnetic spectrum 347
electron beam evaporation 185
electron beam lithography 165, 284, 393, 411
electron cyclotron resonance (ECR) plasma 195
electron density 333
electron field emission tips 238
electron mean free path 371
electron scattering 285, 338, 371
electron tunneling 292

electrophoresis 153
electrophoretic deposition 151
electroplating 144
electrospinning 164
electrostatic fiber processing 164
electrostatic force 34, 207, 314
electrostatic force microscopy 341
electrostatic interaction 314
electrostatic repulsion 35, 38
electrostatic stabilization 38, 152
electrosteric stabilization 47
emulsion polymerizations 98, 260, 261
energy barrier 55, 94, 175
energy dispersive X-ray spectroscopy (EDS) 349
entropic force 33, 34
enzyme immobilization 209
epitaxial aggregation 253
epitaxy 177
equilibrium crystal 21
equilibrium vapor pressure 183
Euler's theorem 231
evanescent wave regime 296
evanescent waves 296
evaporation 183
evaporation-induced self-assembly 265
evaporation–condensation process 112, 119
excimer laser micromachining 321
excimer lasers 282
extinction coefficient 366
extreme UV (EUV) lithography 282

F-face 117
far-field regime 297
fatty acid monolayers 378
Fe_3O_4 nanoparticles 101
ferroelastic 382
ferroelectrics 380
ferroelectric–paraelectric transition 380
ferromagnetic 382
FIB deposition 289
FIB etching 289
field effect transistor (FET) 393
field emitters 404
field evaporation 299, 303
field-assisted diffusion 299
field-gradient induced surface diffusion 305

field-ion microscopy 303
flat panel display 405
flat surfaces 117
flocculation 39
Flory–Huggins theta temperature 44
flow sensors 238
fluorescence 347
focused ion beam (FIB) 289, 393
focused ion beam (FIB) lithography 288
forced hydrolysis 85
formation of cracks 222
Fourier transform infrared spectroscopy (FTIR) 348
Frank–van der Merwe growth 174
Fraunhofer diffraction 282
Fresnel diffraction 282
Fuchs–Sondheimer theory 373
fullerene crystals 232
fullerene solids 230
fullerenes 230
fullerites 232

GaAs nanoparticles 79, 89, 96
GaAs nanowires 142
GaInP$_2$ nanoparticles 78
Ga$_2$O$_3$ nanowires 139
galvanic cell 146
GaN nanoparticles 81
GaN nanowires 140, 375
GaP nanoparticles 78
GaP nanowires 135
gas adsorption isotherm 343
gas impingement flux 180
gas-phase hydrothermal crystallization 408
Ge nanowires 131, 134, 138
gene replacement 395
GeO$_2$ nanowires 139
Gibbs–Thompson relation 28
gold nanoparticle 4, 63, 67, 89, 90, 95, 363, 366, 370, 377, 394, 397, 411
gold–silica core-shell structure 257
good solvent 43
Gouy layer 34
grain boundary scattering 374
graphene 232
graphite 230
gravitational field assisted assembly 319
growth mechanisms of zeolite 252

growth steps 359
growth termination 99
growth-limited process 58

Hall–Petch relationship 268, 360
Hamaker constants 36
Hamaker theory 317
hardness 360
heat mode 403
helical nanostructures 121
hetero-condensation 83
heteroatoms 255
heteroepitaxial growth 125
heteroepitaxy 175, 177
heterogeneous nucleation 93, 174
heterojunction bipolar transistor (HBT) 400
heterojunction materials 406
hexagonal faces 230
hexagonal or cubic packing of cylindrical micelles 240
hexagons 231
hierarchically structured mesoporous materials 245
high-resolution spectral filters 409
highest occupied molecular orbital (HOMO) 345
highly oriented pyrolitic graphite (HOPG) 95
hollow metal tubules 148
holographic lithography 410
homoepitaxy 175, 177
homogeneous nucleation 53
horizontal lifting 216
Hückel equation 154
hydrated antimony oxide 374
hydrogenation of unsaturated hydrocarbons 397
hydrolysis reaction 82
hydrophilicity 207
hydrophobic interactions 314
hydrophobicity 207
hydrothermal growth 126
hydrothermal synthesis 251
hysteresis 383

image-hump model 305
imperfections 358

imprint lithography 393
impurity enrichment 19
incident rate 180
incorporation of organic components 249
indium tin oxide (ITO) 243
inelastic scattering 336, 372
infrared (IR) spectroscopy 345, 347
$InGaO_3(ZnO)_5$ superlattice structure 205
inhomogeneous strains 331
InP nanocrystals 78, 368
InP nanoparticles 78
InP nanowires 135, 141, 370
intercalation method 238
intercalation compounds 266
intermolecular conduction 379
intramolecular conduction 379
ion beam lithography 165
ion bombardment 405
ion exchange 266
ion implantation 268
ion plating 188
ionic spectrometry 350
iron particles 384
island (or Volmer–Weber) growth 95, 174
island-layer (or Stranski–Krastanov) growth 95, 174, 401

Kelvin equation 130, 136, 343
Kelvin probe microscopy 341
kinked surfaces (K-face) 117
Knudsen cells 185
Knudsen diffusion 196
Knudsen number 181
KSV theory 116

laminar flow 182
Langmuir films 215
Langmuir–Blodgett films (LB films) 213, 300
Laplace equation 246
laser ablation 184, 233
laser direct writing 321
laser enhanced or assisted CVD 194
lasers 399
lattice mismatch 177
layer (or Frank–van der Merwe) growth 95, 174

layer-by-layer growth 253
layer-by-layer stacking 410
layer-island growth 95, 174, 401
leaching a phase separated glass 245
lead titanate 355, 381
LIGA 321
light emitting diodes 409
light forces 290
light-absorbing dye 407
liquid chromatography 231
liquid metal ion (LMI) source 288
lithography 278
local oxidation and passivation 303
localized chemical vapor deposition 303
logic functions 393
lowest unoccupied molecular orbital (LUMO) 345
LPCVD (low pressure CVD) 194
luminescence 346, 368

macroporous 238
magnetic force microscopy 295, 341
magnetron sputtering 188
Matthiessen's rule 371
MCM-41 241
MCM-48 241
mean diffusion distance 115
mean free path 179, 185, 371
mechanical properties 357
mechanical strength 358
membrane-based synthesis 260
mesopores 343
mesoporous 238
mesoporous materials 143
metal alloy nanoparticles 74
metal catalyst 235
metal nanoparticles 393
metal-to-semiconductor transition 375
metallic colloidal dispersions 63
metal–polymer core-shell structures 260
micelles 96, 239
microcontact printing 308
microemulsion 96, 121
micromolding in capillaries 310
micropores 343
microporous 238
microtransfer molding 310

Mie theory 362
Miller indices 21
mineralizing agent 251
miniature bull 9
MOCVD (metalorganic CVD) 194
molding 310
molecular beam epitaxy (MBE) 185, 401
molecular density 180
molecular electronics 392
molecular electronics toolbox 394
molecular flow 181
molecular labeling 396
molecular layer epitaxy (MLE) 199
molecular person 9
molecular recognition 393, 396
monolayers 393
mononuclear growth 59
Moore's law 4
multi-pole oscillation 364
multi-wall carbon nanotube (MWCNT) 232

nano-rings 121
nanobelts 120
nanobiotechnology 396
nanobots 6, 394
nanochannel array glass 143
nanocomposites 238, 263, 267
nanocomputers 393
nanocrystals 53, 395, 396
nanoelectronics 392
nanograined materials 267
nanoimprint 310
nanolithography 291
nanomanipulation 291, 298
nanomechanical sensor 403
nanomedicine 5, 394
nanoparticle seeding 125
nanoparticle superlattices 317
nanoparticles 30, 315, 395
nanorobots 5, 394
nanoscience 3
nanosensors 402
nanosurgery 395
nanotechnology 1
nanotwizers 402
nanowires 318
nanowires of the III-V materials 132

near-field coupling 411
near-field photolithography 165
near-field regime 297
near-field scanning optical microscopy 292, 296, 342
negative differential resistance 393
Nernst equation 32, 33, 145, 218
neutral atomic beam lithography 290
Ni nanoparticles 95, 383
noble metal nanoparticles 411
non-adsorbing polymer 44
non-oxide semiconductor nanoparticles 74
nonionic surfactants 239
NSOM 292, 296, 342
numerical aperture 337

oligonucleotides 396
optical absorption 368
optical labeling 395
optical switching and logic devices 400
ordered mesoporous complex metal oxides 242
ordered mesoporous materials 6, 239
order–disorder transition 216
organic aerogels 249
organic–inorganic hybrid fibers 164
organic–inorganic hybrid zeolites 256
organic–inorganic hybrids 223, 263
organometallic vapor phase epitaxy (OMVPE) 195
organosilicon derivatives 208
organosulfur compound 210
ormocers 263
ormosils 263
oscillation 362
osmotic flow 40
Ostwald ripening 11, 24, 25, 30, 65, 74, 76, 86
oxidation of carbon monoxide 397
oxidation of hydrocarbons 397
oxide nanoparticles 81
oxide–polymer structures 261

parallel process 298
paramagnetic 382
PbS nanoparticles 80
$PbTiO_3$ 355, 381

Index

Pd nanoparticles 65, 68
PECVD 188, 194, 233, 237
pentagonal faces 230
pentagons 231
periodic bond chain (PBC) theory 115
perpendicular processes 298
phase masks 283
phase shifters 283
phase-shifting photolithography 283
phosphorescence 347
photoactive polymer 279
photochemical deposition 321
photoelectrochemical cells 406
photolithography 279
photoluminescence 345, 346
photolytical deposition 321
photonic crystals 409
photonic-band-gap 409
photoresist 279
photovoltaic cells 406
physical adsorption 343
physical vapor deposition (PVD) 182
piezoelectrics 380
plasma enhanced chemical vapor deposition (PECVD) 188, 194, 233, 237
plasma etching 188
plasmon bandwidth 366
plasmon oscillation 411
plasmon waveguides 411
point of zero charge (p.z.c.) 32
poly-nuclear growth 59
polyheterocyclic fibris 379
polymer layers 45
polymer nanoparticles 98
polymer nanotubules 150
polymer particles 98, 99
polymer stabilizers 72
polymeric stabilization 42
polymeric stabilizers 63
poor solvent 43
pore volume 343
Porod's law 335
porous nanocrystalline titania 407
porous silicon 143
porous solids 238
powder metallurgy 25
primary minimum 39
projection printing 281

protic solvent 43
proton conductivity 374
proximity printing 281
Pt nanoparticles 65, 70, 72
pulsed electrodeposition 148
purification 238
pyroelectrics 380
pyrolysis 233
pyrolysis growth 237
pyrolytical deposition 321
p–n injection diode 400
p–n junction materials 406

quantum dot heterostructures 401
quantum dot lasers 401
quantum dots 53
quantum resistors 238
quantum well electroabsorption and electro-optic modulators 400
quantum well infrared photodetectors 400
quantum well lasers 399
quantum wells 399

radiation track-etched mica 143
radiation track-etched polymer membranes 143
radiation-track etching 245
Raman scattering 381
Raman spectroscopy 345, 348
random doping fluctuations 11
rate of nucleation 56
Rayleigh instability 355
Rayleigh scattering 349
Rayleigh's equation 281
reactive ion etching (RIE) 188
reactive sputtering 188
reduction of nitrogen oxides 397
reduction reagents 67
reflection high energy electron diffraction (RHEED) 185
relativistic effect 397
replica molding 310
repulsive barrier 39
residence time 115
resist 279
reversible spontaneous polarization 380
Reynolds number 181

RF sputtering 187
Rh nanoparticles 64
rough surface 130
roughening transition 23, 137
Rutherford backscattering spectrometry (RBS) 350

scanning acoustic microscope 295
scanning capacitance microscope 295
scanning capacitance microscopy 341
scanning electron microscopy (SEM) 336
scanning probe microscopy (SPM) 7, 291, 340, 402
scanning probe tip 238
scanning thermal microscopy 341
scanning tunneling microscopy (STM) 7, 292, 340
Schaefer's method 216
Scherrer's formula 331
Schottky barrier 299
secondary ion mass spectrometry (SIMS) 351
secondary minimum 39
sedimentation method 319
seeding nucleation 74
selected-area diffraction (SAD) 339
self assembled (SA) multilayer 210
self assembled monolayer (SAM) 308
self-assembly 205, 249, 257, 262, 314
self-assembly of monodispersed spherical colloids 410
self-limiting growth 199
self-purification 11, 353
semiconductor-to-insulator transition 375
sensors 393
shadow printing 281
shear force assisted assembly 318
Si nanowires 131, 142, 339, 369, 375
silane coupling agents 258
silica colloids 81
silica nanowires 140
silicon pillars 410
single molecular electronics 395
single molecular transistors 393
single-wall carbon nanotube 232
sintering 24

SiO_2 nanoparticles 89
size-selective precipitation 74, 78
sliding process 299
slip casting 159
slip plane 153
small angle X-ray scattering (SAXS) 333
sodium chloride whisker 359
soft lithography 308
soft organic elastomeric polymers 284
sol-gel processing 82, 155, 240, 245, 408
solar cells 406
solution filling 160
solution– liquid–solid (SLS) growth 140
solvent exchange 249
specific surface area 343
specular scattering 372
spin-coating 221
spiral growth 117
SPM 7, 291, 340, 402
SPM-based nanolithography 303
spontaneous growth 111
spontaneous magnetization 383
sputtering 186
static mode 403
static SIMS 351
step-growth theory 116
step surfaces (S-face) 117
steric stabilization 42
Stern layer 34, 152
STM 7, 292, 340
Stöber method 85, 259
Stokes scattering 349
strain energy 176, 178
Stranski–Krastanov growth 95, 174, 401
stress-induced recrystallization 142
structure-directing agent 251, 253
subsequent growth 56
subsequent polymerization 261
supercritical drying 245, 247
supercritical point 247
superlattice 204
superparamagnetics 382
superparamagnetism 353
supersaturation 53
surface adsorption 19
surface atomic density 18

surface charge density 32
surface diffusion coefficient 115
surface energy 11, 54
surface growth limited process 114
surface plasmon resonance 353, 362, 411
surface potential 33
surface relaxation 18
surface restructuring 18
surface roughening 23
surface scattering 353, 365, 372
surfactants 97
SWCNT 232

template 164
template filling 157
Template-assisted assembly 319
template-directed reaction 162
temporally discrete nucleation 74
thermal CVD 237
thermochemical deposition 321
thermometers 394
thermoplastic polymer 312
theta state 44
thiol-gold bonds 396
thiol-stabilized gold nanoparticles 398
Thompson model 372
TiO_2 film 199, 407
TiO_2 nanorods 157
TiO_2 particles 98
tip growth 237
tissue regeneration 395
TO_4 tetrahedra 249
top-down approaches 4, 7
transistors 5
transition metal catalysts 237
transmission electron microscopy (TEM) 338
tunneling conduction 375, 377
tunneling junctions 393
turbulent flow 182
twinned structure 354
two-photon polymerization 9

uniform elastic strain 331

van der Waals attraction force 36, 294, 314
vapor-liquid-solid (VLS) growth 127, 237
vapor phase deposition 257
vapor–solid (VS) process 112
vertical deposition 215
vibrational frequency of adatom 115
vibrational spectroscopy 345
viologen 394
viscous flow 181
visible light scattering 336
VLS growth 127, 237
Volmer–Weber growth 95, 174

waveguides 409
Wulff plot 21, 118
Wulff relationship 354

X-ray diffraction (XRD) 331
X-ray fluorescence 346
X-ray lithography 165, 287
X-ray photoelectron spectroscopy (XPS) 185, 349
xerogel 245

Y_2O_3 nanoparticles 101
Y_2O_3:Eu nanoparticles 87
yield strength 360
Young's equation 95, 175
Young–Laplace equation 26

zeolites 143, 249
zero-point charge (z.p.c.) 32
zeta potential 153
ZnO nanoparticles 88
ZnO nanowires 138
ZnS film 201
ZnS nanoparticles 100, 101
ZrO_2 nanoparticles 100